Mathematical Methods for Engineers and Scientists 3

T0203086

K.T. Tang

Mathematical Methods for Engineers and Scientists 3

Fourier Analysis, Partial Differential Equations and Variational Methods

With 79 Figures and 4 Tables

 Springer

Professor Dr. Kwong-Tin Tang
Pacific Lutheran University
Department of Physics
Tacoma, WA 98447, USA
E-mail: tangka@plu.edu

ISBN 978-3-642-07947-4   e-ISBN 978-3-540-44697-2

Springer is a part of Springer Science+Business Media.

springer.com

© Springer-Verlag Berlin Heidelberg 2010

Cover design: eStudio Calamar Steinen

Preface

For some 30 years, I have taught two "Mathematical Physics" courses. One of them was previously named "Engineering Analysis." There are several textbooks of unquestionable merit for such courses, but I could not find one that fitted our needs. It seemed to me that students might have an easier time if some changes were made in these books. I ended up using class notes. Actually, I felt the same about my own notes, so they got changed again and again. Throughout the years, many students and colleagues have urged me to publish them. I resisted until now, because the topics were not new and I was not sure that my way of presenting them was really much better than others. In recent years, some former students came back to tell me that they still found my notes useful and looked at them from time to time. The fact that they always singled out these courses, among many others I have taught, made me think that besides being kind, they might even mean it. Perhaps, it is worthwhile to share these notes with a wider audience.

It took far more work than expected to transcribe the lecture notes into printed pages. The notes were written in an abbreviated way without much explanation between any two equations, because I was supposed to supply the missing links in person. How much detail I would go into depended on the reaction of the students. Now without them in front of me, I had to decide the appropriate amount of derivation to be included. I chose to err on the side of too much detail rather than too little. As a result, the derivation does not look very elegant, but I also hope it does not leave any gap in students' comprehension.

Precisely stated and elegantly proved theorems looked great to me when I was a young faculty member. But in the later years, I found that elegance in the eyes of the teacher might be stumbling blocks for students. Now I am convinced that before a student can use a mathematical theorem with confidence, he or she must first develop an intuitive feeling. The most effective way to do that is to follow a sufficient number of examples.

This book is written for students who want to learn but need a firm hand-holding. I hope they will find the book readable and easy to learn from.

Learning, as always, has to be done by the student herself or himself. No one can acquire mathematical skill without doing problems, the more the better. However, realistically students have a finite amount of time. They will be overwhelmed if problems are too numerous, and frustrated if problems are too difficult. A common practice in textbooks is to list a large number of problems and let the instructor to choose a few for assignments. It seems to me that is not a confidence building strategy. A self-learning person would not know what to choose. Therefore a moderate number of not overly difficult problems, with answers, are selected at the end of each chapter. Hopefully after the student has successfully solved all of them, he will be encouraged to seek more challenging ones. There are plenty of problems in other books. Of course, an instructor can always assign more problems at levels suitable to the class.

Professor I.I. Rabi used to say "All textbooks are written with the principle of least astonishment." Well, there is a good reason for that. After all, textbooks are supposed to explain away the mysteries and make the profound obvious. This book is no exception. Nevertheless, I still hope the reader will find something in this book exciting.

This set of books is written in the spirit of what Sommerfeld called "physical mathematics." For example, instead of studying the properties of hyperbolic, parabolic, and elliptic partial differential equations, materials on partial differential equations are organized around wave, diffusion and Laplace equations. Physical problems are used as the framework for various mathematical techniques to hang together, rather than as just examples for mathematical theories. In order not to sacrifice the underlying mathematical concepts, these materials are preceded by a chapter on Sturm–Livouville theory in infinite dimensional vector space. It is author's experience that this approach not only stimulates students' intuitive thinking but also increase their confidence in using mathematical tools.

These books are dedicated to my students. I want to thank my A and B students, their diligence and enthusiasm have made teaching enjoyable and worthwhile. I want to thank my C and D students, their difficulties and mistakes made me search for better explanations.

I want to thank Brad Oraw for drawing many figures in this book and Mathew Hacker for helping me to typeset the manuscript.

I want to express my deepest gratitude to Professor S.H. Patil, Indian Institute of Technology, Bombay. He has read the entire manuscript and provided many excellent suggestions. He has also checked the equations and the problems and corrected numerous errors.

The responsibility for remaining errors is, of course, entirely mine. I will greatly appreciate if they are brought to my attention.

Tacoma, Washington   *K.T. Tang*
June 2006

Contents

Part I

Fourier Analysis

1

Fourier Series

One of the most useful tools of mathematical analysis is Fourier series, named after the French mathematical physicist Jean Baptiste Joseph Fourier (1768–1830). Fourier analysis is ubiquitous in almost all fields of physical sciences.

In 1822, Fourier in his work on heat flow made a remarkable assertion that every function $f(x)$ with period 2π can be represented by a trigonometric infinite series of the form

$$f(x) = \frac{1}{2}a_0 + \sum_{n=1}^{\infty}(a_n \cos nx + b_n \sin nx). \tag{1.1}$$

We now know that, with very little restrictions on the function, this is indeed the case. An infinite series of this form is called a Fourier series. The series was originally proposed for the solutions of partial differential equations with boundary (and/or initial) conditions. While it is still one of the most powerful methods for such problems, as we shall see in later chapters, its usefulness has been extended far beyond the problem of heat conduction. Fourier series is now an essential tool for the analysis of all kinds of wave forms, ranging from signal processing to quantum particle waves.

1.1 Fourier Series of Functions with Periodicity 2π

1.1.1 Orthogonality of Trigonotric Functions

To discuss Fourier series, we need the following integrals. If m and n are integers, then

$$\int_{-\pi}^{\pi} \cos mx \, dx = 0, \tag{1.2}$$

$$\int_{-\pi}^{\pi} \sin mx \, dx = 0, \tag{1.3}$$

$$\int_{-\pi}^{\pi} \cos mx \sin nx \, dx = 0, \tag{1.4}$$

$$\int_{-\pi}^{\pi} \cos mx \cos nx \, dx = \begin{cases} 0 & m \neq n, \\ \pi & m = n \neq 0, \\ 2\pi & m = n = 0, \end{cases} \tag{1.5}$$

$$\int_{-\pi}^{\pi} \sin mx \sin nx \, dx = \begin{cases} 0 & m \neq n, \\ \pi & m = n. \end{cases} \tag{1.6}$$

The first two integrals are trivial, either by direct integration or by noting that any trigonometric function integrated over a whole period will give zero since the positive part will cancel the negative part. The rest of the integrals can be shown by using the trigonometry formulas for products and then integrating. An easier way is to use the complex forms

$$\int_{-\pi}^{\pi} \cos mx \sin nx \, dx = \int_{-\pi}^{\pi} \frac{e^{imx} + e^{-imx}}{2} \frac{e^{inx} - e^{-inx}}{2i} \, dx.$$

We can see the results without actually multiplying out. All terms in the product are of the form e^{ikx}, where k is an integer. Since

$$\int_{-\pi}^{\pi} e^{ikx} \, dx = \frac{1}{ik} \left[e^{ikx} \right]_{-\pi}^{\pi} = 0,$$

it follows that all integrals in the product are zero. Similarly

$$\int_{-\pi}^{\pi} \cos mx \cos nx \, dx = \int_{-\pi}^{\pi} \frac{e^{imx} + e^{-imx}}{2} \frac{e^{inx} + e^{-inx}}{2} \, dx$$

is identically zero except $n = m$, in that case

$$\int_{-\pi}^{\pi} \cos mx \cos mx \, dx = \int_{-\pi}^{\pi} \frac{e^{i2mx} + 2 + e^{-i2mx}}{4} \, dx$$

$$= \int_{-\pi}^{\pi} \frac{1}{2} \left[1 + \cos 2mx \right] \, dx = \begin{cases} \pi & m \neq 0, \\ 2\pi & m = 0. \end{cases}$$

In the same way we can show that if $n \neq m$,

$$\int_{-\pi}^{\pi} \sin mx \sin nx \, dx = 0$$

and if $n = m$,

$$\int_{-\pi}^{\pi} \sin mx \sin mx \, dx = \int_{-\pi}^{\pi} \frac{1}{2} \left[1 - \cos 2mx \right] \, dx = \pi.$$

This concludes the proof of (1.2)–(1.6).

In general, if any two members ψ_n, ψ_m of a set of functions $\{\psi_i\}$ satisfy the condition

$$\int_a^b \psi_n(x)\psi_m(x)\mathrm{d}x = 0 \quad \text{if } n \neq m, \tag{1.7}$$

then ψ_n and ψ_m are said to be orthogonal, and (1.7) is known as the orthogonal condition in the interval between a and b. The set $\{\psi_i\}$ is an orthogonal set over the same interval.

Thus if the members of the set of trigonometric functions are

$$1, \ \cos x, \ \sin x, \ \cos 2x, \ \sin 2x, \ \cos 3x, \ \sin 3x, \ldots,$$

then this is an orthogonal set in the interval from $-\pi$ to π.

1.1.2 The Fourier Coefficients

If $f(x)$ is a periodic function of period 2π, i.e.,

$$f(x + 2\pi) = f(x)$$

and it is represented by the Fourier series of the form (1.1), the coefficients a_n and b_n can be found in the following way.

We multiply both sides of (1.1) by $\cos mx$, where m is an positive integer

$$f(x) \cos mx = \frac{1}{2}a_0 \cos mx + \sum_{n=1}^{\infty}(a_n \cos nx \cos mx + b_n \sin nx \cos mx).$$

This series can be integrated term by term

$$\int_{-\pi}^{\pi} f(x) \cos mx \, \mathrm{d}x = \frac{1}{2}a_0 \int_{-\pi}^{\pi} \cos mx \, \mathrm{d}x + \sum_{n=1}^{\infty} a_n \int_{-\pi}^{\pi} \cos nx \cos mx \, \mathrm{d}x$$

$$+ \sum_{n=1}^{\infty} b_n \int_{-\pi}^{\pi} \sin nx \cos mx \, \mathrm{d}x.$$

From the integrals we have discussed, we see that all terms associated with b_n will vanish and all terms associated with a_n will also vanish except the term with $n = m$, and that term is given by

$$\int_{-\pi}^{\pi} f(x) \cos mx \, \mathrm{d}x = \begin{cases} \dfrac{1}{2}a_0 \displaystyle\int_{-\pi}^{\pi} \mathrm{d}x = a_0\pi & \text{for } m = 0, \\[3ex] a_m \displaystyle\int_{-\pi}^{\pi} \cos^2 mx \, \mathrm{d}x = a_m\pi & \text{for } m \neq 0. \end{cases}$$

These relations permit us to calculate any desired coefficient a_m including a_0 when the function $f(x)$ is known.

The coefficients b_m can be similarly obtained. The expansion is multiplied by $\sin mx$ and then integrated term by term. Orthogonality relations yield

$$\int_{-\pi}^{\pi} f(x) \sin mx \, dx = b_m \pi.$$

Since m can be any integer, it follows that a_n (including a_0) and b_n are given by

$$a_n = \frac{1}{\pi} \int_{-\pi}^{\pi} f(x) \cos nx \, dx, \tag{1.8}$$

$$b_n = \frac{1}{\pi} \int_{-\pi}^{\pi} f(x) \sin nx \, dx. \tag{1.9}$$

These coefficients are known as the Euler formulas for Fourier coefficients, or simply as the *Fourier coefficients*.

In essence, Fourier series decomposes the periodic function into cosine and sine waves. From the procedure, it can be observed that:

- The first term $\frac{1}{2}a_0$ represents the average value of $f(x)$ over a period 2π.
- The term $a_n \cos nx$ represents the cosine wave with amplitude a_n. Within one period 2π, there are n complete cosine waves.
- The term $b_n \sin nx$ represents the sine wave with amplitude b_n, and n is the number of complete sine wave in one period 2π.
- In general a_n and b_n can be expected to decrease as n increases.

1.1.3 Expansion of Functions in Fourier Series

Before we discuss the validity of the Fourier series, let us use the following example to show that it is possible to represent a periodic function with period 2π by a Fourier series, provided enough terms are taken.

Suppose we want to expand the square-wave function, shown in Fig. 1.1, into a Fourier series.

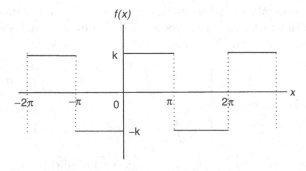

Fig. 1.1. A square-wave function

This function is periodic with period 2π. It can be defined as

$$f(x) = \begin{cases} -k & -\pi < x < 0 \\ k & 0 < x < \pi \end{cases}, \quad f(x + 2\pi) = f(x).$$

To find the coefficients of the Fourier series of this function

$$f(x) = \frac{1}{2}a_0 + \sum_{n=1}^{\infty}(a_n \cos nx + b_n \sin nx)$$

it is always a good idea to calculate a_0 separately, since it is given by simple integral. In this case

$$a_0 = \frac{1}{\pi}\int_{-\pi}^{\pi} f(x)\mathrm{d}x = 0$$

can be seen without integration, since the area under the curve of $f(x)$ between $-\pi$ and π is zero. For the rest of the coefficients, they are given by (1.8) and (1.9). To carry out these integrations, we have to split each of them into two integrals because $f(x)$ is defined by two different formulas on the intervals $(-\pi, 0)$ and $(0, \pi)$. From (1.8)

$$a_n = \frac{1}{\pi}\int_{-\pi}^{\pi} f(x)\cos nx \, \mathrm{d}x = \frac{1}{\pi}\left[\int_{-\pi}^{0}(-k)\cos nx \, \mathrm{d}x + \int_{0}^{\pi} k\cos nx \, \mathrm{d}x\right]$$

$$= \frac{1}{\pi}\left\{\left[-k\frac{\sin nx}{n}\right]_{-\pi}^{0} + \left[k\frac{\sin nx}{n}\right]_{0}^{\pi}\right\} = 0.$$

From (1.9)

$$b_n = \frac{1}{\pi}\int_{-\pi}^{\pi} f(x)\sin nx \, \mathrm{d}x = \frac{1}{\pi}\left[\int_{-\pi}^{0}(-k)\sin nx \, \mathrm{d}x + \int_{0}^{\pi} k\sin nx \, \mathrm{d}x\right]$$

$$= \frac{1}{\pi}\left\{\left[k\frac{\cos nx}{n}\right]_{-\pi}^{0} + \left[-k\frac{\cos nx}{n}\right]_{0}^{\pi}\right\} = \frac{2k}{n\pi}(1 - \cos n\pi)$$

$$= \frac{2k}{n\pi}(1 - (-1)^n) = \begin{cases} \dfrac{4k}{n\pi} & \text{if } n \text{ is odd,} \\ 0 & \text{if } n \text{ is even.} \end{cases}$$

With these coefficients, the Fourier series becomes

$$f(x) = \frac{4k}{\pi}\sum_{n \text{ odd}}\frac{1}{n}\sin nx$$

$$= \frac{4k}{\pi}\left(\sin x + \frac{1}{3}\sin 3x + \frac{1}{5}\sin 5x + \cdots\right). \tag{1.10}$$

Alternatively this series can be written as

$$f(x) = \frac{4k}{\pi}\sum_{n=1}^{\infty}\frac{1}{2n-1}\sin(2n-1)x.$$

To examine the convergence of this series, let us define the partial sums as

$$S_N = \frac{4k}{\pi} \sum_{n=1}^{N} \frac{1}{2n-1} \sin(2n-1)x.$$

In other words, S_N is the sum the first N terms of the Fourier series. S_1 is simply the first term $\frac{4k}{\pi} \sin x$, S_2 is the sum of the first two terms $\frac{4k}{\pi} (\sin x + \frac{1}{3} \sin 3x)$, etc.

In Fig. 1.2a, the first three partial sums are shown in the right column, the individual terms in these sums are shown in the left column. It is seen that S_N gets closer to $f(x)$ as N increases, although the contributions of the

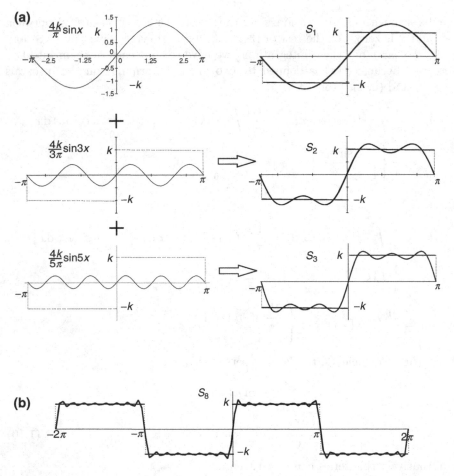

Fig. 1.2. The convergence of a Fourier series expansion of a square-wave function. (a) The first three partial sums are shown in the *right*; the individual terms in these sums are shown in the *left*. (b) The sum of the first eight terms of the Fourier series of the function

individual terms are steadily decreasing as n gets larger. In Fig. 1.2b, we show
the result of S_8. With eight terms, the partial sum already looks very similar
to the square-wave function. We notice that at the points of discontinuity
$x = -\pi, x = 0$, and $x = \pi$, all the partial sums have the value zero, which
is the average of the values of k and $-k$ of the function. Note also that as x
approaches a discontinuity of $f(x)$ from either side, the value of $S_N(x)$ tends
to overshoot the value of $f(x)$, in this case $-k$ or $+k$. As N increases, the
overshoots (about 9% of the discontinuity) are pushed closer to the points of
discontinuity, but they will not disappear even if N goes to infinity. This behavior of a Fourier series near a point of discontinuity of its function is known
as *Gibbs' phenomenon*.

1.2 Convergence of Fourier Series

1.2.1 Dirichlet Conditions

The conditions imposed on $f(x)$ to make (1.1) valid are stated in the following
theorem.

Theorem 1.2.1. *If a periodic function $f(x)$ of period 2π is bounded and piecewise continuous, and has a finite number of maxima and minima in each
period, then the trigonometric series*

$$\frac{1}{2}a_0 + \sum_{n=1}^{\infty}(a_n \cos nx + b_n \sin nx)$$

with

$$a_n = \frac{1}{\pi}\int_{-\pi}^{\pi} f(x)\cos nx\,\mathrm{d}x, \quad n = 0, 1, 2, \ldots$$

$$b_n = \frac{1}{\pi}\int_{-\pi}^{\pi} f(x)\sin nx\,\mathrm{d}x, \quad n = 1, 2, \ldots$$

*converges to $f(x)$ where $f(x)$ is continuous, and it converges to the average
of the left- and right-hand limits of $f(x)$ at points of discontinuity.*

A proof of this theorem may be found in G.P. Tolstov, *Fourier Series*,
Dover, New York, 1976.

As long as $f(t)$ is periodic, the choice of the symmetric upper and
lower integration limits $(-\pi, \pi)$ is not essential. Any interval of 2π, such as
$(x_0, x_0 + 2\pi)$ will give the same result.

The conditions of convergence were first proved by the German mathematician P.G. Lejeune Dirichlet (1805–1859), and therefore known as Dirichlet
conditions. These conditions impose very little restrictions on the function.
Furthermore, these are only sufficient conditions. It is known that certain
function that does not satisfy these conditions can also be represented by the

Fourier series. The minimum necessary conditions for its convergence are not known. In any case, it can be safely assumed that functions of interests in physical problems can all be represented by their Fourier series.

1.2.2 Fourier Series and Delta Function

(For those who have not yet studied complex contour integration, this section can be skipped.)

Instead of proving the convergence theorem, we will use a delta function to explicitly demonstrate that the Fourier series

$$S_\infty(x) = \frac{1}{2}a_0 + \sum_{n=1}^{\infty}(a_n \cos nx + b_n \sin nx)$$

converges to $f(x)$.

With a_n and b_n given by (1.8) and (1.9), $S_\infty(x)$ can be written as

$$S_\infty(x) = \frac{1}{2\pi}\int_{-\pi}^{\pi} f(x')\mathrm{d}x' + \frac{1}{\pi}\sum_{n=1}^{\infty}\left(\int_{-\pi}^{\pi} f(x') \cos nx'\mathrm{d}x'\right) \cos nx$$

$$+\frac{1}{\pi}\sum_{n=1}^{\infty}\left(\int_{-\pi}^{\pi} f(x') \sin nx'\mathrm{d}x'\right) \sin nx$$

$$= \int_{-\pi}^{\pi} f(x')\left[\frac{1}{2\pi} + \frac{1}{\pi}\sum_{n=1}^{\infty}(\cos nx' \cos nx + \sin nx' \sin nx)\right]\mathrm{d}x'$$

$$= \int_{-\pi}^{\pi} f(x')\left[\frac{1}{2\pi} + \frac{1}{\pi}\sum_{n=1}^{\infty}\cos n(x' - x)\right]\mathrm{d}x'.$$

If the cosine series

$$D(x' - x) = \frac{1}{2\pi} + \frac{1}{\pi}\sum_{n=1}^{\infty}\cos n(x' - x)$$

behaves like a delta function $\delta(x' - x)$, then $S_\infty(x) = f(x)$ because

$$\int_{-\pi}^{\pi} f(x')\delta(x' - x)\mathrm{d}x' = f(x) \quad \text{for} \quad -\pi < x < \pi.$$

Recall that the delta function $\delta(x' - x)$ can be defined as

$$\delta(x' - x) = \begin{cases} 0 & x' \neq x \\ \infty & x' = x \end{cases},$$

$$\int_{-\pi}^{\pi} \delta(x' - x)\mathrm{d}x' = 1 \quad \text{for} \quad -\pi < x < \pi.$$

Now we will show that indeed $D(x' - x)$ has these properties. First, to ensure the convergence, we write the cosine series as

$$D(x' - x) = \lim_{\gamma \to 1^-} D_\gamma(x' - x),$$

$$D_\gamma(x' - x) = \frac{1}{\pi} \left[\frac{1}{2} + \sum_{n=1}^{\infty} \gamma^n \cos n(x' - x) \right],$$

where the limit $\gamma \to 1^-$ means that γ approaches one from below, i.e., γ is infinitely close to 1, but is always less than 1. To sum this series, it is advantageous to regard $D_\gamma(x' - x)$ as the real part of the complex series

$$D_\gamma(x' - x) = \mathrm{Re} \left[\frac{1}{\pi} \left(\frac{1}{2} + \sum_{n=1}^{\infty} \gamma^n e^{in(x'-x)} \right) \right].$$

Since

$$\frac{1}{1 - \gamma e^{i(x'-x)}} = 1 + \gamma e^{i(x'-x)} + \gamma^2 e^{i2(x'-x)} + \cdots,$$

$$\frac{\gamma e^{i(x'-x)}}{1 - \gamma e^{i(x'-x)}} = \gamma e^{i(x'-x)} + \gamma^2 e^{i2(x'-x)} + \gamma^3 e^{i3(x'-x)} + \cdots,$$

so

$$\frac{1}{2} + \sum_{n=1}^{\infty} \gamma^n e^{in(x'-x)} = \frac{1}{2} + \frac{\gamma e^{i(x'-x)}}{1 - \gamma e^{i(x'-x)}}$$

$$= \frac{1 + \gamma e^{i(x'-x)}}{2(1 - \gamma e^{i(x'-x)})} = \frac{1 + \gamma e^{i(x'-x)}}{2(1 - \gamma e^{i(x'-x)})} \frac{1 - \gamma e^{-i(x'-x)}}{1 - \gamma e^{-i(x'-x)}}$$

$$= \frac{1 - \gamma^2 + \gamma e^{i(x'-x)} - \gamma e^{-i(x'-x)}}{2[1 - \gamma(e^{i(x'-x)} + e^{-i(x'-x)}) + \gamma^2]} = \frac{1 - \gamma^2 + i2\gamma \sin(x' - x)}{2[1 - 2\gamma \cos(x' - x) + \gamma^2]}.$$

Thus

$$D_\gamma(x' - x) = \mathrm{Re} \left[\frac{1 - \gamma^2 + i2\gamma \sin(x' - x)}{2\pi[1 - 2\gamma \cos(x' - x) + \gamma^2]} \right]$$

$$= \frac{1 - \gamma^2}{2\pi[1 - 2\gamma \cos(x' - x) + \gamma^2]}.$$

Clearly, if $x' \neq x$,

$$D(x' - x) = \lim_{\gamma \to 1} \frac{1 - \gamma^2}{2\pi[1 - 2\gamma \cos(x' - x) + \gamma^2]} = 0.$$

If $x' = x$, then $\cos(x' - x) = 1$, and

$$\frac{1 - \gamma^2}{2\pi[1 - 2\gamma\cos(x' - x) + \gamma^2]} = \frac{1 - \gamma^2}{2\pi[1 - 2\gamma + \gamma^2]}$$

$$= \frac{(1 - \gamma)(1 + \gamma)}{2\pi[1 - \gamma]^2} = \frac{1 + \gamma}{2\pi(1 - \gamma)}.$$

It follows that

$$D(x' - x) = \lim_{\gamma \to 1} \frac{1 + \gamma}{2\pi(1 - \gamma)} \to \infty, \quad x' = x.$$

Furthermore

$$\int_{-\pi}^{\pi} D_\gamma(x' - x)dx' = \frac{1 - \gamma^2}{2\pi} \int_{-\pi}^{\pi} \frac{dx'}{(1 + \gamma^2) - 2\gamma\cos(x' - x)}.$$

We have shown in the chapter on the theory of residue (see Example 3.5.2 of Volume 1) that

$$\oint \frac{d\theta}{a - b\cos\theta} = \frac{2\pi}{\sqrt{a^2 - b^2}}, \quad a > b.$$

With a substitution $x' - x = \theta$,

$$\int_{-\pi}^{\pi} \frac{dx'}{(1 + \gamma^2) - 2\gamma\cos(x' - x)} = \oint \frac{d\theta}{(1 + \gamma^2) - 2\gamma\cos\theta}.$$

As long as γ is not exactly one, $1 + \gamma^2 > 2\gamma$, so

$$\oint \frac{d\theta}{(1 + \gamma^2) - 2\gamma\cos\theta} = \frac{2\pi}{\sqrt{(1 + \gamma^2)^2 - 4\gamma^2}} = \frac{2\pi}{1 - \gamma^2}.$$

Therefore

$$\int_{-\pi}^{\pi} D_\gamma(x' - x)dx' = \frac{1 - \gamma^2}{2\pi} \frac{2\pi}{1 - \gamma^2} = 1.$$

This concludes our proof that $D(x' - x)$ behaves like the delta function $\delta(x' - x)$. Therefore if $f(x)$ is continuous, then the Fourier series converges to $f(x)$,

$$S_\infty(x) = \int_{-\pi}^{\pi} f(x')D(x' - x)dx' = f(x).$$

Suppose that $f(x)$ is discontinuous at some point x, and that $f(x^+)$ and $f(x^-)$ are the limiting values as we approach x from the right and from the left. Then in evaluating the last integral, half of $D(x' - x)$ is multiplied by $f(x^+)$ and half by $f(x^-)$, as shown in the following figure.

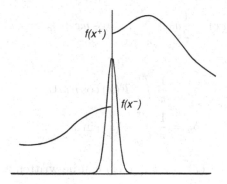

Therefore the last equation becomes

$$S_\infty(x) = \frac{1}{2}[f(x^+) + f(x^-)].$$

Thus at points where $f(x)$ is continuous, the Fourier series gives the value of $f(x)$, and at points where $f(x)$ is discontinuous, the Fourier series gives the mean value of the right and left limits of $f(x)$.

1.3 Fourier Series of Functions of any Period

1.3.1 Change of Interval

So far attention has been restricted to functions of period 2π. This restriction may easily be relaxed. If $f(t)$ is periodic with a period $2L$, we can make a change of variable

$$t = \frac{L}{\pi}x$$

and let

$$f(t) = f\left(\frac{L}{\pi}x\right) \equiv F(x).$$

By this definition,

$$f(t + 2L) = f\left(\frac{L}{\pi}x + 2L\right) = f\left(\frac{L}{\pi}[x + 2\pi]\right) = F(x + 2\pi).$$

Since $f(t)$ is a periodic function with a period $2L$

$$f(t + 2L) = f(t)$$

it follows that:

$$F(x + 2\pi) = F(x).$$

So $F(x)$ is periodic with a period 2π.

We can expand $F(x)$ into a Fourier series, then transform back to a function of t

$$F(x) = \frac{1}{2}a_0 + \sum_{n=1}^{\infty}(a_n \cos nx + b_n \sin nx) \qquad (1.11)$$

with

$$a_n = \frac{1}{\pi}\int_{-\pi}^{\pi} F(x)\cos nx \, \mathrm{d}x,$$

$$b_n = \frac{1}{\pi}\int_{-\pi}^{\pi} F(x)\sin nx \, \mathrm{d}x.$$

Since $x = \frac{\pi}{L}t$ and $F(x) = f(t)$, (1.11) can be written as

$$f(t) = \frac{1}{2}a_0 + \sum_{n=1}^{\infty}\left(a_n \cos \frac{n\pi}{L}t + b_n \sin \frac{n\pi}{L}t\right) \qquad (1.12)$$

and the coefficients can also be expressed as integrals over t. Changing the integration variable from x to t with $\mathrm{d}x = \frac{\pi}{L}\mathrm{d}t$, we have

$$a_n = \frac{1}{L}\int_{-L}^{L} f(t)\cos\left(\frac{n\pi}{L}t\right)\mathrm{d}t, \qquad (1.13)$$

$$b_n = \frac{1}{L}\int_{-L}^{L} f(t)\sin\left(\frac{n\pi}{L}t\right)\mathrm{d}t. \qquad (1.14)$$

Kronecker's method. As a practical matter, very often $f(t)$ is in the form of t^k, $\sin kt$, $\cos kt$, or e^{kt} for various integer values of k. We will have to carry out the integrations of the type

$$\int t^k \cos \frac{n\pi t}{L}\mathrm{d}t, \quad \int \sin kt \cos \frac{n\pi t}{L}\mathrm{d}t.$$

These integrals can be evaluated by repeated integration by parts. The following systematic approach is helpful in reducing the tedious details inherent in such computation. Consider the integral

$$\int f(t)g(t)\mathrm{d}t$$

and let

$$g(t)\mathrm{d}t = \mathrm{d}G(t), \quad \text{then} \quad G(t) = \int g(t)\mathrm{d}t.$$

With integration by parts, one gets

$$\int f(t)g(t)\mathrm{d}t = f(t)G(t) - \int f'(t)G(t)\mathrm{d}t.$$

Continuing this process, with

$$G_1(t) = \int G(t)\mathrm{d}t, \quad G_2(t) = \int G_1(t)\mathrm{d}t, \ldots, G_n(t) = \int G_{n-1}(t)\mathrm{d}t,$$

we have

$$\int f(t)g(t)\mathrm{d}t = f(t)G(t) - f'(t)G_1(t) + \int f''(t)G_1(t)\mathrm{d}t \qquad (1.15)$$

$$= f(t)G(t) - f'(t)G_1(t) + f''(t)G_2(t) - f'''(t)G_3(t) + \cdots . \qquad (1.16)$$

This procedure is known as Kronecker's method.

Now if $f(t) = t^k$, then

$$f'(t) = kt^{k-1}, \ldots, f^k(t) = k!, \ f^{k+1}(t) = 0,$$

the above expression will terminate. Furthermore, if $g(t) = \cos\frac{n\pi t}{L}$, then

$$G(t) = \int \cos\frac{n\pi t}{L}\mathrm{d}t = \left(\frac{L}{n\pi}\right)\sin\frac{n\pi t}{L},$$

$$G_1(t) = \left(\frac{L}{n\pi}\right)\int \sin\frac{n\pi t}{L}\mathrm{d}t = -\left(\frac{L}{n\pi}\right)^2\cos\frac{n\pi t}{L},$$

$$G_2(t) = -\left(\frac{L}{n\pi}\right)^3\sin\frac{n\pi t}{L}, \quad G_3(t) = \left(\frac{L}{n\pi}\right)^4\cos\frac{n\pi t}{L},\ldots.$$

Similarly, if $g(t) = \sin\frac{n\pi t}{L}$, then

$$G(t) = \int \sin\frac{n\pi t}{L}\mathrm{d}t = -\left(\frac{L}{n\pi}\right)\cos\frac{n\pi t}{L}, \quad G_1(t) = -\left(\frac{L}{n\pi}\right)^2\sin\frac{n\pi t}{L},$$

$$G_2(t) = \left(\frac{L}{n\pi}\right)^3\cos\frac{n\pi t}{L}, \quad G_3(t) = \left(\frac{L}{n\pi}\right)^4\sin\frac{n\pi t}{L},\ldots.$$

Thus

$$\int_a^b t^k\cos\frac{n\pi t}{L}\mathrm{d}t = \left[\frac{L}{n\pi}t^k\sin\frac{n\pi t}{L} + \left(\frac{L}{n\pi}\right)^2 kt^{k-1}\cos\frac{n\pi t}{L}\right.$$

$$\left. -\left(\frac{L}{n\pi}\right)^3 k(k-1)t^{k-2}\sin\frac{n\pi t}{L} + \cdots\right]_a^b \qquad (1.17)$$

and

$$\int_a^b t^k\sin\frac{n\pi t}{L}\mathrm{d}t = \left[-\frac{L}{n\pi}t^k\cos\frac{n\pi t}{L} + \left(\frac{L}{n\pi}\right)^2 kt^{k-1}\sin\frac{n\pi t}{L}\right.$$

$$\left. +\left(\frac{L}{n\pi}\right)^3 k(k-1)t^{k-2}\cos\frac{n\pi t}{L} + \cdots\right]_a^b . \qquad (1.18)$$

If $f(t) = \sin kt$, then

$$f'(t) = k\cos kt, \quad f''(t) = -k^2\sin kt.$$

we can use (1.15) to write

$$\int_a^b \sin kt \cos \frac{n\pi}{L} t \, dt = \left[\frac{L}{n\pi} \sin kt \sin \frac{n\pi t}{L} + k \left(\frac{L}{n\pi} \right)^2 \cos kt \cos \frac{n\pi t}{L} \right]_a^b$$

$$+ k^2 \left(\frac{L}{n\pi} \right)^2 \int_a^b \sin kt \cos \frac{n\pi}{L} t \, dt.$$

Combining the last term with the left-hand side, we have

$$\left[1 - k^2 \left(\frac{L}{n\pi} \right)^2 \right] \int_a^b \sin kt \cos \frac{n\pi}{L} t \, dt$$

$$= \left[\frac{L}{n\pi} \sin kt \sin \frac{n\pi t}{L} + k \left(\frac{L}{n\pi} \right)^2 \cos kt \cos \frac{n\pi t}{L} \right]_a^b$$

or

$$\int_a^b \sin kt \cos \frac{n\pi}{L} t \, dt$$

$$= \frac{(n\pi)^2}{(n\pi)^2 - (kL)^2} \left[\frac{L}{n\pi} \sin kt \sin \frac{n\pi t}{L} + k \left(\frac{L}{n\pi} \right)^2 \cos kt \cos \frac{n\pi t}{L} \right]_a^b.$$

Clearly, integrals such as

$$\int_a^b \sin kt \sin \frac{n\pi}{L} t \, dt, \quad \int_a^b \cos kt \cos \frac{n\pi}{L} t \, dt, \quad \int_a^b \cos kt \sin \frac{n\pi}{L} t \, dt,$$

$$\int_a^b e^{kt} \cos \frac{n\pi}{L} t \, dt, \quad \int_a^b e^{kt} \sin \frac{n\pi}{L} t \, dt$$

can similarly be integrated.

Example 1.3.1. Find the Fourier series for $f(t)$ which is defined as

$$f(t) = t \quad \text{for} \ -L < t \le L, \quad \text{and} \quad f(t + 2L) = f(t).$$

Solution 1.3.1.

$$f(t) = \frac{1}{2} a_0 + \sum_{n=1}^{\infty} \left(a_n \cos \frac{n\pi t}{L} + b_n \sin \frac{n\pi t}{L} \right),$$

$$a_0 = \frac{1}{L} \int_{-L}^{L} t \, dt = 0,$$

$$a_n = \frac{1}{L} \int_{-L}^{L} t \cos \frac{n\pi t}{L} dt = \frac{1}{L} \left[\frac{L}{n\pi} t \sin \frac{n\pi t}{L} + \left(\frac{L}{n\pi} \right)^2 \cos \frac{n\pi t}{L} \right]_{-L}^{L} = 0,$$

$$b_n = \frac{1}{L} \int_{-L}^{L} t \sin \frac{n\pi t}{L} dt$$

$$= \frac{1}{L} \left[-\frac{L}{n\pi} t \cos \frac{n\pi t}{L} + \left(\frac{L}{n\pi}\right)^2 \sin \frac{n\pi t}{L} \right]_{-L}^{L} = -\frac{2L}{n\pi} \cos n\pi.$$

Thus

$$f(t) = \frac{2L}{\pi} \sum_{n=1}^{\infty} -\frac{1}{n} \cos n\pi \sin \frac{n\pi t}{L} = \frac{2L}{\pi} \sum_{n=1}^{\infty} \frac{(-1)^{n+1}}{n} \sin \frac{n\pi t}{L}$$

$$= \frac{2L}{\pi} \left(\sin \frac{\pi t}{L} - \frac{1}{2} \sin \frac{2\pi t}{L} + \frac{1}{3} \sin \frac{3\pi t}{L} - \cdots \right). \tag{1.19}$$

The convergence of this series is shown in Fig. 1.3, where S_N is the partial sum defined as

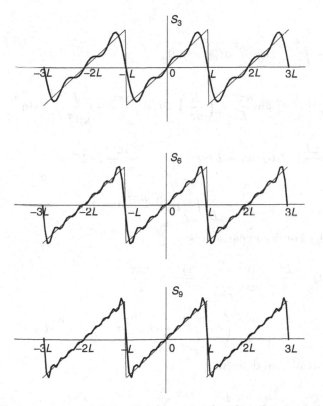

Fig. 1.3. The convergence of the Fourier series for the periodic function whose definition in one period is $f(t) = t$, $-L < t < L$. The first N terms approximations are shown as S_N

$$S_N = \frac{2L}{\pi} \sum_{n=1}^{N} \frac{(-1)^{n+1}}{n} \sin \frac{n\pi t}{L}.$$

Note the increasing accuracy with which the terms approximate the function. With three terms, S_3 already looks like the function. Except for the Gibbs' phenomenon, a very good approximation is obtained with S_9.

Example 1.3.2. Find the Fourier series of the periodic function whose definition in one period is

$$f(t) = t^2 \text{ for } -L < t \le L, \quad \text{and} \quad f(t+2L) = f(t).$$

Solution 1.3.2. The Fourier coefficients are given by

$$a_0 = \frac{1}{L} \int_{-L}^{L} t^2 \mathrm{d}t = \frac{1}{L}\frac{1}{3}[L^3 - (-L)^3] = \frac{2}{3}L^2.$$

$$a_n = \frac{1}{L} \int_{-L}^{L} t^2 \cos \frac{n\pi t}{L} \mathrm{d}t, \quad n \ne 0$$

$$= \frac{1}{L}\left[\frac{L}{n\pi}t^2 \sin\frac{n\pi t}{L} + \left(\frac{L}{n\pi}\right)^2 2t\cos\frac{n\pi t}{L} - \left(\frac{L}{n\pi}\right)^3 2\sin\frac{n\pi t}{L}\right]_{-L}^{L}$$

$$= \frac{2L}{(n\pi)^2}\left[L\cos n\pi + L\cos(-n\pi)\right] = \frac{4L^2}{n^2\pi^2}(-1)^n.$$

$$b_n = \frac{1}{L}\int_{-L}^{L} t^2 \sin\frac{n\pi t}{L}\mathrm{d}t = 0.$$

Therefore the Fourier expansion is

$$f(t) = \frac{L^2}{3} + \frac{4L^2}{\pi^2}\sum_{n=1}^{\infty}\frac{(-1)^n}{n^2}\cos\frac{n\pi t}{L}$$

$$= \frac{L^2}{3} - \frac{4L^2}{\pi^2}\left(\cos\frac{\pi}{L}t - \frac{1}{4}\cos\frac{2\pi}{L}t + \frac{1}{9}\cos\frac{3\pi}{L}t + \cdots\right). \quad (1.20)$$

With the partial sum defined as

$$S_N = \frac{L^2}{3} + \frac{4L^2}{\pi^2}\sum_{n=1}^{N}\frac{(-1)^n}{n^2}\cos\frac{n\pi t}{L},$$

we compare S_3 and S_6 with $f(t)$ in Fig. 1.4.

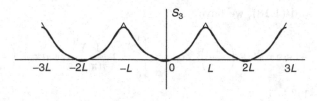

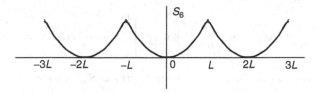

Fig. 1.4. The convergence of the Fourier expansion of the periodic function whose definition in one period is $f(t) = t^2, -L < t \leq L$. The partial sum of S_3 is already a very good approximation

It is seen that S_3 is already a very good approximation of $f(t)$. The difference between S_6 and $f(t)$ is hardly noticeable. This Fourier series converges much faster than that of the previous example. The difference is that $f(t)$ in this problem is continuous not only within the period but also in the extended range, whereas $f(t)$ in the previous example is discontinuous in the extended range.

Example 1.3.3. Find the Fourier series of the periodic function whose definition in one period is

$$f(t) = \begin{cases} 0 & -1 < t < 0 \\ t & 0 < t < 1 \end{cases}, \quad f(t+2) = f(t). \tag{1.21}$$

Solution 1.3.3. The periodicity $2L$ of this function is 2, so $L = 1$, and the Fourier series is given by

$$f(t) = \frac{1}{2}a_0 + \sum_{n=1}^{\infty}[a_n \cos(n\pi t) + b_n \sin(n\pi t)]$$

with

$$a_0 = \int_{-1}^{1} f(t)\mathrm{d}t = \int_{0}^{1} t\,\mathrm{d}t = \frac{1}{2},$$

$$a_n = \int_{-1}^{1} f(t)\cos(n\pi t)\mathrm{d}t = \int_{0}^{1} t\cos(n\pi t)\mathrm{d}t,$$

$$b_n = \int_{-1}^{1} f(t)\sin(n\pi t)\mathrm{d}t = \int_{0}^{1} t\sin(n\pi t)\mathrm{d}t.$$

Using (1.17) and (1.18), we have

$$a_n = \left[\frac{1}{n\pi}t\sin n\pi t + \left(\frac{1}{n\pi}\right)^2 \cos n\pi t\right]_0^1 = \left(\frac{1}{n\pi}\right)^2 \cos n\pi - \left(\frac{1}{n\pi}\right)^2$$

$$= \frac{(-1)^n - 1}{(n\pi)^2},$$

$$b_n = \left[-\frac{1}{n\pi}t\cos n\pi t + \left(\frac{1}{n\pi}\right)^2 \sin n\pi t\right]_0^1 = -\frac{1}{n\pi}\cos n\pi = -\frac{(-1)^n}{n\pi}.$$

Thus the Fourier series for this function is $f(t) = S_\infty$, where

$$S_N = \frac{1}{4} + \sum_{n=1}^{N}\left[\frac{(-1)^n - 1}{(n\pi)^2}\cos n\pi t - \frac{(-1)^n}{n\pi}\sin n\pi t\right].$$

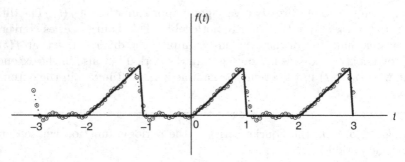

Fig. 1.5. The periodic function of (1.21) is shown together with the partial sum S_5 of its Fourier series. The function is shown as the *solid line* and S_5 as a *line of circles*

In Fig. 1.5 this function (shown as the solid line) is approximated with S_5 which is given by

$$S_5 = \frac{1}{4} - \frac{2}{\pi^2}\cos \pi t - \frac{2}{9\pi^2}\cos 3\pi t - \frac{2}{25\pi^2}\cos 5\pi t$$

$$+ \frac{1}{\pi}\sin \pi t - \frac{1}{2\pi}\sin 2\pi t + \frac{1}{3\pi}\sin 3\pi t - \frac{1}{4\pi}\sin 4\pi t + \frac{1}{5\pi}\sin 5\pi t.$$

While the convergence in this case is not very fast, but it is clear that with sufficient number of terms, the Fourier series can give an accurate representation of this function.

1.3.2 Fourier Series of Even and Odd Functions

If $f(t)$ is a even function, such that

$$f(-t) = f(t),$$

then its Fourier series contains cosine terms only. This can be seen as follows. The b_n coefficients can be written as

$$b_n = \frac{1}{L} \int_{-L}^{0} f(s) \sin\left(\frac{n\pi}{L}s\right) ds + \frac{1}{L} \int_{0}^{L} f(t) \sin\left(\frac{n\pi}{L}t\right) dt. \qquad (1.22)$$

If we make a change of variable and let $s = -t$, the first integral on the right-hand side becomes

$$\frac{1}{L} \int_{-L}^{0} f(s) \sin\left(\frac{n\pi}{L}s\right) ds = \frac{1}{L} \int_{L}^{0} f(-t) \sin\left(-\frac{n\pi}{L}t\right) d(-t)$$

$$= \frac{1}{L} \int_{L}^{0} f(t) \sin\left(\frac{n\pi}{L}t\right) dt,$$

since $\sin(-x) = -\sin(x)$ and $f(-x) = f(x)$. But

$$\frac{1}{L} \int_{L}^{0} f(t) \sin\left(\frac{n\pi}{L}t\right) dt = -\frac{1}{L} \int_{0}^{L} f(t) \sin\left(\frac{n\pi}{L}t\right) dt,$$

which is the negative of the second integral on the right-hand side of (1.22). Therefore $b_n = 0$ for all n.

Following the same procedure and using the fact that $\cos(-x) = \cos(x)$, we find

$$a_n = \frac{1}{L} \int_{-L}^{0} f(s) \cos\left(\frac{n\pi}{L}s\right) ds + \frac{1}{L} \int_{0}^{L} f(t) \cos\left(\frac{n\pi}{L}t\right) dt$$

$$= \frac{1}{L} \int_{L}^{0} f(-t) \cos\left(-\frac{n\pi}{L}\right) d(-t) + \frac{1}{L} \int_{0}^{L} f(t) \cos\left(\frac{n\pi}{L}t\right) dt$$

$$= -\frac{1}{L} \int_{L}^{0} f(t) \cos\left(\frac{n\pi}{L}t\right) dt + \frac{1}{L} \int_{0}^{L} f(t) \cos\left(\frac{n\pi}{L}t\right) dt$$

$$= \frac{2}{L} \int_{0}^{L} f(t) \cos\left(\frac{n\pi}{L}t\right) dt. \qquad (1.23)$$

Hence

$$f(t) = \frac{1}{L} \int_{0}^{L} f(t')dt' + \sum_{n=1}^{\infty} \left[\frac{2}{L} \int_{0}^{L} f(t') \cos\left(\frac{n\pi}{L}t'\right) dt' \right] \cos\frac{n\pi}{L}t. \qquad (1.24)$$

Similarly, if $f(t)$ is an odd function

$$f(-t) = -f(t),$$

then

$$f(t) = \sum_{n=1}^{\infty} \left[\frac{2}{L} \int_0^L f(t') \sin\left(\frac{n\pi}{L}t'\right) dt' \right] \sin\frac{n\pi}{L}t. \qquad (1.25)$$

In the previous examples, the periodic function in Fig. 1.3 is an odd function, therefore its Fourier expansion is a sine series. In Fig. 1.4, the function is an even function, so its Fourier series is a cosine series. In Fig. 1.5, the periodic function has no symmetry, therefore its Fourier series contains both cosine and sine terms.

Example 1.3.4. Find the Fourier series of the function shown in Fig. 1.6.

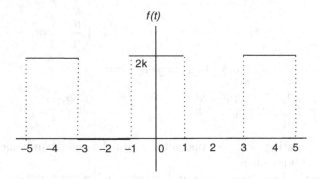

Fig. 1.6. An even square-wave function

Solution 1.3.4. The function shown in Fig. 1.6 can be defined as

$$f(t) = \begin{cases} 0 & \text{if } -2 < t < -1 \\ 2k & \text{if } -1 < t < 1 \\ 0 & \text{if } 1 < t < 2 \end{cases}, \quad f(t) = f(t+4).$$

The period of the function $2L$ is equal to 4, therefore $L = 2$. Furthermore, the function is even, so the Fourier expansion is a cosine series, all coefficients for the sine terms are equal to zero

$$b_n = 0.$$

The coefficients for the cosine series are given by

$$a_0 = \frac{2}{2} \int_0^2 f(t)dt = \int_0^1 2k\,dt = 2k,$$

$$a_n = \frac{2}{2} \int_0^2 f(t) \cos \frac{n\pi t}{2} dt = \int_0^1 2k \cos \frac{n\pi t}{2} dt = \frac{4k}{n\pi} \sin \frac{n\pi}{2}.$$

Thus the Fourier series of $f(t)$ is

$$f(t) = k + \frac{4k}{\pi} \left(\cos \frac{\pi}{2} t - \frac{1}{3} \cos \frac{3\pi}{2} t + \frac{1}{5} \cos \frac{5\pi}{2} t - \cdots \right). \qquad (1.26)$$

It is instructive to compare Fig. 1.6 with Fig. 1.1. Figure 1.6 represents an even function whose Fourier expansion is a cosine series, whereas the function associated with Fig. 1.1 is an odd function and its Fourier series contains only sine terms. Yet they are clearly related. The two figures can be brought to coincide with each other if (a) we move y-axis in Fig. 1.6 one unit to the left (from $t = 0$ to $t = -1$), (b) make a change of variable so that the periodicity is changed from 4 to 2π, (c) shift Fig. 1.6 downward by an amount of k.

The changes in the Fourier series due to these operations are as follows. First let $t' = t + 1$, so that $t = t' - 1$ in (1.26),

$$f(t) = k + \frac{4k}{\pi} \left(\cos \frac{\pi}{2}(t'-1) - \frac{1}{3} \cos \frac{3\pi}{2}(t'-1) + \frac{1}{5} \cos \frac{5\pi}{2}(t'-1) - \cdots \right).$$

Since

$$\cos \frac{n\pi}{2}(t'-1) = \cos \left(\frac{n\pi}{2}t' - \frac{n\pi}{2} \right) = \begin{cases} \sin \dfrac{n\pi}{2} t' & n = 1, 5, 9, \ldots \\ -\sin \dfrac{n\pi}{2} t' & n = 3, 7, 11, \ldots \end{cases},$$

$f(t)$ expressed in terms of t' becomes

$$f(t) = k + \frac{4k}{\pi} \left(\sin \frac{\pi}{2}t' + \frac{1}{3} \sin \frac{3\pi}{2}t' + \frac{1}{5} \sin \frac{5\pi}{2}t' - \cdots \right) = g(t').$$

We call this expression $g(t')$, it still has a periodicity of 4. Next let us make a change of variable $t' = 2x/\pi$, so that the function expressed in terms of x will have a period of 2π,

$$g(t') = k + \frac{4k}{\pi} \left(\sin \frac{\pi}{2} \left(\frac{2x}{\pi} \right) + \frac{1}{3} \sin \frac{3\pi}{2} \left(\frac{2x}{\pi} \right) + \frac{1}{5} \sin \frac{5\pi}{2} \left(\frac{2x}{\pi} \right) - \cdots \right)$$

$$= k + \frac{4k}{\pi} \left(\sin x + \frac{1}{3} \sin 3x + \frac{1}{5} \sin 5x - \cdots \right) = h(x).$$

Finally, shifting it down by k, we have

$$h(x) - k = \frac{4k}{\pi} \left(\sin x + \frac{1}{3} \sin 3x + \frac{1}{5} \sin 5x - \cdots \right).$$

This is the Fourier series (1.10) for the odd function shown in Fig. 1.1.

1.4 Fourier Series of Nonperiodic Functions in Limited Range

So far we have considered only periodic functions extending from $-\infty$ to $+\infty$. In physical applications, often we are interested in the values of a function only in a limited interval. Within that interval the function may not be periodic. For example, in the study of a vibrating string fixed at both ends. There is no condition of periodicity as far as the physical problem is concerned, but there is also no interest in the function beyond the length of the string. Fourier analysis can still be applied to such problem, since we may continue the function outside the desired range so as to make it periodic.

Suppose that the interval of interest in the the function $f(t)$ shown in Fig. 1.7a is between 0 and L. We can extend the function between $-L$ and 0 any way we want. If we extend it first symmetrically as in part (b), then to the entire real line by the periodicity condition $f(t + 2L) = f(t)$, a Fourier series consisting of only cosine terms can be found for the even function. An extension as in part (c) will enable us to find a Fourier sine series for the odd function. Both series would converge to the given $f(t)$ in the interval from 0 to L. Such series expansions are known as half-range expansions. The following examples will illustrate such expansions.

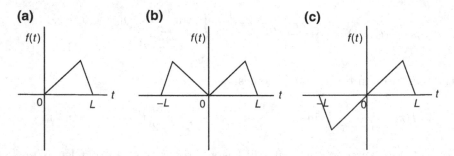

Fig. 1.7. Extension of a function. (a) The function is defined only between 0 and L. (b) A symmetrical extension yields an even function with a periodicity of $2L$. (c) An antisymmetrical extension yields an odd function with a periodicity of $2L$

Example 1.4.1. The function $f(t)$ is defined only over the range $0 < t < 1$ to be

$$f(t) = t - t^2.$$

Find the half-range cosine and sine Fourier expansions of $f(t)$.

Solution 1.4.1. (a) Let the interval $(0,1)$ be half period of the symmetrically extended function, so that $2L = 2$ or $L = 1$. A half-range expansion of this even function is a cosine series

$$f(t) = \frac{1}{2}a_0 + \sum_{n=1} a_n \cos n\pi t$$

with

$$a_0 = 2 \int_0^1 (t - t^2)dt = \frac{1}{3},$$

$$a_n = 2 \int_0^1 (t - t^2) \cos n\pi t\, dt, \quad n \neq 0.$$

Using the Kronecker's method, we have

$$\int_0^1 t \cos n\pi t\, dt - \left[\frac{1}{n\pi} t \sin n\pi t + \left(\frac{1}{n\pi} \right)^2 \cos n\pi t \right]_0^1$$

$$= \left(\frac{1}{n\pi} \right)^2 (\cos n\pi - 1),$$

$$\int_0^1 t^2 \cos n\pi t\, dt = \left[\frac{1}{n\pi} t^2 \sin n\pi t + \left(\frac{1}{n\pi} \right)^2 2t \cos n\pi t - \left(\frac{1}{n\pi} \right)^3 2 \sin n\pi t \right]_0^1$$

$$= 2 \left(\frac{1}{n\pi} \right)^2 \cos n\pi,$$

so

$$a_n = 2 \int_0^1 (t - t^2) \cos n\pi t\, dt = -2 \left(\frac{1}{n\pi} \right)^2 (\cos n\pi + 1).$$

With these coefficients, the half-range Fourier cosine expansion is given by S_∞^{even}, where

$$S_N^{even} = \frac{1}{6} - \frac{2}{\pi^2} \sum_{n=1}^N \frac{(\cos n\pi + 1)}{n^2} \cos n\pi t$$

$$= \frac{1}{6} - \frac{1}{\pi^2} \left(\cos 2\pi t + \frac{1}{4} \cos 4\pi t + \frac{1}{9} \cos 6\pi t + \cdots \right).$$

The convergence of this series is shown in Fig. 1.8a.

(b) A half-range sine expansion would be found by forming an anti-symmetric extension. Since it is an odd function, the Fourier expansion is a sine series

$$f(t) = \sum_{n=1} b_n \sin \pi t$$

with

$$b_n = 2 \int_0^1 (t - t^2) \sin n\pi t\, dt.$$

(a)

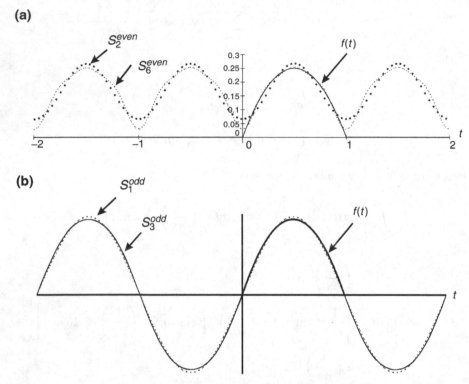

(b)

Fig. 1.8. Convergence of the half-range expansion series. The function $f(t) = t - t^2$ is given between 0 and 1. Both cosine and sine series converge to the function within this range. But outside this range, cosine series converges to an even function shown in (**a**) and sine series converges to an odd function shown in (**b**). S_2^{even} and S_6^{even} are two- and four-term approximations of the cosine series. S_1^{odd} and S_3^{odd} are one- and two-term approximations of the sine series

Now

$$\int_0^1 t \sin n\pi t \, dt = \left[-\frac{1}{n\pi} t \cos n\pi t + \left(\frac{1}{n\pi} \right)^2 \sin n\pi t \right]_0^1 = -\frac{1}{n\pi} \cos n\pi,$$

$$\int_0^1 t^2 \sin n\pi t \, dt = \left[-\frac{1}{n\pi} t^2 \cos n\pi t + \left(\frac{1}{n\pi} \right)^2 2t \sin n\pi t + \left(\frac{1}{n\pi} \right)^3 2 \cos n\pi t \right]_0^1$$

$$= -\frac{1}{n\pi} \cos n\pi + 2 \left(\frac{1}{n\pi} \right)^3 \cos n\pi - 2 \left(\frac{1}{n\pi} \right)^3,$$

so

$$b_n = 2 \int_0^1 (t - t^2) \sin n\pi t \, dt = 4 \left(\frac{1}{n\pi} \right)^3 (1 - \cos n\pi).$$

Therefore the half-range sine expansion is given by S_∞^{odd}, with

$$S_N^{\text{odd}} = \frac{4}{\pi^3} \sum_{n=1}^{N} \frac{(1 - \cos n\pi)}{n^3} \sin n\pi t$$

$$= \frac{8}{\pi^3} \left(\sin \pi t + \frac{1}{27} \sin 3\pi t + \frac{1}{125} \sin 5\pi t + \cdots \right).$$

The convergence of this series is shown in Fig. 1.8b.

It is seen that both the cosine and sine series converge to $t - t^2$ in the range between 0 and 1. Outside this range, the cosine series converges to an even function, and the sine series converges to an odd function. The rate of convergence is also different. For the sine series in (b), with only one term, S_1^{odd} is already very close to $f(t)$. With only two terms, S_3^{odd} (three terms if we include the $n = 2$ term that is equal to zero) is indistinguishable from $f(t)$ in the range of interest. The convergence of the cosine series in (a) is much slower. Although the four-term approximation S_6^{even} is much closer to $f(t)$ than the two-term approximation S_2^{even}, the difference between S_6^{even} and $f(t)$ in the range of interest is still noticeable.

This is generally the case that if we make extension smooth, greater accuracy results for a particular number of terms.

Example 1.4.2. A function $f(t)$ is defined only over the range $0 \leq t \leq 2$ to be $f(t) = t$. Find a Fourier series with only sine terms for this function.

Solution 1.4.2. One can obtain a half-range sine expansion by antisymmetrically extending the function. Such a function is described by

$$f(t) = t \text{ for } -2 < t \leq 2, \quad \text{and} \quad f(t + 4) = f(t).$$

The Fourier series for this function is given by (1.19) with $L = 2$

$$f(t) = \frac{4}{\pi} \sum_{n=1}^{\infty} \frac{(-1)^{n+1}}{n} \sin \frac{n\pi t}{2}.$$

However, this series does not converge to 2, the value of the function at $t = 2$. It converges to 0, the average value of the right- and left-hand limit of the function at $t = 2$, as shown in Fig. 1.3.

We can find a Fourier sine series that converges to the correct value at the end points, if we consider the function

$$f(t) = \begin{cases} t & \text{for } 0 < t \leq 2, \\ 4 - t & \text{for } 2 < t \leq 4. \end{cases}$$

An antisymmetrical extension will give us an odd function with a periodicity of 8 ($2L = 8$, $L = 4$). The Fourier expansion for this function is a sine series

$$f(t) = \sum_{n=1}^{\infty} b_n \sin \frac{n\pi t}{4}$$

with

$$b_n = \frac{2}{4} \int_0^4 f(t) \sin \frac{n\pi t}{4} dt$$

$$= \frac{2}{4} \int_0^2 t \sin \frac{n\pi t}{4} dt + \frac{2}{4} \int_2^4 (4-t) \sin \frac{n\pi t}{4} dt.$$

Using the Kronecker's method, we have

$$b_n = \frac{1}{2} \left[-\frac{4}{n\pi} t \cos \frac{n\pi t}{4} + \left(\frac{4}{n\pi}\right)^2 \sin \frac{n\pi t}{4} \right]_0^2 + 2 \left[-\frac{4}{n\pi} \cos \frac{n\pi t}{4} \right]_2^4$$

$$- \frac{1}{2} \left[-\frac{4}{n\pi} t \cos \frac{n\pi t}{4} + \left(\frac{4}{n\pi}\right)^2 \sin \frac{n\pi t}{4} \right]_2^4$$

$$= \left(\frac{4}{n\pi}\right)^2 \sin \frac{n\pi}{2}.$$

Thus

$$f(t) = \sum_{n=1}^{\infty} \left(\frac{4}{n\pi}\right)^2 \sin \frac{n\pi}{2} \sin \frac{n\pi t}{4}$$

$$= \frac{16}{\pi^2} \left[\sin \frac{\pi t}{4} - \frac{1}{9} \sin \frac{3\pi t}{4} + \frac{1}{25} \sin \frac{5\pi t}{4} - \cdots \right]. \qquad (1.27)$$

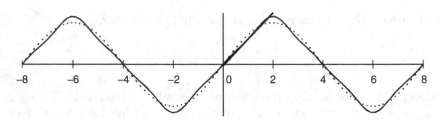

Fig. 1.9. Fourier series for a function defined in a limited range. Within the range $0 \leq t \leq 2$, the series (1.27) converges to $f(t) = t$. Outside this range the series converges to a odd periodic function with a periodicity of 8

Within the range of $0 \leq t \leq 2$, this sine series converges to $f(t) = t$. Outside this range, this series converges to an odd periodic function shown in Fig. 1.9. It converges much faster than the series in (1.19). The first term, shown as dashed line, already provides a reasonable approximation. The difference between the three-term approximation and the given function is hardly noticeable.

As we have seen, for a function that is defined only in a limited range, it is possible to have many different Fourier series. They all converge to the function in the given range, although their rate of convergence may be different. Fortunately, in physical applications, the question of which series we should use for the description the function is usually determined automatically by the boundary conditions.

From all the examples so far, we make the following observations:

- If the function is discontinuous at some point, the Fourier coefficients are decreasing as $1/n$.
- If the function is continuous but its first derivative is discontinuous at some point, the Fourier coefficients are decreasing as $1/n^2$.
- If the function and its first derivative are continuous, the Fourier coefficients are decreasing as $1/n^3$.

Although these comments are based on a few examples, they are generally valid (see the Method of Jumps for the Fourier Coefficients). It is useful to keep them in mind when calculating Fourier coefficients.

1.5 Complex Fourier Series

The Fourier series

$$f(t) = \frac{1}{2}a_0 + \sum_{n=1}^{\infty} \left(a_n \cos \frac{n\pi}{p}t + b_n \sin \frac{n\pi}{p}t \right)$$

can be put in the complex form. Since

$$\cos \frac{n\pi}{p}t = \frac{1}{2} \left(e^{i(n\pi/p)t} + e^{-i(n\pi/p)t} \right),$$

$$\sin \frac{n\pi}{p}t = \frac{1}{2i} \left(e^{i(n\pi/p)t} - e^{-i(n\pi/p)t} \right),$$

it follows:

$$f(t) = \frac{1}{2}a_0 + \sum_{n=1}^{\infty} \left[\left(\frac{1}{2}a_n + \frac{1}{2i}b_n \right) e^{i(n\pi/p)t} + \left(\frac{1}{2}a_n - \frac{1}{2i}b_n \right) e^{-i(n\pi/p)t} \right].$$

Now if we define c_n as

$$c_n = \frac{1}{2}a_n + \frac{1}{2i}b_n$$

$$= \frac{1}{2}\frac{1}{p} \int_{-p}^{p} f(t) \cos \left(\frac{n\pi}{p}t \right) dt + \frac{1}{2i}\frac{1}{p} \int_{-p}^{p} f(t) \sin \left(\frac{n\pi}{p}t \right) dt$$

$$= \frac{1}{2p} \int_{-p}^{p} f(t) \left[\cos\left(\frac{n\pi}{p}t\right) - i\sin\left(\frac{n\pi}{p}t\right) \right] dt$$

$$= \frac{1}{2p} \int_{-p}^{p} f(t) e^{-i(n\pi/p)t} dt,$$

$$c_{-n} = \frac{1}{2}a_n - \frac{1}{2i}b_n$$

$$= \frac{1}{2}\frac{1}{p} \int_{-p}^{p} f(t) \cos\left(\frac{n\pi}{p}t\right) dt - \frac{1}{2i}\frac{1}{p} \int_{-p}^{p} f(t) \sin\left(\frac{n\pi}{p}t\right) dt$$

$$= \frac{1}{2p} \int_{-p}^{p} f(t) e^{i(n\pi/p)t} dt$$

and

$$c_0 = \frac{1}{2}a_0 = \frac{1}{2}\frac{1}{p} \int_{-p}^{p} f(t) dt,$$

then the series can be written as

$$f(t) = c_0 + \sum_{n=1}^{\infty} \left[c_n e^{i(n\pi/p)t} + c_{-n} e^{i(n\pi/p)t} \right]$$

$$= \sum_{n=-\infty}^{\infty} c_n e^{i(n\pi/p)t} \tag{1.28}$$

with

$$c_n = \frac{1}{2p} \int_{-p}^{p} f(t) e^{-i(n\pi/p)t} dt \tag{1.29}$$

for positive n, negative n, or $n = 0$.

Now the Fourier series appears in complex form. If $f(t)$ is a complex function of real variable t, then the complex Fourier series is a natural one. If $f(t)$ is a real function, it can still be represented by the complex series (1.28). In that case, c_{-n} is the complex conjugate of c_n ($c_{-n} = c_n^*$).

Since

$$c_n = \frac{1}{2}(a_n - ib_n), \quad c_{-n} = \frac{1}{2}(a_n + ib_n),$$

if follows that:

$$a_n = c_n + c_{-n}, \quad b_n = i(c_n - c_{-n}).$$

Thus if $f(t)$ is an even function, then $c_{-n} = c_n$. If $f(t)$ is an odd function, then $c_{-n} = -c_n$.

Example 1.5.1. Find the complex Fourier series of the function

$$f(t) = \begin{cases} 0 & -\pi < t < 0, \\ 1 & 0 < t < \pi. \end{cases}$$

Solution 1.5.1. Since the period is 2π, so $p = \pi$, and the complex Fourier series is given by

$$f(t) = \sum_{n=-\infty}^{\infty} c_n e^{int}$$

with

$$c_0 = \frac{1}{2\pi} \int_0^{\pi} dt = \frac{1}{2},$$

$$c_n = \frac{1}{2\pi} \int_0^{\pi} e^{-int} dt = \frac{1 - e^{-in\pi}}{2\pi ni} = \begin{cases} 0 & n = \text{even}, \\ \frac{1}{\pi ni} & n = \text{odd}. \end{cases}$$

Therefore the complex series is

$$f(t) = \frac{1}{2} + \frac{1}{i\pi}\left(\cdots - \frac{1}{3}e^{-i3t} - e^{-it} + e^{it} + \frac{1}{3}e^{i3t} + \cdots\right).$$

It is clear that

$$c_{-n} = \frac{1}{\pi(-n)i} = \frac{1}{\pi n(-i)} = c_n^*$$

as we expect, sine $f(t)$ is real. Furthermore, since

$$e^{int} - e^{-int} = 2i \sin nt,$$

the Fourier series can be written as

$$f(t) = \frac{1}{2} + \frac{2}{\pi}\left(\sin t + \frac{1}{3}\sin 3t + \frac{1}{5}\sin 5t + \cdots\right).$$

This is also what we expected, since $f(t) - \frac{1}{2}$ is an odd function, and

$$a_n = c_n + c_{-n} = \frac{1}{\pi ni} + \frac{1}{\pi(-n)i} = 0,$$

$$b_n = i(c_n - c_{-n}) = i\left(\frac{1}{\pi ni} - \frac{1}{\pi(-n)i}\right) = \frac{2}{\pi n}.$$

Example 1.5.2. Find the Fourier series of the function defined as

$$f(t) = e^t \quad \text{for} \quad -\pi < t < \pi, \quad f(t + 2\pi) = f(t).$$

Solution 1.5.2. This periodic function has a period of 2π. We can express it as the Fourier series

$$f(t) = \frac{1}{2}a_0 + \sum_{n=1}^{\infty}(a_n \cos nt + b_n \sin nt).$$

However, the complex Fourier coefficients are easier to compute, so we first express it as a complex Fourier series

$$f(t) = \sum_{n=-\infty}^{\infty} c_n e^{int}$$

with

$$c_n = \frac{1}{2\pi} \int_{-\pi}^{\pi} e^t e^{-int} dt = \frac{1}{2\pi} \left[\frac{1}{1-in} e^{(1-in)t} \right]_{-\pi}^{\pi}.$$

Since

$$e^{(1-in)\pi} = e^\pi e^{-in\pi} = (-1)^n e^\pi,$$

$$e^{-(1-in)\pi} = e^{-\pi} e^{in\pi} = (-1)^n e^{-\pi},$$

$$e^\pi - e^{-\pi} = 2\sinh\pi,$$

so

$$c_n = \frac{(-1)^n}{2\pi(1-in)} (e^\pi - e^{-\pi}) = \frac{(-1)^n}{\pi} \frac{1+in}{1+n^2} \sinh\pi.$$

Now

$$a_n = c_n + c_{-n} = \frac{(-1)^n}{\pi} \frac{2}{1+n^2} \sinh\pi,$$

$$b_n = i(c_n - c_{-n}) = -\frac{(-1)^n}{\pi} \frac{2n}{1+n^2} \sinh\pi.$$

Thus, the Fourier series is given by

$$e^x = \frac{\sinh\pi}{\pi} + \frac{2\sinh\pi}{\pi} \sum_{n=1}^{\infty} \frac{(-1)^n}{1+n^2} (\cos nt - n\sin nt).$$

1.6 The Method of Jumps

There is an effective way of computing the Fourier coefficients, known as the method of jumps. As long as the given function is piecewise continuous, this method enables us to find Fourier coefficients by graphical techniques.

Suppose that $f(t)$, shown in Fig. 1.10, is a periodic function with a period $2p$. It is piecewise continuous. The locations of the discontinuity are at $t_1, t_2, \ldots, t_{N-1}$, counting from left to right. The two end points t_0 and t_N may or may not be points of discontinuity. Let $f(t_i^+)$ be the right-hand limit of the function as t approaches t_i from the right, and $f(t_i^-)$, the left-hand limit. At each discontinuity t_i, except at two end points t_0 and $t_N = t_0 + 2p$, we define a jump J_i as

$$J_i = f(t_i^+) - f(t_i^-).$$

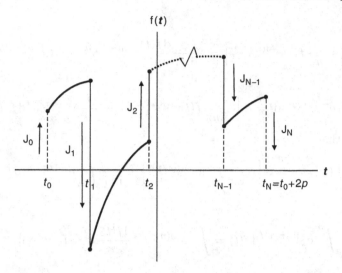

Fig. 1.10. One period of a periodic piecewise continuous function $f(t)$ with period $2p$

At t_0, the jump J_0 is defined as

$$J_0 = f(t_0^+) - 0 = f(t_0^+)$$

and at t_N, the jump J_N is

$$J_N = 0 - f(t_N^-) = -f(t_N^-).$$

These jumps are indicated by the arrows in Fig. 1.10. It is seen that J_i will be positive if the jump at t_i is up and negative if the jump is down. Note that at t_0, the jump is from zero to $f\left(t_0^+\right)$, and at t_N, the jump is from $f\left(t_N^-\right)$ to zero.

We will now show that the coefficients of the Fourier series can be expressed in terms of these jumps.

The coefficients of the complex Fourier series, as seen in (1.29), is given by

$$c_n = \frac{1}{2p} \int_{-p}^{p} f(t) e^{-i(n\pi/p)t} dt.$$

Let us define the integral as

$$\int_{-p}^{p} f(t) e^{-i(n\pi/p)t} dt = I_n[f(t)].$$

So $c_n = \frac{1}{2p} I_n[f(t)]$.

Since

$$\frac{d}{dt}\left[-\frac{p}{in\pi}f(t)e^{-i(n\pi/p)t}\right] = -\frac{p}{in\pi}\frac{df(t)}{dt}e^{-i(n\pi/p)t} + f(t)e^{-i(n\pi/p)t},$$

so

$$f(t)e^{-i(n\pi/p)t}dt = d\left[-\frac{p}{in\pi}f(t)e^{-i(n\pi/p)t}\right] + \frac{p}{in\pi}e^{-i(n\pi/p)t}df(t),$$

it follows that:

$$I_n\left[f(t)\right] = \int_{-p}^{p}d\left[-\frac{p}{in\pi}f(t)e^{-i(n\pi/p)t}\right] + \frac{p}{in\pi}\int_{-p}^{p}e^{-i(n\pi/p)t}df(t).$$

Note that

$$\int_{-p}^{p}e^{-i(n\pi/p)t}df(t) = \int_{-p}^{p}e^{-i(n\pi/p)t}\frac{df(t)}{dt}dt = I_n\left[f'(t)\right],$$

and

$$\int_{-p}^{p}d\left[-\frac{p}{in\pi}f(t)e^{-i(n\pi/p)t}\right] = -\frac{p}{in\pi}\left[\int_{t_0}^{t_1} + \int_{t_1}^{t_2} + \cdots + \int_{t_{N-1}}^{t_N}\right]$$
$$\times d\left[f(t)e^{-i(n\pi/p)t}\right].$$

Since

$$\int_{t_0}^{t_1}d\left[f(t)e^{-i(n\pi/p)t}\right] = f(t_1^-)e^{-i(n\pi/p)t_1} - f(t_0^+)e^{-i(n\pi/p)t_0},$$

$$\int_{t_1}^{t_2}d\left[f(t)e^{-i(n\pi/p)t}\right] = f(t_2^-)e^{-i(n\pi/p)t_2} - f(t_1^+)e^{-i(n\pi/p)t_1},$$

$$\int_{t_{N-1}}^{t_N}d\left[f(t)e^{-i(n\pi/p)t}\right] = f(t_N^-)e^{-i(n\pi/p)t_N} - f(t_{N-1}^+)e^{-i(n\pi/p)t_{N-1}},$$

we have

$$\int_{-p}^{p}d\left[-\frac{p}{in\pi}f(t)e^{-i(n\pi/p)t}\right] = \frac{p}{in\pi}f(t_0^+)e^{-i(n\pi/p)t_0}$$

$$+ \frac{p}{in\pi}[f(t_1^+) - f(t_1^-)]e^{-i(n\pi/p)t_1}$$

$$+ \cdots - \frac{p}{in\pi}f(t_N^-)e^{-i(n\pi/p)t_N} = \frac{p}{in\pi}\sum_{k=0}^{k=N}J_k e^{-i(n\pi/p)t_k}.$$

Thus

$$I_n[f(t)] = \frac{p}{in\pi}\sum_{k=0}^{k=N}J_k e^{-i(n\pi/p)t_k} + \frac{p}{in\pi}I_n[f'(t)].$$

Clearly, $I_n[f'(t)]$ can be evaluated similarly as $I_n[f(t)]$. This formula can be used iteratively to find the Fourier coefficient c_n for nonzero n, since $c_n = I_n[f(t)]/2p$. Together with c_0, which is given by a simple integral, these coefficients determine all terms of the Fourier series. For many practical functions, their Fourier series can be simply obtained from the jumps at the points of discontinuity. The following examples will illustrate how quickly this can be done with the sketches of the function and its derivatives.

Example 1.6.1. Use the method of jumps to find the Fourier series of the periodic function $f(t)$, one of its periods is defined on the interval of $-\pi < t < \pi$ as

$$f(t) = \begin{cases} k & \text{for } -\pi < t < 0 \\ -k & \text{for } 0 < t < \pi \end{cases} .$$

Solution 1.6.1. The sketch of this function is

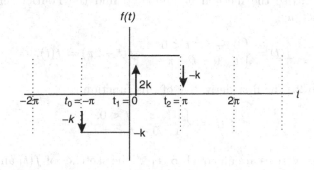

The period of this function is 2π, therefore $p = \pi$. It is clear that all derivatives of this function are equal to zero, thus we have

$$c_n = \frac{1}{2\pi} I_n[f(t)] = \frac{1}{i2\pi n} \sum_{k=0}^{2} J_k e^{-i(n\pi/p)t_k}, \quad n \neq 0,$$

where

$$t_0 = -\pi, \quad t_1 = 0, \quad t_2 = \pi$$

and

$$J_0 = -k, \quad J_1 = 2k, \quad J_2 = -k.$$

Hence

$$c_n = \frac{1}{i2\pi n}[-ke^{in\pi} + 2k - ke^{-in\pi}]$$

$$= \frac{k}{i2\pi n}[2 - 2\cos(n\pi)] = \begin{cases} 0 & n = \text{even} \\ \frac{2k}{in\pi} & n = \text{odd} \end{cases} .$$

It follows that:

$$a_n = c_n + c_{-n} = 0,$$

$$b_n = i(c_n - c_{-n}) = \begin{cases} 0 & n = \text{even} \\ \frac{4k}{n\pi} & n = \text{odd} \end{cases}.$$

Furthermore,

$$c_0 = \frac{1}{2\pi} \int_{-\pi}^{\pi} f(t)dt = 0.$$

Therefore the Fourier series is given by

$$f(t) = \frac{4k}{\pi}\left(\sin t + \frac{1}{3}\sin 3t + \frac{1}{5}\sin 5t + \cdots\right).$$

Example 1.6.2. Use the method of jumps to find the Fourier series of the following function:

$$f(t) = \begin{cases} 0 & -\pi < t < 0 \\ t & 0 < t < \pi \end{cases}, \quad f(t + 2\pi) = f(t).$$

Solution 1.6.2. The first derivative of this function is

$$f'(t) = \begin{cases} 0 & -\pi < t < 0, \\ 1 & 0 < t < \pi \end{cases}$$

and higher derivatives are all equal to zero. The sketches of $f(t)$ and $f'(t)$ are shown as follows:

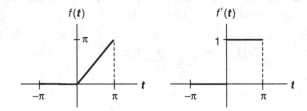

In this case

$$p = \pi, \quad t_0 = -\pi, \quad t_1 = 0, \quad t_2 = \pi.$$

Thus

$$I_n[f(t)] = \frac{1}{in}\sum_{k=0}^{2} J_k e^{-int_k} + \frac{1}{in}I_n[f'(t)],$$

where

$$J_0 = 0, \quad J_1 = 0, \quad J_2 = -\pi,$$

and

$$I_n[f'(t)] = \frac{1}{\mathrm{i}n} \sum_{k=0}^{2} J_k' \mathrm{e}^{-\mathrm{i}nt_k}$$

with

$$J_0' = 0, \quad J_1' = 1, \quad J_2' = -1.$$

It follows that:

$$I_n[f(t)] = \frac{1}{\mathrm{i}n}(-\pi)\mathrm{e}^{-\mathrm{i}n\pi} + \frac{1}{\mathrm{i}n}\left[\frac{1}{\mathrm{i}n}(1 - \mathrm{e}^{-\mathrm{i}n\pi})\right]$$

and

$$c_n = \frac{1}{2\pi} I_n[f(t)] = -\frac{1}{\mathrm{i}2n}\mathrm{e}^{-\mathrm{i}n\pi} - \frac{1}{2\pi n^2}(1 - \mathrm{e}^{-\mathrm{i}n\pi}), \quad n \neq 0.$$

In addition

$$c_0 = \frac{1}{2\pi} \int_0^\pi t\, \mathrm{d}t = \frac{\pi}{4}.$$

Therefore the Fourier coefficients a_n and b_n are given by

$$a_n = c_n + c_{-n} = \frac{1}{\mathrm{i}2n}(-\mathrm{e}^{-\mathrm{i}n\pi} + \mathrm{e}^{\mathrm{i}n\pi}) + \frac{1}{2\pi n^2}(\mathrm{e}^{-\mathrm{i}n\pi} + \mathrm{e}^{\mathrm{i}n\pi}) - \frac{1}{\pi n^2}$$

$$= \frac{1}{n}\sin n\pi + \frac{1}{\pi n^2}\cos n\pi - \frac{1}{\pi n^2} = \begin{cases} -\frac{2}{\pi n^2} & n = \text{odd} \\ 0 & n = \text{even} \end{cases},$$

$$b_n = \mathrm{i}(c_n - c_{-n}) = \mathrm{i}\left[-\frac{1}{\mathrm{i}2n}(\mathrm{e}^{-\mathrm{i}n\pi} + \mathrm{e}^{\mathrm{i}n\pi}) + \frac{1}{2\pi n^2}(\mathrm{e}^{-\mathrm{i}n\pi} - \mathrm{e}^{\mathrm{i}n\pi})\right]$$

$$= -\frac{1}{n}\cos n\pi + \frac{1}{\pi n^2}\sin n\pi = \begin{cases} \frac{1}{n} & n = \text{odd} \\ -\frac{1}{n} & n = \text{even} \end{cases}.$$

So the Fourier series can be written as

$$f(t) = \frac{\pi}{4} - \frac{2}{\pi}\sum_{n=1}^{\infty} \frac{1}{(2n-1)^2}\cos(2n-1)t - \sum_{n=1}^{\infty} \frac{(-1)^n}{n}\sin nt.$$

1.7 Properties of Fourier Series

1.7.1 Parseval's Theorem

If the periodicity of a periodic function $f(t)$ is $2p$, the Parseval's theorem states that

$$\frac{1}{2p} \int_{-p}^{p} [f(t)]^2 \mathrm{d}t = \frac{1}{4}a_0^2 + \frac{1}{2}\sum_{n=1}^{\infty}(a_n^2 + b_n^2),$$

where a_n and b_n are the Fourier coefficients. This theorem can be proved by expressed $f(t)$ as the Fourier series

$$f(t) = \frac{1}{2}a_0 + \sum_{n=1}^{\infty}\left(a_n \cos\frac{n\pi t}{p} + b_n \sin\frac{n\pi t}{p}\right),$$

and carrying out the integration. However, the computation is simpler if we first work with the complex Fourier series

$$f(t) = \sum_{n=-\infty}^{\infty} c_n e^{i(n\pi/p)t},$$

$$c_n = \frac{1}{2p}\int_{-p}^{p} f(t)e^{-i(n\pi/p)t}.$$

With these expressions, the integral can be written as

$$\frac{1}{2p}\int_{-p}^{p} [f(t)]^2 dt = \frac{1}{2p}\int_{-p}^{p} f(t)\sum_{n=-\infty}^{\infty} c_n e^{i(n\pi/p)t} dt$$

$$= \sum_{n=-\infty}^{\infty} c_n \frac{1}{2p}\int_{-p}^{p} f(t)e^{i(n\pi/p)t} dt.$$

Since

$$c_{-n} = \frac{1}{2p}\int_{-p}^{p} f(t)e^{-i((-n)\pi/p)t} = \frac{1}{2p}\int_{-p}^{p} f(t)e^{i(n\pi/p)t} dt,$$

it follows that:

$$\frac{1}{2p}\int_{-p}^{p} [f(t)]^2 dt = \sum_{n=-\infty}^{\infty} c_n c_{-n} = c_0^2 + 2\sum_{n=1}^{\infty} c_n c_{-n}.$$

If $f(t)$ is a real function, then $c_{-n} = c_n^*$. Since

$$c_n = \frac{1}{2}(a_n - ib_n), \quad c_n^* = \frac{1}{2}(a_n + ib_n),$$

so

$$c_n c_{-n} = c_n c_n^* = \frac{1}{4}\left[a_n^2 - (ib_n)^2\right] = \frac{1}{4}(a_n^2 + b_n^2).$$

Therefore

$$\frac{1}{2p}\int_{-p}^{p} [f(t)]^2 dt = c_0^2 + 2\sum_{n=1}^{\infty} c_n c_{-n} = \left(\frac{1}{2}a_0\right)^2 + \frac{1}{2}\sum_{n=1}^{\infty}(a_n^2 + b_n^2).$$

This theorem has an interesting and important interpretation. In physics we learnt that the energy in a wave is proportional to the square of its amplitude. For the wave represented by $f(t)$, the energy in one period will be proportional to $\int_{-p}^{p}[f(t)]^2 dt$. Since $a_n \cos \frac{n\pi t}{p}$ also represents a wave, so the energy in this pure cosine wave is proportional to

$$\int_{-p}^{p} \left(a_n \cos \frac{n\pi}{p} t \right)^2 dt = a_n^2 \int_{-p}^{p} \cos^2 \frac{n\pi t}{p} dt = pa_n^2$$

so the energy in the pure sine wave is

$$\int_{-p}^{p} \left(b_n \sin \frac{n\pi}{p} t \right)^2 dt = b_n^2 \int_{-p}^{p} \sin^2 \frac{n\pi t}{p} dt = pb_n^2.$$

From the Parseval's theorem, we have

$$\int_{-p}^{p} [f(t)]^2 dt = p\frac{1}{2}a_0^2 + p\sum_{n=1}^{\infty}(a_n^2 + b_n^2).$$

This says that the total energy in a wave is just the sum of the energies in all the Fourier components. For this reason, Parseval's theorem is also called "energy theorem."

1.7.2 Sums of Reciprocal Powers of Integers

An interesting application of Fourier series is that it can be used to sum up a series of reciprocal powers of integers. For example, we have shown that the Fourier series of the square-wave

$$f(x) = \begin{cases} -k & -\pi < x < 0 \\ k & 0 < x < \pi \end{cases}, \quad f(x + 2\pi) = f(x)$$

is given by

$$f(x) = \frac{4k}{\pi} \left(\sin x + \frac{1}{3} \sin 3x + \frac{1}{5} \sin 5x + \cdots \right).$$

At $x = \pi/2$, we have

$$f\left(\frac{\pi}{2}\right) = k = \frac{4k}{\pi} \left(1 - \frac{1}{3} + \frac{1}{5} - \frac{1}{7} + \cdots \right),$$

thus

$$\frac{\pi}{4} = 1 - \frac{1}{3} + \frac{1}{5} - \frac{1}{7} + \cdots = \sum_{n=1}^{\infty} \frac{(-1)^{n+1}}{2n-1}.$$

This is a famous result obtained by Leibniz in 1673 from geometrical considerations. It became well known because it was the first series involving π ever discovered.

The Parseval's theorem can also be used to give additional results. In this problem,

$$[f(t)]^2 = k^2, \quad a_n = 0, \quad b_n = \begin{cases} \frac{4k}{\pi n} & n = \text{odd} \\ 0 & n = \text{even} \end{cases},$$

$$\frac{1}{2\pi} \int_{-\pi}^{\pi} [f(t)]^2 dt = k^2 = \frac{1}{2} \sum_{n=1}^{\infty} b_n^2 = \frac{1}{2} \left(\frac{4k}{\pi} \right)^2 \left(1 + \frac{1}{3^2} + \frac{1}{5^2} + \cdots \right).$$

So we have

$$\frac{\pi^2}{8} = 1 + \frac{1}{3^2} + \frac{1}{5^2} + \cdots = \sum_{n=1}^{\infty} \frac{1}{(2n-1)^2}.$$

In the following example, we will demonstrate that a number of such sums can be obtained with one Fourier series.

Example 1.7.1. Use the Fourier series for the function whose definition is

$$f(x) = x^2 \text{ for } -1 < x < 1, \quad \text{and} \quad f(x+2) = f(x),$$

to show that

(a) $\sum_{n=1}^{\infty} \frac{(-1)^{n+1}}{n^2} = \frac{\pi^2}{12}$, (b) $\sum_{n=1}^{\infty} \frac{1}{n^2} = \frac{\pi^2}{6}$,

(c) $\sum_{n=1}^{\infty} \frac{(-1)^{n+1}}{(2n-1)^3} = \frac{\pi^3}{32}$, (d) $\sum_{n=1}^{\infty} \frac{1}{n^4} = \frac{\pi^4}{90}$.

Solution 1.7.1. The Fourier series for the function is given by (1.20) with $L = 1$:

$$x^2 = \frac{1}{3} + \frac{4}{\pi^2} \sum_{n=1}^{\infty} \frac{(-1)^n}{n^2} \cos n\pi x.$$

(a) Set $x = 0$, so we have

$$x^2 = 0, \quad \cos n\pi x = 1.$$

Thus

$$0 = \frac{1}{3} + \frac{4}{\pi^2} \sum_{n=1}^{\infty} \frac{(-1)^n}{n^2}$$

or

$$-\frac{4}{\pi^2} \sum_{n=1}^{\infty} \frac{(-1)^n}{n^2} = \frac{1}{3}.$$

It follows that:

$$1 - \frac{1}{2^2} + \frac{1}{3^2} - \frac{1}{4^2} + \cdots = \frac{\pi^2}{12}.$$

(b) With $x = 1$, the series becomes

$$1 = \frac{1}{3} + \frac{4}{\pi^2} \sum_{n=1}^{\infty} \frac{(-1)^n}{n^2} \cos n\pi.$$

Since $\cos n\pi = (-1)^n$, we have

$$1 - \frac{1}{3} = \frac{4}{\pi^2} \sum_{n=1}^{\infty} \frac{(-1)^{2n}}{n^2}$$

or

$$\frac{\pi^2}{6} = 1 + \frac{1}{2^2} + \frac{1}{3^2} + \frac{1}{4^2} + \cdots.$$

(c) Integrating both sides from 0 to 1/2,

$$\int_0^{1/2} x^2 \mathrm{d}x = \int_0^{1/2} \left[\frac{1}{3} + \frac{4}{\pi^2} \sum_{n=1}^{\infty} \frac{(-1)^n}{n^2} \cos n\pi x \right] \mathrm{d}x$$

we get

$$\frac{1}{3} \left(\frac{1}{2} \right)^3 = \frac{1}{3} \left(\frac{1}{2} \right) + \frac{4}{\pi^2} \sum_{n=1}^{\infty} \frac{(-1)^n}{n^2} \frac{1}{n\pi} \sin \frac{n\pi}{2}$$

or

$$-\frac{1}{8} = \frac{4}{\pi^3} \sum_{n=1}^{\infty} \frac{(-1)^n}{n^3} \sin \frac{n\pi}{2}.$$

Since

$$\sin \frac{n\pi}{2} = \begin{cases} 0 & n = \text{even}, \\ 1 & n = 1, 5, 9, \ldots, \\ -1 & n = 3, 7, 11, \ldots \end{cases}$$

the sum can be written as

$$-\frac{1}{8} = -\frac{4}{\pi^3} \left(1 - \frac{1}{3^3} + \frac{1}{5^3} - \frac{1}{7^3} + \cdots \right).$$

It follows that:

$$\frac{\pi^3}{32} = \sum_{n=1}^{\infty} \frac{(-1)^{n+1}}{(2n-1)^3}.$$

(d) Using the Parseval's theorem, we have

$$\frac{1}{2} \int_{-1}^{1} \left(x^2 \right)^2 \mathrm{d}x = \left(\frac{1}{3} \right)^2 + \frac{1}{2} \sum_{n=1}^{\infty} \left[\frac{4}{\pi^2} \frac{(-1)^n}{n^2} \right]^2.$$

Thus

$$\frac{1}{5} = \frac{1}{9} + \frac{8}{\pi^4} \sum_{n=1}^{\infty} \frac{1}{n^4}.$$

It follows that:

$$\frac{\pi^4}{90} = \sum_{n=1}^{\infty} \frac{1}{n^4}.$$

This last series played an important role in the theory of black-body radiation, which was crucial in the development of quantum mechanics.

1.7.3 Integration of Fourier Series

If a Fourier series of $f(x)$ is integrated term-by-term, a factor of $1/n$ is introduced into the series. This has the effect of enhancing the convergence. Therefore we expect the series resulting from term-by-term integration will converge to the integral of $f(x)$. For example, we have shown that the Fourier series for the odd function $f(t) = t$ of period $2L$ is given by

$$t = \frac{2L}{\pi} \sum_{n=1}^{\infty} \frac{(-1)^{n+1}}{n} \sin \frac{n\pi}{L} t.$$

We expect a term-by-term integration of the right-hand side of this equation to converge to the integral of t. That is

$$\int_0^t x \, dx = \frac{2L}{\pi} \sum_{n=1}^{\infty} \frac{(-1)^{n+1}}{n} \int_0^t \sin \frac{n\pi}{L} x \, dx.$$

The result of this integration is

$$\frac{1}{2} t^2 = \frac{2L}{\pi} \sum_{n=1}^{\infty} \frac{(-1)^{n+1}}{n} \left[-\frac{L}{n\pi} \cos \frac{n\pi}{L} x \right]_0^t$$

or

$$t^2 = \frac{4L^2}{\pi^2} \sum_{n=1}^{\infty} \frac{(-1)^{n+1}}{n^2} - \frac{4L^2}{\pi^2} \sum_{n=1}^{\infty} \frac{(-1)^{n+1}}{n^2} \cos \frac{n\pi}{L} t.$$

Since

$$\sum_{n=1}^{\infty} \frac{(-1)^{n+1}}{n^2} = \frac{\pi^2}{12},$$

we obtain

$$t^2 = \frac{L^2}{3} + \frac{4L^2}{\pi^2} \sum_{n=1}^{\infty} \frac{(-1)^n}{n^2} \cos \frac{n\pi}{L} t.$$

This is indeed the correct Fourier series converging to t^2 of period $2L$, as seen in (1.20).

Example 1.7.2. Find the Fourier series of the function whose definition in one period is

$$f(t) = t^3, \quad -L < t < L.$$

Solution 1.7.2. Integrating the Fourier series for t^2 in the required range term-by-term

$$\int t^2 dt = \int \left[\frac{L^2}{3} + \frac{4L^2}{\pi^2} \sum_{n=1}^{\infty} \frac{(-1)^n}{n^2} \cos \frac{n\pi}{L} t \right] dt,$$

we obtain

$$\frac{1}{3} t^3 = \frac{L^2}{3} t + \frac{4L^2}{\pi^2} \sum_{n=1}^{\infty} \frac{(-1)^n}{n^2} \frac{L}{n\pi} \sin \frac{n\pi}{L} t + C.$$

We can find the integration constant C by looking at the values of both sides of this equation at $t = 0$. Clearly $C = 0$. Furthermore, since in the range of $-L < t < L$,

$$t = \frac{2L}{\pi} \sum_{n=1}^{\infty} \frac{(-1)^{n+1}}{n} \sin \frac{n\pi}{L} t,$$

therefore the Fourier series of t^3 in the required range is

$$t^3 = \frac{2L^3}{\pi} \sum_{n=1}^{\infty} \frac{(-1)^{n+1}}{n} \sin \frac{n\pi}{L} t + \frac{12L^3}{\pi^3} \sum_{n=1}^{\infty} \frac{(-1)^n}{n^3} \sin \frac{n\pi}{L} t.$$

1.7.4 Differentiation of Fourier Series

In differentiating a Fourier series term-by-term, we have to be more careful. A term-by-term differentiation will cause the coefficients a_n and b_n to be multiplied by a factor n. Since it grows linearly, the resulting series may not even converge. Take, for example

$$t = \frac{2L}{\pi} \sum_{n=1}^{\infty} \frac{(-1)^{n+1}}{n} \sin \frac{n\pi}{L} t.$$

This equation is valid in the range of $-L < t < L$, as seen in (1.19). The derivative of t is of course equal to 1. However, a term-by-term differentiation of the Fourier series on the right-hand side

$$\frac{d}{dt} \left[\frac{2L}{\pi} \sum_{n=1}^{\infty} \frac{(-1)^{n+1}}{n} \sin \frac{n\pi}{L} t \right] = 2 \sum_{n=1}^{\infty} (-1)^{n+1} \cos \frac{n\pi}{L} t$$

does not even converge, let alone equal to 1.

In order to see under what conditions, if any, that the Fourier series of the function $f(t)$

$$f(t) = \frac{1}{2}a_0 + \sum_{n=1}^{\infty} \left(a_n \cos\frac{n\pi}{L}t + b_n \sin\frac{n\pi}{L}t\right)$$

can be differentiated term-by-term, let us first assume that $f(t)$ is continuous within the range $-L < t < L$, and the derivative of the function $f'(t)$ can be expanded in another Fourier series

$$f'(t) = \frac{1}{2}a_0' + \sum_{n=1}^{\infty} \left(a_n' \cos\frac{n\pi}{L}t + b_n' \sin\frac{n\pi}{L}t\right).$$

The coefficients a_n' are given by

$$a_n' = \frac{1}{L}\int_{-L}^{L} f'(t)\cos\frac{n\pi}{L}t \, dt$$

$$= \frac{1}{L}\left[f(t)\cos\frac{n\pi}{L}t\right]_{-L}^{L} + \frac{n\pi}{L^2}\int_{-L}^{L} f(t)\sin\frac{n\pi}{L}t \, dt$$

$$= \frac{1}{L}[f(L) - f(-L)]\cos n\pi + \frac{n\pi}{L}b_n. \tag{1.30}$$

Similarly

$$b_n' = \frac{1}{L}[f(L) - f(-L)]\sin n\pi - \frac{n\pi}{L}na_n. \tag{1.31}$$

On the other hand, differentiating the Fourier series of the function term-by-term, we get

$$\frac{d}{dt}\left[\frac{1}{2}a_0 + \sum_{n=1}^{\infty}\left(a_n\cos\frac{n\pi}{L}t + b_n\sin\frac{n\pi}{L}t\right)\right]$$

$$= \sum_{n=1}^{\infty}\left(-a_n\frac{n\pi}{L}\sin\frac{n\pi}{L}t + b_n\frac{n\pi}{L}\cos\frac{n\pi}{L}t\right).$$

This would simply give coefficients

$$a_n' = \frac{n\pi}{L}b_n, \quad b_n' = -\frac{n\pi}{L}a_n. \tag{1.32}$$

Thus we see that the derivative of a function is not, in general, given by differentiating the Fourier series of the function term-by-term. However, if the function satisfies the condition

$$f(L) = f(-L), \tag{1.33}$$

then a_n' and b_n' given by (1.30) and (1.31) are identical to those given by (1.32). We call (1.33) the "head equals tail" condition. Once this condition is satisfied, a term-by-term differentiation of the Fourier series of the function will converge to the derivative of the function. Note that if the periodic function $f(t)$ is continuous everywhere, this condition is automatically satisfied.

Now it is clear why (1.19) cannot be differentiated term-by-term. For this function

$$f(L) = L \neq -L = f(-L),$$

the "head equals tail" condition is not satisfied. In the following example, the function satisfies this condition. Its derivative is indeed given by the result of the term-by-term differentiation.

Example 1.7.3. The fourier series for t^2 in the range $-L < t < L$ is given by (1.20)

$$\frac{L^2}{3} + \frac{4L^2}{\pi^2} \sum_{n=1}^{\infty} \frac{(-1)^n}{n^2} \cos \frac{n\pi}{L} t = t^2.$$

It satisfies the "head equals tail" condition, as shown in Fig. 1.4. Show that a term-by-term differentiation of this series is equal to $2t$.

Solution 1.7.3.

$$\frac{d}{dt}\left[\frac{L^2}{3} + \frac{4L^2}{\pi^2} \sum_{n=1}^{\infty} \frac{(-1)^n}{n^2} \cos \frac{n\pi}{L} t\right] = \frac{4L^2}{\pi^2} \sum_{n=1}^{\infty} \frac{(-1)^n}{n^2} \frac{d}{dt} \cos \frac{n\pi}{L} t$$

$$= \frac{4L}{\pi} \sum_{n=1}^{\infty} \frac{(-1)^{n+1}}{n} \sin \frac{n\pi}{L} t$$

which is the Fourier series of $2t$ in the required range, as seen in (1.19) .

1.8 Fourier Series and Differential Equations

Fourier series play an important role in solving partial differential equations, as we shall see in many examples in later chapters. In this section, we shall confine ourselves with some applications of Fourier series in solving nonhomogeneous ordinary differential equations.

1.8.1 Differential Equation with Boundary Conditions

Let us consider the following nonhomogeneous differential equation:

$$\frac{d^2 x}{dt^2} + 4x = 4t,$$

$$x(0) = 0, \quad x(1) = 0.$$

We want to find the solution between $t = 0$ and $t = 1$. Previously we have learned that the general solution of this equation is the sum of the complementary function x_c and the particular solution x_p. That is

$$x = x_c + x_p,$$

where x_c is the solution of the homogeneous equation

$$\frac{d^2 x_c}{dt^2} + 4x_c = 0$$

with two arbitrary constants, and x_p is the particular solution of

$$\frac{d^2 x_p}{dt^2} + 4x_p = 4t$$

with no arbitrary constant. It can be easily verified that in this case

$$x_c = A \cos 2t + B \sin 2t,$$
$$x_p = t.$$

Therefore the general solution is

$$x(t) = A \cos 2t + B \sin 2t + t.$$

The two constants A and B are determined by the boundary conditions. Since

$$x(0) = A = 0,$$
$$x(1) = A \cos 2 + B \sin 2 + 1 = 0,$$

Thus

$$A = 0, \quad B = -\frac{1}{\sin 2}.$$

Therefore the exact solution that satisfies the boundary conditions is given by

$$x(t) = t - \frac{1}{\sin 2} \sin 2t.$$

This function in the range of $0 \le t \le 1$ can be expanded into a half-range Fourier sine series

$$x(t) = \sum_{n=1}^{\infty} b_n \sin n\pi t,$$

where

$$b_n = 2 \int_0^1 \left(t - \frac{1}{\sin 2} \sin 2t \right) \sin n\pi t \, dt.$$

We have already shown that

$$\int_0^1 t \sin n\pi t \, dt = \frac{(-1)^{n+1}}{n\pi}.$$

With integration by parts twice, we find

$$\int_0^1 \sin 2t \sin n\pi t \, dt = \left[-\frac{1}{n\pi} \sin 2t \cos n\pi t + \frac{2}{(n\pi)^2} \cos 2t \sin n\pi t \right]_0^1$$

$$+ \frac{4}{(n\pi)^2} \int_0^1 \sin 2t \sin n\pi t \, dt.$$

Combining the last term with left-hand side and putting in the limits, we get

$$\int_0^1 \sin 2t \sin n\pi t \, dt = \frac{(-1)^{n+1} n\pi}{[(n\pi)^2 - 4]} \sin 2.$$

It follows that:

$$b_n = 2 \left[\frac{(-1)^{n+1}}{n\pi} - \frac{1}{\sin 2} \frac{(-1)^{n+1} n\pi}{[(n\pi)^2 - 4]} \sin 2 \right] = (-1)^{n+1} \frac{8}{n\pi [4 - (n\pi)^2]}. \quad (1.34)$$

Therefore the solution that satisfies the boundary conditions can be written as

$$x(t) = \frac{8}{\pi} \sum_{n=1}^{\infty} \frac{(-1)^{n+1}}{n[4 - (n\pi)^2]} \sin n\pi t.$$

Now we shall show that this result can be obtained directly from the following Fourier series method. First we expand the solution, whatever it is, into a half-range Fourier sine series

$$x(t) = \sum_{n=1}^{\infty} b_n \sin n\pi t.$$

This is a valid procedure because no matter what the solution is, we can always antisymmetrically extend it to the interval $-1 < t < 0$ and then to the entire real line by the periodicity condition $x(t + 2) = x(t)$. The Fourier series representing this odd function with a periodicity of 2 is given by the above expression. This function is continuous everywhere, therefore it can be differentiated term-by-term. Furthermore, the boundary conditions, $x(0) = 0$ and $x(1) = 1$, are automatically satisfied by this series.

When we put this series into the differential equation, the result is

$$\sum_{n=1}^{\infty} \left[-(n\pi)^2 + 4 \right] b_n \sin n\pi t = 4t.$$

This equation can be regarded as the function $4t$ expressed in a Fourier sine series. The coefficients $[-(n\pi)^2 + 4]b_n$ are given by

$$[-(n\pi)^2 + 4] \, b_n = 2 \int_0^1 4t \sin n\pi t \, dt = 8 \frac{(-1)^{n+1}}{n\pi}.$$

It follows that:

$$b_n = \frac{8(-1)^{n+1}}{n\pi[4 - (n\pi)^2]},$$

which is identical to (1.34). Therefore we will get the exactly same result as before.

This shows that the Fourier series method is convenient and direct. Not every boundary value problem can be handled in this way, but many of them can. When the problem is solved by the Fourier series method, often the solution is actually in a more useful form.

Example 1.8.1. A horizontal beam of length L, supported at each end is uniformly loaded. The deflection of the beam $y(x)$ is known to satisfy the equation

$$\frac{d^4 y}{dx^4} = \frac{w}{EI},$$

where w, E, and I are constants (w is load per unit length, E is the Young's modulus, I is the moment of inertia). Furthermore, $y(t)$ satisfies the following four boundary conditions

$$y(0) = 0, \quad y(L) = 0,$$

$$y''(0) = 0, \quad y''(L) = 0.$$

(This is because there is no deflection and no moment at either end.) Find the deflection curve of the beam $y(x)$.

Solution 1.8.1. The function may be conveniently expanded in a Fourier sine series

$$y(x) = \sum_{n=1}^{\infty} b_n \sin \frac{n\pi}{L} x.$$

The four boundary conditions are automatically satisfied. This series and its derivatives are continuous, therefore it can be repeatedly term-by-term differentiated. Putting it in the equation, we have

$$\sum_{n=1}^{\infty} b_n \left(\frac{n\pi}{L}\right)^4 \sin \frac{n\pi}{L} x = \frac{w}{EI}.$$

This means that $b_n (n\pi/L)^4$ is the coefficients of the Fourier sine series of w/EI. Therefore

$$b_n \left(\frac{n\pi}{L}\right)^4 = \frac{2}{L} \int_0^L \frac{w}{EI} \sin \frac{n\pi}{L} x \, dx = -\frac{2}{L} \frac{w}{EI} \frac{L}{n\pi} (\cos n\pi - 1).$$

It follows that:

$$b_n = \begin{cases} \dfrac{4wL^4}{EI}\dfrac{1}{(n\pi)^5} & n = \text{odd} \\ 0 & n = \text{even} \end{cases}.$$

Therefore

$$y(x) = \frac{4wL^4}{EI\pi^5} \sum_{n=1}^{\infty} \frac{1}{(2n-1)^5} \sin \frac{(2n-1)n\pi x}{L}.$$

This series is rapidly convergent due to the fifth power of n in the denominator.

1.8.2 Periodically Driven Oscillator

Consider a damped spring–mass system driven by an external periodic forcing function. The differential equation describing this motion is

$$m\frac{d^2x}{dt} + c\frac{dx}{dt} + kx = F(t). \tag{1.35}$$

We recall that if the external forcing function $F(t)$ is a sine or cosine function, then the steady state solution of the system is an oscillatory motion with the same frequency of the input function. For example, if

$$F(t) = F_0 \sin \omega t,$$

then

$$x_p(t) = \frac{F_0}{\sqrt{(k - m\omega^2)^2 + (c\omega)^2}} \sin(\omega t - \alpha), \tag{1.36}$$

where

$$\alpha = \tan^{-1} \frac{c\omega}{k - m\omega^2}.$$

However, if $F(t)$ is periodic with frequency ω, but is not a sine or cosine function, then the steady state solution will contain not only a term with the input frequency ω, but also other terms of multiples of this frequency. Suppose that the input forcing function is given by a square-wave

$$F(t) = \begin{cases} 1 & 0 < t < L \\ -1 & -L < t < 0 \end{cases}, \quad F(t + 2L) = F(t). \tag{1.37}$$

This square-wave repeats itself in the time interval of $2L$. The number of times that it repeats itself in $1\,\text{s}$ is called frequency ν. Clearly $\nu = 1/(2L)$. Recall that the angular frequency ω is defined as $2\pi\nu$. Therefore

$$\omega = 2\pi\frac{1}{2L} = \frac{\pi}{L}.$$

Often ω is just referred to as frequency.

Now as we have shown, the Fourier series expansion of $F(t)$ is given by

$$F(t) = \sum_{n=1}^{\infty} b_n \sin \frac{n\pi}{L} t,$$

$$b_n = \begin{cases} \dfrac{4}{n\pi} & n = \text{odd}, \\ 0 & n = \text{even}. \end{cases}$$

It is seen that the first term is a pure sine wave with the same frequency as the input square-wave. We called it the fundamental frequency $\omega_1 (\omega_1 = \omega)$. The other terms in the Fourier series have frequencies of multiples of the fundamental frequency. They are called harmonics (or overtones). For example, the second and third harmonics have, respectively, frequencies of $\omega_2 = 2\pi/L = 2\omega$ and $\omega_3 = 3\pi/L = 3\omega$. (In this terminology, there is no first harmonic.)

With the input square-wave $F(t)$ expressed in terms of its Fourier series in (1.35), the response of the system is also a superposition of the harmonics, since (1.35) is a linear differential equation. That is, if x_n is the particular solution of

$$m\frac{\mathrm{d}^2 x_n}{\mathrm{d}t} + c\frac{\mathrm{d}x_n}{\mathrm{d}t} + kx_n = b_n \sin \omega_n t,$$

then the solution to (1.35) is

$$x_{\mathrm{p}} = \sum_{n=1}^{\infty} x_n.$$

Thus it follows from (1.36) that with the input forcing function given by the square-wave, the steady state solution of the spring–mass system is given by

$$x_{\mathrm{p}} = \sum_{n=1}^{\infty} \frac{b_n \sin(\omega_n t - \alpha_n)}{\sqrt{(k - m\omega_n^2)^2 + (c\omega_n)^2}},$$

where

$$\omega_n = \frac{n\pi}{L} = n\omega, \qquad \alpha_n = \tan^{-1} \frac{c\omega_n}{k - m\omega_n^2}.$$

This solution contains not only a term with the same input frequency ω, but also other terms with multiples of this frequency. If one of these higher frequencies is close to the natural frequency of the system ω_0 $(\omega_0 = \sqrt{k/m})$, then the particular term containing that frequency may play the dominant role in the system response. This is an important problem in vibration analysis. The input frequency may be considerably lower than the natural frequency of the system, yet if that input is not purely sinusoidal, it could still lead to resonance. This is best illustrated with a specific example.

Example 1.8.2. Suppose that in some consistent set of units, $m = 1$, $c = 0.2$, $k = 9$, and $\omega = 1$, and the input $F(t)$ is given by (1.37). Find the steady state solution $x_p(t)$ of the spring–mass system.

Solution 1.8.2. Since $\omega = \pi/L = 1$, so $L = \pi$ and $\omega_n = n$. As we have shown, the Fourier series of $F(t)$ is

$$F(t) = \frac{4}{\pi}\left(\sin t + \frac{1}{3}\sin 3t + \frac{1}{5}\sin 5t + \cdots\right).$$

The steady-state solution is therefore given by

$$x_p(t) = \frac{4}{\pi}\sum_{n=\text{odd}}^{\infty}\frac{1}{n}\frac{\sin(nt - \alpha_n)}{\sqrt{(9 - n^2)^2 + (0.2n)^2}},$$

$$\alpha_n = \tan^{-1}\frac{0.2n}{9 - n^2}, \quad 0 \leq \alpha_n \leq \pi.$$

Carrying out the calculation, we find

$$x_p(t) = 0.1591\sin(t - 0.0250) + 0.7073\sin(3t - 1.5708)$$
$$+0.0159\sin(5t - 3.0792) + \cdots.$$

The following figure shows $x_p(t)$ in comparison with the input force function. In order to have the same dimension of distance, the input force is expressed in terms of the "static distance" $F(t)/k$. The term $0.7073\sin(3t - 1.5708)$ is shown as the dotted line. It is seen that this term dominates the response of the system. This is because the term with $n = 3$ in the Fourier series of $F(t)$

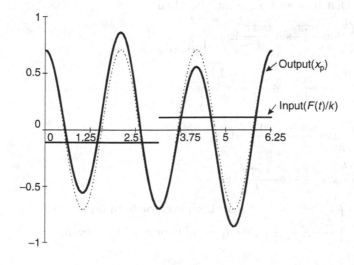

has the same frequency as the natural frequency of the system ($\sqrt{k/m} = 3$). Thus near resonance vibrations occur, with the mass completing essentially three oscillations for every single oscillation of the external input force.

An interesting demonstration of this phenomenon on a piano is given in the Feynman Lecture on Physics, Vol. I, Chap. 50.

Let us label the two successive Cs near the middle of the keyboard by C, C', and the Gs just above by G, G'. The fundamentals will have relative frequencies as follows:

$$C - 2 \quad G - 3$$
$$C' - 4 \quad G' - 6$$

These harmonic relationships can be demonstrated in the following way. Suppose we press C' slowly – so that it does not sound but we cause the damper to be lifted. If we sound C, it will produce its own fundamental and some harmonics. The second harmonic will set the strings of C' into vibration. If we now release C (keeping C' pressed) the damper will stop the vibration of the C strings, and we can hear (*softly*) the note of C' as it dies away. In a similar way, the third harmonic of C can cause a vibration of G'.

This phenomenon is as interesting as important. In a mechanical or electrical system that is forced with a periodic function having a frequency smaller than the natural frequency of the system, as long as the forcing function is not purely sinusoidal, one of its overtones may resonate with the system. To avoid the occurrence of abnormally large and destructive resonance vibrations, one must not allow any overtone of the input function to dominate the response of the system.

Exercises

1. Show that if m and n are integers then

(a) $\displaystyle\int_0^L \sin\frac{n\pi x}{L} \sin\frac{m\pi x}{L}\,\mathrm{d}x = \begin{cases} \dfrac{L}{2} & n = m, \\ 0 & n \neq m. \end{cases}$

(b) $\displaystyle\int_0^L \cos\frac{n\pi x}{L} \cos\frac{m\pi x}{L}\,\mathrm{d}x = \begin{cases} \dfrac{L}{2} & n = m, \\ 0 & n \neq m. \end{cases}$

(c) $\displaystyle\int_{-L}^L \sin\frac{n\pi x}{L} \cos\frac{m\pi x}{L}\,\mathrm{d}x = 0, \quad$ all $\quad n, m.$

(d) $\displaystyle\int_0^L \sin\frac{n\pi x}{L} \cos\frac{m\pi x}{L}\,\mathrm{d}x$

$$= \begin{cases} 0 & n, m \text{ both even or both odd,} \\ \dfrac{L}{\pi}\dfrac{2n}{n^2 - m^2} & n \text{ even, } m \text{ odd; or } n \text{ odd, } m \text{ even.} \end{cases}$$

2. Find the Fourier series of the following functions:

(a) $f(x) = \begin{cases} 0 & -\pi < x < 0 \\ 2 & 0 < x < \pi \end{cases}$, $f(x + 2\pi) = f(x)$,

(b) $f(x) = \begin{cases} 1 & -\dfrac{\pi}{2} < x < \dfrac{\pi}{2} \\ -1 & \dfrac{\pi}{2} < x < \dfrac{3\pi}{2} \end{cases}$, $f(x + 2\pi) = f(x)$,

(c) $f(x) = \begin{cases} 0 & -\pi < x < 0 \\ \sin x & 0 < x < \pi \end{cases}$, $f(x + 2\pi) = f(x)$.

Ans. (a) $f(x) = 1 + \dfrac{4}{\pi}\left[\sin x + \dfrac{1}{3}\sin 3x + \dfrac{1}{5}\sin 5x - \cdots\right]$,

(b) $f(x) = \dfrac{4}{\pi}\left[\cos x - \dfrac{1}{3}\cos 3x + \dfrac{1}{5}\cos 5x - \cdots\right]$,

(c) $f(x) = \dfrac{1}{\pi} - \dfrac{2}{\pi}\left[\dfrac{1}{3}\cos 2x + \dfrac{1}{15}\cos 4x + \dfrac{1}{35}\cos 6x - \cdots\right]$
$+ \dfrac{1}{2}\sin x.$

3. Find the Fourier series of the following functions:

(a) $f(t) = \begin{cases} -1 & -2 < t < 0 \\ 1 & 0 < t < 2 \end{cases}$, $f(t + 4) = f(t)$,

(b) $f(t) = t^2$, $0 < t < 2$, $f(t + 2) = f(t)$.

Ans. (a) $f(t) = \dfrac{4}{\pi}\left[\sin\dfrac{\pi t}{2} + \dfrac{1}{3}\sin\dfrac{3\pi t}{2} + \dfrac{1}{5}\sin\dfrac{5\pi t}{2} \cdots\right]$,

(b) $f(t) = \dfrac{4}{3} + \dfrac{4}{\pi^2}\sum\dfrac{1}{n^2}\cos n\pi t + \dfrac{4}{\pi}\sum\dfrac{1}{n}\sin n\pi t.$

4. Find the half-range Fourier cosine and sine expansions of the following functions:

(a) $f(t) = 1$, $0 < t < 2$.
(b) $f(t) = t$, $0 < t < 1$.
(c) $f(t) = t^2$, $0 < t < 3$.

Ans. (a) 1; $\dfrac{4}{\pi}\displaystyle\sum \dfrac{1}{2n-1}\sin\dfrac{(2n-1)\pi t}{2}$,

(b) $\dfrac{1}{2}-\dfrac{4}{\pi^2}\displaystyle\sum\dfrac{1}{(2n-1)^2}\cos(2n-1)\pi t$; $\dfrac{2}{\pi}\displaystyle\sum\dfrac{(-1)^{n+1}}{n}\sin n\pi t$,

(c) $3+\dfrac{36}{\pi^2}\displaystyle\sum\dfrac{(-1)^n}{n^2}\cos\dfrac{n\pi t}{3};\dfrac{18}{\pi^3}\left[\left(\dfrac{\pi^2}{1}-\dfrac{4}{1^3}\right)\sin\dfrac{\pi t}{3}\right.$

$-\dfrac{\pi^2}{2}\sin\dfrac{2\pi t}{3}+\left(\dfrac{\pi^2}{3}-\dfrac{4}{3^3}\right)\sin\dfrac{3\pi t}{3}$

$\left.-\dfrac{\pi^2}{4}\sin\dfrac{4\pi t}{3}+\cdots\right]$.

5. The output from an electronic oscillator takes the form of a sine wave $f(t)=\sin t$ for $0<t\le\pi/2$, it then drops to zero and starts again. Find the complex Fourier series of this wave form.
   Ans.

$$\sum_{n=-\infty}^{\infty}\dfrac{2}{\pi}\dfrac{4ni-1}{16n^2-1}e^{i4nt}.$$

6. Use the method of jumps to find the half-range cosine series of the function $g(t)=\sin t$ defined in the interval of $0<t<\pi$.

Hint: For a cosine series, we need an even extension of the function. Let

$$f(t)=\begin{cases}g(t)=\sin t & 0<t<\pi,\\ g(-t)=-\sin t & -\pi<t<0.\end{cases}$$

Its derivatives are

$$f'(t)=\begin{cases}\cos t & 0<t<\pi\\ -\cos t & -\pi<t<0\end{cases},\qquad f''(t)=-f(t).$$

The sketches of the function and its derivatives are shown as follows:

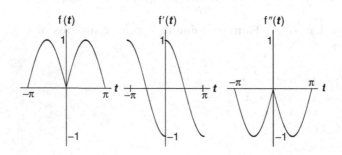

Ans.

$$f(t) = \frac{2}{\pi} - \frac{4}{\pi}\left(\frac{1}{3}\cos 2t + \frac{1}{15}\cos 4t + \frac{1}{35}\cos 6t + \cdots\right).$$

7. Use the method of jumps to find the half range (a) cosine and (b) sine Fourier expansions of $g(t)$, which is defined only over the range $0 < t < 1$ as

$$g(t) = t - t^2, \quad 0 < t < 1.$$

Hint: (a) For the half-range cosine expansion, the function must be symmetrically extended to negative t. That is, we have to expand into a Fourier series the even function $f(t)$ defined as

$$f(t) = \begin{cases} g(t) = t - t^2 & 0 < t < 1, \\ g(-t) = -t - t^2 & -1 < t < 0. \end{cases}$$

The first and second derivatives of this function are given by

$$f'(t) = \begin{cases} 1 - 2t & 0 < t < 1 \\ -1 - 2t & -1 < t < 0 \end{cases}, \quad f''(t) = -2$$

and all higher derivatives are zero. The sketches of this function and its derivatives are as follows:

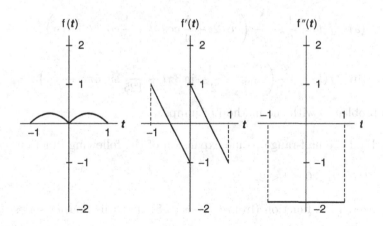

(b) For the half-range sine expansion, an antisymmetric extension of $g(t)$ to negative t is needed. Let

$$f(t) = \begin{cases} g(t) = t - t^2 & 0 < t < 1, \\ -g(-t) = t + t^2 & -1 < t < 0. \end{cases}$$

The first and second derivatives of this function are given by

$$f'(t) = \begin{cases} 1 - 2t & 0 < t < 1, \\ 1 + 2t & -1 < t < 0, \end{cases} \qquad f''(t) = \begin{cases} -2 & 0 < t < 1, \\ 2 & -1 < t < 0 \end{cases}$$

and all higher derivatives are zero. The sketches of these functions are shown below

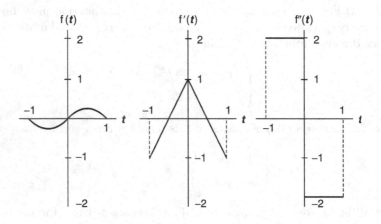

Ans. (a) $f(t) = \dfrac{1}{6} - \dfrac{1}{\pi^2}\left(\cos 2t + \dfrac{1}{4}\cos 4t + \dfrac{1}{9}\cos 6t + \cdots\right).$

(b) $f(t) = \dfrac{8}{\pi^3}\left(\sin \pi t + \dfrac{1}{27}\sin 3\pi t + \dfrac{1}{125}\sin 5\pi t + \cdots\right).$

8. Do problem 3 with the method of jumps.

9. (a) Find the half-range cosine expansion of the following function:

$$f(t) = t, \quad 0 < t < 2.$$

(b) Sketch the function (from $t = -8$ to 8) that this Fourier series represents.

(c) What is the periodicity of this function.

Ans.

$$f(t) = 1 + \frac{4}{\pi^2}\sum_{1}^{\infty}\frac{1}{n^2}(\cos n\pi - 1)\cos\frac{n\pi}{2}t; \quad \text{period} = 4.$$

10. (a) Find the half-range cosine expansion of the following function:

$$f(t) = \begin{cases} t & 0 < t \le 2 \\ 4 - t & 2 \le t < 4 \end{cases}.$$

(b) Sketch the function (from $t = -8$ to 8) this Fourier series represents.

(c) What is the periodicity of this function.

Ans.

$$f(t) = 1 - \frac{8}{\pi^2} \sum_1^\infty \frac{1}{n^2}(1 + \cos n\pi - 2\cos \frac{n\pi}{2}) \cos \frac{n\pi}{4}t; \quad \text{period} = 8.$$

11. (a) Show that the Fourier series in the two preceding problems are identical to each other.

(b) Compare the two sketches to find out the reason why this is so.
Ans. Since they represent the same function, both Fourier series can be expressed as

$$f(t) = 1 - \frac{8}{\pi^2}(\cos \frac{\pi t}{2} + \frac{1}{9}\cos \frac{3\pi t}{2} + \frac{1}{25}\cos \frac{5\pi t}{2} + \cdots).$$

12. Use the Fourier series for

$$f(t) = t \quad \text{for} \quad -1 < t < 1, \quad \text{and} \quad f(t+2) = f(t)$$

to show that

(a) $1 - \frac{1}{3} + \frac{1}{5} - \frac{1}{7} + \cdots = \frac{\pi}{4},$

(b) $1 + \frac{1}{2^2} + \frac{1}{3^2} + \frac{1}{4^2} + \cdots = \frac{\pi^2}{6}.$

13. Use the Fourier series shown in Fig. 1.5 to show that

(a) $1 + \frac{1}{3^2} + \frac{1}{5^2} + \frac{1}{7^2} + \cdots = \frac{\pi^2}{8},$

(b) $1 + \frac{1}{3^4} + \frac{1}{5^4} + \frac{1}{7^4} + \cdots = \frac{\pi^4}{96}.$

Hint: (a) Set $t = 0$. (b) Use Parseval's theorem and $\sum 1/n^2 = \pi^2/6$.

14. Use

$$\sum_{n=1}^{\infty} \frac{1}{n^4} = \frac{\pi^4}{90} \quad \text{and} \quad \sum_{n=1}^{\infty} \frac{1}{(2n-1)^4} = \frac{\pi^4}{96}$$

to show that

$$1 - \frac{1}{2^4} + \frac{1}{3^4} - \frac{1}{4^4} + \cdots = \frac{7\pi^4}{720}.$$

15. An odd function $f(t)$ of period of 2π is to be approximated by a Fourier series having only N terms. The so called "square deviation" is defined to be

$$\varepsilon = \int_{-\pi}^{\pi} \left[f(t) - \sum_{n=1}^{N} b_n \sin nt \right]^2 dt.$$

It is a measure of the error of this approximation. Show that for ε to be minimum, b_n must be given by the Fourier coefficient

$$b_n = \frac{1}{\pi} \int_{-\pi}^{\pi} f(t) \sin nt \ dt.$$

Hint: Set $\dfrac{\partial \varepsilon}{\partial b_n} = 0$.

16. Show that for $-\pi \le x \le \pi$

(a) $\cos kx = \dfrac{\sin k\pi}{k\pi} + \sum_{n=1}^{\infty} (-1)^n \dfrac{2k \sin k\pi}{\pi(k^2 - n^2)} \cos nx,$

(b) $\cot k\pi = \dfrac{1}{\pi} \left(\dfrac{1}{k} - \sum_{n=1}^{\infty} \dfrac{2k}{n^2 - k^2} \right).$

17. Find the steady-state solution of

$$\frac{d^2 x}{dt^2} + 2\frac{dx}{dt} + 3x = f(t),$$

where $f(t) = t, \quad -\pi \le t < \pi$, and $f(t + 2\pi) = f(t)$.

Ans.

$$x_p = \sum \frac{(-1)^n 2(n^2 - 3)}{n(n^4 - 2n^2 + 9)} \sin nt + \sum \frac{(-1)^n 4}{n^4 - 2n^2 + 9} \cos nt.$$

18. Use the Fourier series method to solve the following boundary value problem

$$\frac{d^4 y}{dx^4} = \frac{Px}{EIL}$$

$$y(0) = 0, \qquad y(L) = 0,$$

$$y''(0) = 0, \qquad y''(L) = 0.$$

($y(x)$ is the deflection of a beam bearing a linearly increasing load given by Px/L)

Ans.

$$y(x) = \frac{2PL^4}{\pi^4 EI} \sum \frac{(-1)^{n+1}}{n^5} \sin \frac{n\pi x}{L}.$$

19. Find the Fourier series for

    (a) $f(t) = t$ for $-\pi < t < \pi$, and $f(t+2\pi) = f(t)$,
    (b) $f(t) = |t|$ for $-\pi < t < \pi$, and $f(t+2\pi) = f(t)$.

Show that the series resulting from a term-by-term differentiation of the series in (a) does not converge to $f'(t)$, whereas the series resulting from a term-by-term differentiation of the series in (b) converges to $f'(t)$. Why?

2

Fourier Transforms

Fourier transform is a generalization of Fourier series. It provides representations, in terms of a superposition of sinusoidal waves, for functions defined over an infinite interval with no particular periodicity. It is an indispensable mathematical tool in the study of waves, which in one form or another, consist of most of physics and modern technology.

Like Laplace transform, Fourier transform is a member of a class of representations known as integral transforms. As such, it is useful in solving differential equations. But the importance of Fourier transforms far exceeds just being able to solve differential equations. In quantum mechanics, it enables us to look at the wave functions either in the coordinate space or in the momentum space. In information theory, it allows one to examine a wave form from the perspective of both the time and frequency domains. For these reasons, Fourier transform has become a cornerstone of diverse fields ranging from signal processing technology to quantum description of matter waves.

2.1 Fourier Integral as a Limit of a Fourier Series

As we have seen, Fourier series is useful in representing either periodic functions or functions confined in limited range of interest. However, in many problems, the function of interests, such as a single unrepeated pulse of force or voltage, is nonperiodic over an infinite range. In such a case, we can still imagine that the function is periodic with the period approaching infinity. In this limit, the Fourier series becomes the Fourier integral.

To extend the concept of Fourier series to nonperiodic functions, let us first consider a function which repeats itself after an interval of $2p$

$$f(t) = \sum_{n=0}^{\infty} \left(a_n \cos \frac{n\pi}{p} t + b_n \sin \frac{n\pi}{p} t \right),$$

where

$$a_0 = \frac{1}{2p} \int_{-p}^{p} f(t) dt,$$

$$a_n = \frac{1}{p} \int_{-p}^{p} f(t) \cos \frac{n\pi}{p} t\, dt, \quad n = 1, 2, \ldots,$$

$$b_n = \frac{1}{p} \int_{-p}^{p} f(t) \sin \frac{n\pi}{p} t\, dt, \quad n = 0, 1, 2, \ldots.$$

Note that each individual term $\cos \frac{n\pi}{p} t$ or $\sin \frac{n\pi}{p} t$ is a periodic function. Its period T_n is determined by the relation that when t is increased by T_n, the function returns to its previous value,

$$\cos \frac{n\pi}{p} (t + T_n) = \cos \left(\frac{n\pi}{p} t + \frac{n\pi}{p} T_n \right) = \cos \frac{n\pi}{p} t.$$

Thus,

$$\frac{n\pi}{p} T_n = 2\pi \text{ and } T_n = \frac{2p}{n}.$$

The frequency ν is defined as the number of oscillations in one second. Therefore, each term is associated with a frequency ν_n,

$$\nu_n = \frac{1}{T_n} = \frac{n}{2p}.$$

Now if t stands for time, then ν_n is just the usual temporal frequency. If the variable is x, standing for distance, ν_n is simply the spatial frequency. The distribution of the set of all of the frequencies $\left\{ \frac{n}{2p} \right\}$ is called the frequency spectrum. To see what happens to the frequency spectra as p increases, consider the cases where $p = 1, 2$, and 10. The corresponding frequencies of the spectra are as follows:

$$p = 1, \quad \nu_n = 0,\ 0.50,\ 1.0,\ 1.50,\ 2.0, \ldots$$

$$p = 2, \quad \nu_n = 0,\ 0.25,\ 0.5,\ 0.75,\ 1.0, \ldots$$

$$p = 10, \quad \nu_n = 0,\ 0.05,\ 0.1,\ 0.15,\ 0.2, \ldots.$$

It is seen that as p increases, the discrete spectrum becomes more and more dense. It will approach a continuous spectrum as $p \to \infty$, and the Fourier series appears to be an integral. This is indeed the case, if $f(t)$ is absolutely integrable over the infinite range.

Often the angular frequency, defined as $\omega_n = 2\pi\nu_n$, is used to simplify the writing. Since

$$\omega_n = 2\pi\nu_n = 2\pi \frac{n}{2p} = \frac{n\pi}{p},$$

the Fourier series can be written as

$$f(t) = \frac{1}{2p} \int_{-p}^{p} f(t)\mathrm{d}t + \sum_{n=1}^{\infty} (a_n \cos \omega_n t + b_n \sin \omega_n t).$$

As $f(t)$ is absolutely integrable over the infinite range, this means that the integral $\int_{-p}^{p} |f(t)|\,\mathrm{d}t$ exists even when $p \to \infty$. Therefore

$$\lim_{p \to \infty} \frac{1}{2p} \int_{-p}^{p} f(t)\mathrm{d}t = 0.$$

Hence,

$$f(t) = \sum_{n=1}^{\infty} (a_n \cos \omega_n t + b_n \sin \omega_n t),$$

where

$$a_n = \frac{1}{p} \int_{-p}^{p} f(t) \cos \omega_n t\, \mathrm{d}t,$$

$$b_n = \frac{1}{p} \int_{-p}^{p} f(t) \sin \omega_n t\, \mathrm{d}t.$$

Furthermore, we can define

$$\Delta\omega = \omega_{n+1} - \omega_n = \frac{(n+1)\pi}{p} - \frac{n\pi}{p} = \frac{\pi}{p}.$$

Therefore

$$f(t) = \sum_{n=1}^{\infty} \left[\frac{\Delta\omega}{\pi} \int_{-p}^{p} f(t) \cos \omega_n t\, \mathrm{d}t \right] \cos \omega_n t$$

$$+ \sum_{n=1}^{\infty} \left[\frac{\Delta\omega}{\pi} \int_{-p}^{p} f(t) \sin \omega_n t\, \mathrm{d}t \right] \sin \omega_n t.$$

If we write the series as

$$f(t) = \sum_{n=1}^{\infty} [A_p(\omega_n) \cos \omega_n t + B_p(\omega_n) \sin \omega_n t]\, \Delta\omega,$$

then

$$A_p(\omega_n) = \frac{1}{\pi} \int_{-p}^{p} f(t) \cos \omega_n t\, \mathrm{d}t,$$

$$B_p(\omega_n) = \frac{1}{\pi} \int_{-p}^{p} f(t) \sin \omega_n t\, \mathrm{d}t.$$

Now if we let $p \to \infty$, then $\Delta\omega \to 0$ and ω_n becomes a continuous variable. Furthermore, let

$$A(\omega) = \lim_{p \to \infty} A_p(\omega_n) = \frac{1}{\pi} \int_{-\infty}^{\infty} f(t) \cos \omega t \, dt,$$

$$B(\omega) = \lim_{p \to \infty} B_p(\omega_n) = \frac{1}{\pi} \int_{-\infty}^{\infty} f(t) \sin \omega t \, dt.$$

Then the infinite series becomes a Riemann sum of an integral

$$f(t) = \int_0^{\infty} [A(\omega) \cos \omega t + B(\omega) \sin \omega t] \, d\omega.$$

This integral is known as Fourier integral. This development is purely formal. However, it can be made rigorous provided (1) $f(t)$ is piecewise continuous and differentiable and (2) it is absolutely integrable in the infinite range, as we have assumed.

This integral will converge to $f(t)$ where $f(t)$ is continuous, and it converges to the average of the left- and right-hand limits of $f(t)$ at points of discontinuity, just like a Fourier series.

Example 2.1.1. (a) Find the Fourier integral of

$$f(t) = \begin{cases} 1 & \text{if } -1 < t < 1, \\ 0 & \text{otherwise.} \end{cases}$$

(b) Show that

$$\int_0^{\infty} \frac{\cos \omega t \sin \omega}{\omega} d\omega = \begin{cases} \frac{\pi}{2} & if \quad -1 < t < 1, \\ \frac{\pi}{4} & if \quad |t| = 1, \\ 0 & if \quad |t| > 1. \end{cases}$$

(c) Show that

$$\int_0^{\infty} \frac{\sin \omega}{\omega} d\omega = \frac{\pi}{2}.$$

Solution 2.1.1. (a)

$$A(\omega) = \frac{1}{\pi} \int_{-\infty}^{\infty} f(t) \cos \omega t \, dt = \frac{1}{\pi} \int_{-1}^{1} \cos \omega t \, dt = \frac{2 \sin \omega}{\pi \omega}.$$

Since $f(t)$ is an even function

$$B(\omega) = \frac{1}{\pi} \int_{-\infty}^{\infty} f(t) \sin \omega t \, dt = 0.$$

Therefore the Fourier integral is given by

$$f(t) = \frac{2}{\pi} \int_0^\infty \frac{\sin \omega}{\omega} \cos \omega t \, d\omega.$$

(b) In the range of $-1 < t < 1$, $f(t) = 1$, therefore

$$\int_0^\infty \frac{\sin \omega}{\omega} \cos \omega t \, d\omega = \frac{\pi}{2}, \quad \text{for} \quad -1 < t < 1.$$

At $|t| = 1$, it is a point of discontinuity, the Fourier integral converges to the average of 1 and 0, which is $\frac{1}{2}$. Therefore

$$\frac{1}{2} = \frac{2}{\pi} \int_0^\infty \frac{\sin \omega}{\omega} \cos \omega \, d\omega$$

or

$$\int_0^\infty \frac{\sin \omega}{\omega} \cos \omega \, d\omega = \frac{\pi}{4}.$$

For $|x| > 1$, $f(t) = 0$. Thus

$$\int_0^\infty \frac{\sin \omega}{\omega} \cos \omega t \, d\omega = 0, \quad for \quad |t| > 1.$$

(c) In particular at $t = 0$,

$$\int_0^\infty \frac{\sin \omega}{\omega} \cos \omega t \, d\omega = \int_0^\infty \frac{\sin \omega}{\omega} d\omega.$$

At $t = 0$, $f(0) = 1$, therefore

$$\int_0^\infty \frac{\sin \omega}{\omega} d\omega = \frac{\pi}{2}.$$

2.1.1 Fourier Cosine and Sine Integrals

If $f(t)$ is a even function, then

$$A(\omega) = \frac{1}{\pi} \int_{-\infty}^\infty f(t) \cos \omega t \, dt = \frac{2}{\pi} \int_0^\infty f(t) \cos \omega t \, dt,$$

$$B(\omega) = \frac{1}{\pi} \int_{-\infty}^\infty f(t) \sin \omega t \, dt = 0$$

and

$$f(t) = \int_0^\infty A(\omega) \cos \omega t \, d\omega.$$

This is known as Fourier cosine integral.

If $f(t)$ is an odd function, then

$$A(\omega) = \frac{1}{\pi} \int_{-\infty}^{\infty} f(t) \cos \omega t \, dt = 0,$$

$$B(\omega) = \frac{1}{\pi} \int_{-\infty}^{\infty} f(t) \sin \omega t \, dt = \frac{2}{\pi} \int_{0}^{\infty} f(t) \sin \omega t \, dt$$

and

$$f(t) = \int_{0}^{\infty} B(\omega) \sin \omega t \, d\omega.$$

This is known as Fourier sine integral.

Note that the function is supposed to be defined from $-\infty$ to $+\infty$, but because of the parity of the function, to define the transform, we only need the function from 0 to ∞. This also means that if we are only interested in the range of 0 to ∞, we can define the function from $-\infty$ to 0 any way we want, then we can have either cosine integral or sine integral by extending the function into the negative range either in an even or odd form. In this sense, Fourier cosine and sine integrals are equivalent to the half-range expansion of Fourier series.

Example 2.1.2. Find the Fourier cosine and sine integrals of

$$f(t) = e^{-st}, \quad t > 0, \quad s > 0.$$

Solution 2.1.2. For the Fourier cosine integral, we can imagine $f(t)$ is an even function with respect to $t = 0$. Thus

$$A(\omega) = \frac{2}{\pi} \int_{0}^{\infty} e^{-st} \cos \omega t \, dt.$$

This integral can be evaluated with integration by parts twice. Better still, we recognize that the integral is just the Laplace transform of $\cos \omega t$. So

$$A(\omega) = \frac{2}{\pi} \frac{s}{s^2 + \omega^2} .$$

It follows that the Fourier cosine integral is given by:

$$f(t) = \int_{0}^{\infty} A(\omega) \cos \omega t \, dt = \frac{2s}{\pi} \int_{0}^{\infty} \frac{\cos \omega t}{s^2 + \omega^2} dt.$$

Since $f(t) = e^{-st}$, a byproduct of this cosine integral is

$$\int_{0}^{\infty} \frac{\cos \omega t}{s^2 + \omega^2} d\omega = \frac{\pi}{2s} e^{-st}$$

a formula we have obtained before by contour integration. In particular, for $t = 0$, we have

$$\int_0^\infty \frac{1}{s^2 + \omega^2} d\omega = \frac{\pi}{2s}.$$

Similarly, for Fourier sine integral, we can imagine $f(t)$ is an odd function. In this case

$$B(\omega) = \frac{2}{\pi} \int_0^\infty e^{-st} \sin \omega t \, dt = \frac{2}{\pi} \frac{\omega}{s^2 + \omega^2}$$

as the integral is just a Laplace transform of $\sin \omega t$. Thus, the Fourier sine integral is given by

$$f(t) = e^{-st} = \frac{2}{\pi} \int_0^\infty \frac{\omega}{s^2 + \omega^2} \sin \omega t \, d\omega.$$

From this, we can obtain another integration formula

$$\int_0^\infty \frac{\omega \sin \omega t}{s^2 + \omega^2} d\omega = \frac{\pi}{2} e^{-st}.$$

Example 2.1.3. Find $f(t)$, if $f(t)$ is an even function and

$$\int_0^\infty f(t) \cos at \, dt = \begin{cases} 1 - a & \text{if } 0 \le a \le 1, \\ 0 & \text{if } \quad a > 1. \end{cases}$$

Solution 2.1.3. We can use Fourier cosine integral to solve this integral equation. Let

$$A(\omega) = \frac{2}{\pi} \int_0^\infty f(t) \cos \omega t \, dt = \begin{cases} \frac{2}{\pi}(1 - \omega) & \text{if } 0 \le a \le 1, \\ 0 & \text{if } \quad a > 1, \end{cases}$$

then

$$f(t) = \int_0^\infty A(\omega) \cos \omega t \, d\omega = \int_0^1 \frac{2}{\pi}(1 - \omega) \cos \omega t \, d\omega$$

$$= \frac{2}{\pi} \frac{1}{t^2}(1 - \cos t).$$

2.1.2 Fourier Cosine and Sine Transforms

If $f(t)$ is an even function, we have just seen that it can be expressed as a Fourier integral

$$f(t) = \int_0^\infty A(\omega) \cos \omega t \, d\omega, \tag{2.1}$$

$$A(\omega) = \frac{2}{\pi} \int_0^\infty f(t) \cos \omega t \, dt. \tag{2.2}$$

Now if we define a function

$$\widehat{f}_{\mathrm{c}}(\omega) = \sqrt{\frac{\pi}{2}} A(\omega) = \sqrt{\frac{2}{\pi}} \int_0^\infty f(t) \cos \omega t \, dt, \qquad (2.3)$$

then

$$A(\omega) = \sqrt{\frac{2}{\pi}} \widehat{f}_{\mathrm{c}}(\omega).$$

Putting it into (2.1), we have

$$f(t) = \sqrt{\frac{2}{\pi}} \int_0^\infty \widehat{f}_{\mathrm{c}}(\omega) \cos \omega t \, d\omega. \qquad (2.4)$$

The symmetry between (2.3) and (2.4) is unmistakable. They form what is known as the Fourier cosine transform pair. The function $\widehat{f}_{\mathrm{c}}(\omega)$ is known as the Fourier cosine transform. Formula (2.4) gives us back $f(t)$ from $\widehat{f}_{\mathrm{c}}(\omega)$, therefore it is called the inverse Fourier cosine transform of $\widehat{f}_{\mathrm{c}}(\omega)$. The process of obtaining the transform $\widehat{f}_{\mathrm{c}}(\omega)$ from a given function $f(t)$ is also called Fourier cosine transform and is denoted by $F_{\mathrm{c}}\{f(t)\}$, that is, when F_{c} operates on $f(t)$, it gives us $\widehat{f}_{\mathrm{c}}(\omega)$,

$$F_{\mathrm{c}}\{f(t)\} = \sqrt{\frac{2}{\pi}} \int_0^\infty f(t) \cos \omega t \, dt = \widehat{f}_{\mathrm{c}}(\omega).$$

The inverse operation is called inverse Fourier cosine transform and is denoted as $F_{\mathrm{c}}^{-1}\left\{\widehat{f}_{\mathrm{c}}(\omega)\right\}$,

$$F_{\mathrm{c}}^{-1}\left\{\widehat{f}_{\mathrm{c}}(\omega)\right\} = \sqrt{\frac{2}{\pi}} \int_0^\infty \widehat{f}_{\mathrm{c}}(\omega) \cos \omega t \, d\omega = f(t).$$

Similarly, if $f(t)$ is an odd function, we have the Fourier sine transform pair

$$F_{\mathrm{s}}\{f(t)\} = \sqrt{\frac{2}{\pi}} \int_0^\infty f(t) \sin \omega t \, dt = \widehat{f}_{\mathrm{s}}(\omega),$$

$$F_{\mathrm{s}}^{-1}\left\{\widehat{f}_{\mathrm{s}}(\omega)\right\} = \sqrt{\frac{2}{\pi}} \int_0^\infty \widehat{f}_{\mathrm{s}}(\omega) \sin \omega t \, d\omega = f(t).$$

Note that Fourier integral and Fourier transform are essentially the same. The modification of the multiplicative constant is of minor significance. It can be easily shown that if we define

$$\widehat{f}_{\mathrm{c}}(\omega) = \alpha \int_0^\infty f(t) \cos \omega t \, dt, \qquad (2.5)$$

then

$$f(t) = \beta \int_0^\infty \widehat{f_c}(\omega) \cos \omega t \, d\omega, \tag{2.6}$$

where

$$\beta = \frac{2}{\pi} \frac{1}{\alpha}.$$

Therefore as long as

$$\alpha\beta = \frac{2}{\pi},$$

where α can be assigned any number, (2.5) and (2.6) are still a Fourier cosine transform pair. As a matter of fact, in the literature, there are several different conventions in defining Fourier transforms. The differences are where to put the factor $\frac{2}{\pi}$. Using a Fourier transform table, one needs to pay attention to where that factor is in the definition.

Then why should we have two different names for essentially the same thing. This is because we have two different perspectives of looking at it. In Fourier integral, $f(t)$ is being described by a continuum of cosine (or sine) waves and $A(\omega)$ is just the amplitude of the harmonic components of $f(t)$ in the time domain. Whereas in Fourier transform, $\widehat{f_c}(\omega)$ is regarded as a function in the frequency domain. This frequency domain function describes the same entity as the time domain function $f(t)$. There are many reasons why sometimes we would like to work with the transform of the function. For example, in the frequency domain we may easily perform relatively difficult mathematical operations such as differentiation and integration via simple multiplication and division.

Example 2.1.4. Show that

$$F_c\{f'(t)\} = \omega F_s\{f(t)\} - \sqrt{\frac{2}{\pi}} f(0),$$

$$F_s\{f'(t)\} = -\omega F_c\{f(t)\},$$

$$F_c\{f''(t)\} = -\omega^2 F_c\{f(t)\} - \sqrt{\frac{2}{\pi}} f'(0),$$

$$F_s\{f''(t)\} = -\omega^2 F_s\{f(t)\} + \sqrt{\frac{2}{\pi}} \omega f(0).$$

Solution 2.1.4. Since $f(t)$ is absolutely integrable, we assume

$$f(t) \to 0 \quad \text{as} \quad t \to \infty.$$

With the integration by parts, we can evaluate the transform of derivatives

$$F_c\{f'(t)\} = \sqrt{\frac{2}{\pi}} \int_0^\infty \frac{df}{dt} \cos \omega t \, dt$$

$$= \sqrt{\frac{2}{\pi}} \left[f(t) \cos \omega t \Big|_0^\infty - \int_0^\infty f(t) \frac{d}{dt} \cos \omega t \, dt \right]$$

$$= \sqrt{\frac{2}{\pi}} \left[-f(0) + \omega \int_0^\infty f(t) \sin \omega t \, dt \right] = \omega F_s\{f(t)\} - \sqrt{\frac{2}{\pi}} f(0).$$

$$F_s\{f'(t)\} = \sqrt{\frac{2}{\pi}} \int_0^\infty \frac{df}{dt} \sin \omega t \, dt$$

$$= \sqrt{\frac{2}{\pi}} \left[f(t) \sin \omega t \Big|_0^\infty - \int_0^\infty f(t) \frac{d}{dt} \sin \omega t \, dt \right]$$

$$= \sqrt{\frac{2}{\pi}} \left[-\omega \int_0^\infty f(t) \cos \omega t \, dt \right] = -\omega F_c\{f(t)\}.$$

$$F_c\{f''(t)\} = F_c\{[f'(t)]'\} = \omega F_s\{f'(t)\} - \sqrt{\frac{2}{\pi}} f'(0)$$

$$= \omega \left[-\omega F_c\{f(t)\} \right] - \sqrt{\frac{2}{\pi}} f'(0) = -\omega^2 F_c\{f(t)\} - \sqrt{\frac{2}{\pi}} f'(0).$$

$$F_s\{f''(t)\} = F_s\{[f'(t)]'\} = -\omega F_c\{f'(t)\}$$

$$= -\omega \left[\omega F_s\{f(t)\} - \sqrt{\frac{2}{\pi}} f(0) \right] = -\omega^2 F_s\{f(t)\} + \omega \sqrt{\frac{2}{\pi}} f(0).$$

Example 2.1.5. Use the transform of derivatives to show

$$F_s\{e^{-at}\} = \sqrt{\frac{2}{\pi}} \frac{\omega}{a^2 + \omega^2}.$$

Solution 2.1.5. Let $f(t) = e^{-at}$, so $f(0) = 1$ and

$$f'(t) = -a\,e^{-at}, \quad f''(t) = a^2\,e^{-at} = a^2 f(t).$$

Thus

$$F_s\{f''(t)\} = F_s\{a^2 f(t)\} = a^2 F_s\{f(t)\}.$$

But

$$F_s\{f''(t)\} = -\omega^2 F_s\{f(t)\} + \omega \sqrt{\frac{2}{\pi}} f(0)$$

it follows that:

$$-\omega^2 F_s\{f(t)\} + \omega\sqrt{\frac{2}{\pi}} = a^2 F_s\{f(t)\}$$

or

$$(a^2 + \omega^2)F_s\{f(t)\} = \omega\sqrt{\frac{2}{\pi}}.$$

Thus,

$$F_s\{f(t)\} = F_s\{e^{-at}\} = \sqrt{\frac{2}{\pi}}\frac{\omega}{a^2 + \omega^2}.$$

Example 2.1.6. Use the Fourier sine transform to solve the following differential equation:

$$y''(t) - 9y(t) = 50e^{-2t},$$
$$y(0) = y_0.$$

Solution 2.1.6. Since we are interested in positive + region, we can take $y(t)$ to be an odd function and take Fourier sine transforms. It is clear from its definition that Fourier transform is linear

$$F_s\{af_1(t) + bf_2(t)\} = aF_s\{f_1(t)\} + bF_s\{f_2(t)\}.$$

Using this property and taking Fourier transform of both sides of the differential equation, we have

$$F_s\{y''(t)\} - 9F_s\{y(t)\} = 50F_s\{e^{-2t}\}.$$

Since

$$F_s\{y''(t)\} = -\omega^2 F_s\{y(t)\} + \omega\sqrt{\frac{2}{\pi}}y(0),$$

so

$$-\omega^2 F_s\{y(t)\} + \omega\sqrt{\frac{2}{\pi}}y_0 - 9F_s\{y(t)\} = 50F_s\{e^{-2t}\},$$

which, after collecting terms, becomes

$$(\omega^2 + 9)F_s\{y(t)\} = -50\sqrt{\frac{2}{\pi}}\frac{\omega}{\omega^2 + 4} + \omega\sqrt{\frac{2}{\pi}}y_0.$$

Thus

$$F_s\{y(t)\} = -50\sqrt{\frac{2}{\pi}}\frac{\omega}{\omega^2 + 4}\frac{1}{(\omega^2 + 9)} + \sqrt{\frac{2}{\pi}}y_0\frac{\omega}{(\omega^2 + 9)}.$$

With partial fraction of

$$\frac{1}{(\omega^2 + 4)(\omega^2 + 9)} = \frac{1}{5}\frac{1}{\omega^2 + 4} - \frac{1}{5}\frac{1}{\omega^2 + 9}$$

we have

$$F_s\{y(t)\} = \sqrt{\frac{2}{\pi}}10\frac{\omega}{\omega^2+9} - \sqrt{\frac{2}{\pi}}10\frac{\omega}{\omega^2+4} + \sqrt{\frac{2}{\pi}}y_0\frac{\omega}{(\omega^2+9)}$$

$$= (10+y_0)\sqrt{\frac{2}{\pi}}\frac{\omega}{\omega^2+9} - 10\sqrt{\frac{2}{\pi}}\frac{\omega}{\omega^2+4}$$

$$= (10+y_0)F_s\{e^{-3t}\} - 10F_s\{e^{-2t}\}.$$

Taking the inverse transform, we get the solution

$$y(t) = (10+y_0)e^{-3t} - 10e^{-2t}.$$

2.2 Tables of Transforms

There are extensive tables of Fourier transforms (For example, A. Erdélyi, W. Magnus, F. Oberhettinger, and F. Tricomi: "Tables of Integral Transforms," vol. 1, McGraw-Hill Book Company, New York, 1954). A short list of some simple Fourier cosine and sine transforms is given in Tables 2.1 and 2.2, respectively. A short table of Fourier transform, which we will explain in the Sect. 2.3, is given in Table 2.3.

2.3 The Fourier Transform

As we have seen in (1.28) and (1.29) that the Fourier series of a function repeating itself in the interval of $2p$, can also be written in the complex form

$$f(t) = \sum_{n=-\infty}^{\infty} c_n e^{i\frac{n\pi}{p}t}, \quad c_n = \frac{1}{2p}\int_{-p}^{p} f(t)e^{-i\frac{n\pi}{p}t}dt,$$

so

$$f(t) = \sum_{n=-\infty}^{\infty}\left[\frac{1}{2p}\int_{-p}^{p} f(t)e^{-i\frac{n\pi}{p}t}dt\right]e^{i\frac{n\pi}{p}t}.$$

Again let us define

$$\omega_n = \frac{n\pi}{p}$$

and

$$\Delta\omega = \omega_{n+1} - \omega_n = \frac{\pi}{p}$$

Table 2.1. A short table of Fourier cosine transforms

$f(t) = \sqrt{\dfrac{2}{\pi}} \int_0^\infty \widehat{f}_c(\omega) \cos \omega t\, d\omega$	$\widehat{f}_c(\omega) = \sqrt{\dfrac{2}{\pi}} \int_0^\infty f(t) \cos \omega t\, dt$
$\begin{cases} 1 & \text{if } 0 < t < a \\ 0 & \text{otherwise} \end{cases}$	$\sqrt{\dfrac{2}{\pi}} \dfrac{\sin \omega}{\omega}$
$t^{a-1} \quad (0 < a < 1)$	$\sqrt{\dfrac{2}{\pi}} \dfrac{\Gamma(a)}{\omega^a} \cos \dfrac{a\pi}{2}$
$e^{-at} \quad (a > 0)$	$\sqrt{\dfrac{2}{\pi}} \dfrac{a}{a^2 + \omega^2}$
$e^{-at^2} \quad (a > 0)$	$\dfrac{1}{\sqrt{2a}} e^{-\omega^2/4a}$
$\dfrac{1}{t^2 + a^2} \quad (a > 0)$	$\sqrt{\dfrac{\pi}{2}} \dfrac{1}{a} e^{-a\omega}$
$t^n e^{-at} \quad (a > 0)$	$\sqrt{\dfrac{2}{\pi}} \dfrac{n!}{(a^2 + \omega^2)^{n+1}} \operatorname{Re}(a + i\omega)^{n+1}$
$\begin{cases} \cos t & \text{if } 0 < t < a \\ 0 & \text{otherwise} \end{cases}$	$\sqrt{\dfrac{1}{2\pi}} \left[\dfrac{\sin a(1-\omega)}{1-\omega} + \dfrac{\sin a(1+\omega)}{1+\omega} \right]$
$\cos at^2 \quad (a > 0)$	$\sqrt{\dfrac{1}{2a}} \cos \left(\dfrac{\omega^2}{4a} - \dfrac{\pi}{4} \right)$
$\sin at^2 \quad (a > 0)$	$\sqrt{\dfrac{1}{2a}} \cos \left(\dfrac{\omega^2}{4a} + \dfrac{\pi}{4} \right)$
$\dfrac{\sin at}{t} \quad (a > 0)$	$\sqrt{\dfrac{\pi}{2}} u(a - \omega)$
Linearity of transform and inverse: $\alpha f(t) + \beta g(t)$	$\alpha \widehat{f}_c(\omega) + \beta \widehat{g}_c(\omega)$
Transform of derivatives: $f'(t)$	$\omega \widehat{f}_s(\omega) - \sqrt{\dfrac{2}{\pi}} f(0)$
$f''(t)$	$-\omega^2 \widehat{f}_c(\omega) - \sqrt{\dfrac{2}{\pi}} f'(0)$
Convolution theorem: $\frac{1}{2} \int_0^\infty [f(|t - x|) + f(|t + x|)] g(x)\, dx$	$\widehat{f}_c(\omega) \widehat{g}_c(\omega)$

Table 2.2. A short table of Fourier sine transforms

$f(t) = \sqrt{\dfrac{2}{\pi}} \int_0^\infty \widehat{f}_{\mathrm{s}}(\omega) \sin \omega t \, d\omega$	$\widehat{f}_{\mathrm{s}}(\omega) = \sqrt{\dfrac{2}{\pi}} \int_0^\infty f(t) \sin \omega t \, dt$
$\begin{cases} 1 & \text{if } 0 < t < a \\ 0 & \text{otherwise} \end{cases}$	$\sqrt{\dfrac{2}{\pi}} \dfrac{1 - \cos a\omega}{\omega}$
$\dfrac{1}{\sqrt{t}}$	$\dfrac{1}{\sqrt{\omega}}$
$t^{a-1} \quad (0 < a < 1)$	$\sqrt{\dfrac{2}{\pi}} \dfrac{\Gamma(a)}{\omega^a} \sin \dfrac{a\pi}{2}$
e^{-t}	$\sqrt{\dfrac{2}{\pi}} \dfrac{\omega}{1 + \omega^2}$
$\dfrac{t}{t^2 + a^2} \quad (a > 0)$	$\sqrt{\dfrac{\pi}{2}} e^{-a\omega}$
$t^n e^{-at} \quad (a > 0)$	$\sqrt{\dfrac{2}{\pi}} \dfrac{n!}{(a^2 + \omega^2)^{n+1}} \operatorname{Im}(a + i\omega)^{n+1}$
$t e^{-at^2} \quad (a > 0)$	$\dfrac{\omega}{(2a)^{3/2}} e^{-\omega^2/4a}$
$\begin{cases} \sin t & \text{if } 0 < t < a \\ 0 & \text{otherwise} \end{cases}$	$\sqrt{\dfrac{1}{2\pi}} \left[\dfrac{\sin a(1 - \omega)}{1 - \omega} - \dfrac{\sin a(1 + \omega)}{1 + \omega} \right]$
$\dfrac{\cos at}{t} \quad (a > 0)$	$\sqrt{\dfrac{\pi}{2}} u(\omega - a)$
Linearity of transform and inverse: $\alpha f(t) + \beta g(t)$	$\alpha \widehat{f}_{\mathrm{s}}(\omega) + \beta \widehat{g}_{\mathrm{s}}(\omega)$
Transform of derivatives: $f'(t)$	$-\omega \widehat{f}_{\mathrm{c}}(\omega)$
$f''(t)$	$-\omega^2 \widehat{f}_{\mathrm{s}}(\omega) - \sqrt{\dfrac{2}{\pi}} \omega f(0)$
Convolution theorem: $\frac{1}{2} \int_0^\infty [f(|t - x|) - f(|t + x|)] g(x) dx$	$\widehat{f}_{\mathrm{c}}(\omega) \widehat{g}_{\mathrm{s}}(\omega)$

Table 2.3. A short table of Fourier transforms: u is the Heaviside step function

$f(t) = \frac{1}{2\pi}\int_{-\infty}^{\infty} e^{i\omega t}\widehat{f}(\omega)d\omega$	$\widehat{f}(\omega) = \int_{-\infty}^{\infty} e^{-i\omega t} f(t)\,dt$
$\dfrac{1}{t^2 + a^2}$ $(a > 0)$	$\dfrac{\pi}{a}e^{-a|\omega|}$
$u(t)e^{-at}$	$\dfrac{1}{a + i\omega}$
$u(-t)e^{at}$	$\dfrac{1}{a - i\omega}$
$e^{-a|t|}$ $(a > 0)$	$\dfrac{2a}{a^2 + \omega^2}$
e^{-t^2}	$\sqrt{\pi}e^{-\omega^2/4}$
$\dfrac{1}{2a\sqrt{\pi}}e^{-t^2/(2a)^2}$ $(a > 0)$	$e^{-a^2\omega^2}$
$\dfrac{1}{\sqrt{|t|}}$	$\sqrt{\dfrac{2\pi}{|\omega|}}$
$u(t + a) - u(t - a)$	$\dfrac{2\sin\omega a}{\omega}$
$\delta(t - a)$	$e^{-i\omega a}$
$f(at + b)$ $(a > 0)$	$\dfrac{1}{a}e^{ib\omega/a}\widehat{f}(\dfrac{\omega}{a})$
Linearity of transform and inverse: $\alpha f(t) + \beta g(t)$	$\alpha\widehat{f}(\omega) + \beta\widehat{g}$
Transform of derivative: $f^{(n)}(t)$	$(i\omega)^n\widehat{f}(\omega)$
Transform of integral: $f(t) = \int_{-\infty}^{t} g(x)dx$	$\widehat{f}(\omega) = \dfrac{1}{i\omega}\widehat{g}(\omega)$
Convolution theorems: $f(t) * g(t) = \int_{-\infty}^{\infty} f(t - x)g(x)dx$	$\widehat{f}(\omega)\widehat{g}(\omega)$
$f(t)g(t)$	$\dfrac{1}{2\pi}\widehat{f}(\omega) * \widehat{g}(\omega)$

and write the series as

$$f(t) = \sum_{n=-\infty}^{\infty} \left[\frac{1}{2\pi} \int_{-p}^{p} f(t)e^{-i\omega_n t}dt \right] e^{i\omega_n t} \Delta\omega \qquad (2.7)$$

$$= \sum_{n=-\infty}^{\infty} \frac{1}{2\pi} \widehat{f_p}(\omega_n)e^{i\omega_n t} \Delta\omega$$

with

$$\widehat{f_p}(\omega_n) = \int_{-p}^{p} f(t)e^{-i\omega_n t}dt.$$

Now if we let $p \to \infty$, then $\Delta\omega \to 0$ and ω_n becomes a continuous variable. Furthermore

$$\widehat{f}(\omega) = \lim_{p \to \infty} \widehat{f_p}(\omega_n) = \int_{-\infty}^{\infty} f(t)e^{-i\omega t} \, dt \qquad (2.8)$$

and the infinite sum of (2.7) becomes an integral

$$f(t) = \frac{1}{2\pi} \int_{-\infty}^{\infty} \widehat{f}(\omega)e^{i\omega t}d\omega. \qquad (2.9)$$

This integral is known as Fourier integral.

The coefficient function $\widehat{f}(\omega)$ is known as the Fourier transform of $f(t)$. The process of transforming the function $f(t)$ in the time domain into the same function $\widehat{f}(\omega)$ in the frequency domain is expressed as $\mathcal{F}\{f(t)\}$,

$$\mathcal{F}\{f(t)\} = \int_{-\infty}^{\infty} f(t)e^{-i\omega t} \, dt = \widehat{f}(\omega). \qquad (2.10)$$

The process of getting back to $f(t)$ from $\widehat{f}(\omega)$ is known as inverse Fourier transform $\mathcal{F}^{-1}\{\widehat{f}(\omega)\}$,

$$\mathcal{F}^{-1}\{\widehat{f}(\omega)\} = \frac{1}{2\pi} \int_{-\infty}^{\infty} \widehat{f}(\omega)e^{i\omega t} \, d\omega = f(t). \qquad (2.11)$$

We have "derived" this pair of Fourier transforms with the same heuristic arguments as we introduced the Fourier cosine transform. Comments there are also applicable here. Formulas (2.10) and (2.11) can be established rigorously provided (1) $f(t)$ is piecewise continuous and differentiable and (2) it is absolutely integrable, that is, $\int_{-\infty}^{\infty} |f(t)| \, dt$ is finite.

The multiplicative factor in front of the integral is somewhat arbitrary. If $\widehat{f}(\omega)$ is defined as

$$\mathcal{F}\{f(t)\} = \alpha \int_{-\infty}^{\infty} f(t)e^{-i\omega t} \, dt = \widehat{f}(\omega),$$

then $\mathcal{F}^{-1}\{\widehat{f}(\omega)\}$ becomes

$$\mathcal{F}^{-1}\{\widehat{f}(\omega)\} = \beta \int_{-\infty}^{\infty} \widehat{f}(\omega)e^{i\omega t}\,d\omega = f(t),$$

where

$$\alpha\beta = \frac{1}{2\pi}.$$

Some authors chose $\alpha = \beta = \sqrt{\dfrac{1}{2\pi}}$, so that the Fourier pair is symmetrical. Others chose $\alpha = \dfrac{1}{2\pi}$, $\beta = 1$. In (2.10) and (2.11), α is chosen to be 1 and β to be $\dfrac{1}{2\pi}$.

Another convention, that is common in spectral analysis, is to use frequency ν, instead of angular frequency ω in defining the Fourier transforms. Since $\omega = 2\pi\nu$, (2.10) can be written as

$$\mathcal{F}\{f(t)\} - \int_{-\infty}^{\infty} f(t)e^{-i2\pi\nu t}\,dt = \widehat{f}(\nu) \tag{2.12}$$

and (2.11) becomes

$$\mathcal{F}^{-1}\{\widehat{f}(\nu)\} = \int_{-\infty}^{\infty} \widehat{f}(\nu)e^{i2\pi\nu t}\,d\nu = f(t). \tag{2.13}$$

Note that in this pair of equations, the factor 2π is no longer there. Besides, frequency is a well-defined concept and no one actually measures angular frequency. These are good reasons to use (2.12) and (2.13) as the definition of Fourier transforms. However, for historic reasons, most books in engineering and physics use ω. Therefore we will continue to use (2.10) and (2.11) as the definition of the Fourier transforms.

The function $f(t)$ in the Fourier transform may or may not have any even or odd parity. However, if it is an even function, it can be easily shown that it reduces to the Fourier cosine transform. If it is an odd function, it reduces to the Fourier sine transform.

Example 2.3.1. Find the Fourier transform of

$$f(t) = \begin{cases} e^{-\alpha t} & t > 0, \\ 0 & t < 0. \end{cases}$$

Solution 2.3.1.

$$\mathcal{F}\{f(t)\} = \int_{-\infty}^{\infty} f(t)e^{-i\omega t}\,dt = \int_{0}^{\infty} e^{-(\alpha+i\omega)t}\,dt$$

$$= -\frac{1}{\alpha+i\omega}e^{-(\alpha+i\omega)t}\Big|_{0}^{\infty} = \frac{1}{\alpha+i\omega}.$$

This result can, of course, be expressed as a real part plus an imaginary part,

$$\frac{1}{\alpha + i\omega} = \frac{1}{\alpha + i\omega}\frac{\alpha - i\omega}{\alpha - i\omega} = \frac{\alpha}{\alpha^2 + \omega^2} - i\frac{\omega}{\alpha^2 + \omega^2}.$$

Example 2.3.2. Find the inverse Fourier transform of

$$\widehat{f}(\omega) = \frac{1}{\alpha + i\omega}.$$

(This problem can be skipped for those who have not yet studied the complex contour integration.)

Solution 2.3.2.

$$\mathcal{F}^{-1}\{\widehat{f}(\omega)\} = \frac{1}{2\pi}\int_{-\infty}^{\infty}\frac{1}{\alpha + i\omega}e^{i\omega t}\,d\omega = \frac{1}{2\pi i}\int_{-\infty}^{\infty}\frac{1}{\omega - i\alpha}e^{i\omega t}\,d\omega.$$

This integrals can be evaluated with contour integration. For $t > 0$, the contour can be closed counterclockwise in the upper half plane as shown in Fig. 2.1a.

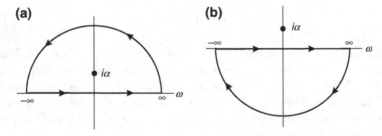

Fig. 2.1. Contour integration for inverse Fourier transform. **(a)** The contour is closed in the upper half plane. **(b)** The contour is closed in the lower half plane

$$\frac{1}{2\pi i}\int_{-\infty}^{\infty}\frac{1}{\omega - i\alpha}e^{i\omega t}\,d\omega = \frac{1}{2\pi i}\oint_{u.h.p.}\frac{1}{\omega - i\alpha}e^{i\omega t}\,d\omega$$

$$= \lim_{\omega \to i\alpha} e^{i\omega t} = e^{-\alpha t}.$$

It follows that for $t > 0$:

$$\mathcal{F}^{-1}\{\widehat{f}(\omega)\} = e^{-\alpha t}.$$

For $t < 0$, the contour can be closed clockwise in the lower half plane as shown in Fig. 2.1b. Since there is no singular point in the lower half plane

$$\frac{1}{2\pi i}\int_{-\infty}^{\infty}\frac{1}{\omega - i\alpha}e^{i\omega t}\,d\omega = \frac{1}{2\pi i}\oint_{l.h.p.}\frac{1}{\omega - i\alpha}e^{i\omega t}\,d\omega = 0.$$

Thus, for $t < 0$,

$$\mathcal{F}^{-1}\{\widehat{f}(\omega)\} = 0.$$

With the Heaviside step function

$$u(t) = \begin{cases} 1 & \text{for } t > 0, \\ 0 & \text{for } t < 0, \end{cases}$$

we can combine the results for $t > 0$ and for $t < 0$ as

$$\mathcal{F}^{-1}\{\widehat{f}(\omega)\} = u(t)e^{-\alpha t}.$$

It is seen that the inverse transform is indeed equal to $f(t)$ of the previous problem.

2.4 Fourier Transform and Delta Function

2.4.1 Orthogonality

If we put $\widehat{f}(\omega)$ of (2.8) back in the Fourier integral of (2.9), the Fourier representation of $f(t)$ takes the form

$$f(t) = \frac{1}{2\pi} \int_{-\infty}^{\infty} \left[\int_{-\infty}^{\infty} f(t')e^{-i\omega t'}\, dt' \right] e^{i\omega t}\, d\omega$$

which, after reversing the order of integration, can be written as

$$f(t) = \int_{-\infty}^{\infty} f(t') \left[\frac{1}{2\pi} \int_{-\infty}^{\infty} e^{i\omega(t-t')}d\omega \right] dt'.$$

Recall that the Dirac delta function $\delta(t - t')$ is defined by the relation

$$f(t) = \int_{-\infty}^{\infty} f(t')\delta(t - t')dt'.$$

Comparing the last two equations, we see that $\delta(t - t')$ can be written as

$$\delta(t - t') = \frac{1}{2\pi} \int_{-\infty}^{\infty} e^{i\omega(t-t')}d\omega. \tag{2.14}$$

Interchange the variables gives the inverted form

$$\delta(\omega - \omega') = \frac{1}{2\pi} \int_{-\infty}^{\infty} e^{i(\omega-\omega')t}dt.$$

The last two equations are known as the orthogonality conditions. A function $e^{i\omega t}$ is orthogonal to all other functions in the form of $e^{-i\omega' t}$ when integrated over all t, as long as $\omega' \neq \omega$.

Since $\delta(x) = \delta(-x)$, (2.14) can also written as

$$\delta(t - t') = \frac{1}{2\pi} \int_{-\infty}^{\infty} e^{-i\omega(t-t')} d\omega.$$

These formulas are very useful representations of delta functions. The derivation of many transform pairs are greatly simplified with the use of delta functions. Although they are not proper mathematical functions, their use can be justified by the distribution theory.

2.4.2 Fourier Transforms Involving Delta Functions

Dirac Delta Function. Consider the function

$$f(t) = K\delta(t),$$

where K is a constant. The Fourier transform of $f(t)$ is easily derived using the definition of the delta function

$$\mathcal{F}\{f(t)\} = \int_{-\infty}^{\infty} K\delta(t)e^{-i\omega t}\, dt = Ke^0 = K.$$

The inverse Fourier transform is given by

$$\mathcal{F}^{-1}\{\widehat{f}(\omega)\} = \frac{1}{2\pi} \int_{-\infty}^{\infty} Ke^{i\omega t}\, dt = K\delta(t).$$

Similarly, the Fourier transform of a constant function K is

$$\mathcal{F}\{K\} = 2\pi K\delta(\omega)$$

and its inverse is

$$\mathcal{F}^{-1}\{2\pi K\delta(\omega)\} = K.$$

These Fourier transform pairs are illustrated in Fig. 2.2.

Periodic Functions. To illustrate the Fourier transform of a periodic function, consider

$$f(t) = A \cos \omega_0 t.$$

The Fourier transform is given by

$$\mathcal{F}\{A \cos \omega_0 t\} = \int_{-\infty}^{\infty} A \cos(\omega_0 t)e^{-i\omega t}\, dt.$$

Since

$$\cos \omega_0 t = \frac{1}{2}\left(e^{i\omega_0 t} + e^{-i\omega_0 t}\right),$$

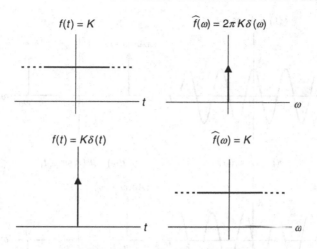

Fig. 2.2. The Fourier transform pair of constant and delta functions. The Fourier transform of constant function is a delta function. The Fourier transform of a delta function is a constant function

so

$$\mathcal{F}\{A\cos\omega_0 t\} = \frac{A}{2}\int_{-\infty}^{\infty}\left[e^{-\mathrm{i}(\omega-\omega_0)t} + e^{-\mathrm{i}(\omega+\omega_0)t}\right]\mathrm{d}t.$$

Using (2.14), we have

$$\mathcal{F}\{A\cos\omega_0 t\} = \pi A\delta(\omega+\omega_0) + \pi A\delta(\omega-\omega_0). \qquad (2.15)$$

Similarly,

$$\mathcal{F}\{A\sin\omega_0 t\} = \mathrm{i}\pi A\delta(\omega+\omega_0) - \mathrm{i}\pi A\delta(\omega-\omega_0). \qquad (2.16)$$

Note that the Fourier transform of a sine function is imaginary.

These Fourier transform pairs are shown in Fig. 2.3, leaving out the factor of i in (2.16).

2.4.3 Three-Dimensional Fourier Transform Pair

So far we have used as variables t and ω, representing time and angular frequency, respectively. Mathematics will, of course, be exactly the same if we change the names of these variables. In describing the spatial variations of a wave, it is more natural to use either r or x, y, and z to represent distances. In a function of time, the period T is the time interval after which the function repeats itself. In a function of distance, the corresponding quantity is called wavelength λ, which is the increase in distance that the function will repeat itself. Therefore, if t is replaced by r, then the angular frequency ω, which is equal to $2\pi/T$, should be replaced by a quantity equal to $2\pi/\lambda$, which is known as the wave number k.

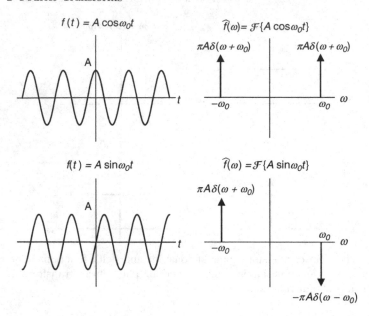

Fig. 2.3. Fourier transform pair of cosine and sine functions

Thus, corresponding to (2.14), we have

$$\delta(x - x') = \frac{1}{2\pi} \int_{-\infty}^{\infty} e^{ik_1(x-x')} dk_1,$$

$$\delta(y - y') = \frac{1}{2\pi} \int_{-\infty}^{\infty} e^{ik_2(y-y')} dk_2,$$

$$\delta(z - z') = \frac{1}{2\pi} \int_{-\infty}^{\infty} e^{ik_3(z-z')} dk_3.$$

Therefore in three-dimensional space, the delta function is given by

$$\delta(\mathbf{r} - \mathbf{r}') = \delta(x - x')\delta(y - y')\delta(z - z')$$

$$= \frac{1}{2\pi} \int_{-\infty}^{\infty} e^{ik_1(x-x')} dk_1 \frac{1}{2\pi} \int_{-\infty}^{\infty} e^{ik_2(y-y')} dk_2 \frac{1}{2\pi} \int_{-\infty}^{\infty} e^{ik_3(z-z')} dk_3$$

$$= \frac{1}{(2\pi)^3} \iiint_{-\infty}^{\infty} e^{i[k_1(x-x')+k_2(y-y')+k_3(z-z')]} dk_1\, dk_2\, dk_3.$$

A convenient notation is to introduce a wave vector $\mathbf{k}$,

$$\mathbf{k} = k_1\widehat{\mathbf{i}} + k_2\widehat{\mathbf{j}} + k_3\widehat{\mathbf{k}}.$$

Together with

$$\mathbf{r} - \mathbf{r}' = (x - x')\widehat{\mathbf{i}} + (y - y')\widehat{\mathbf{j}} + (z - z')\widehat{\mathbf{k}}$$

the three-dimensional delta function can be written as

$$\delta(\mathbf{r} - \mathbf{r}') = \frac{1}{(2\pi)^3} \iiint_{-\infty}^{\infty} e^{i\mathbf{k}\cdot(\mathbf{r}-\mathbf{r}')} d^3k.$$

Now by definition of the delta function

$$f(\mathbf{r}) = \iiint_{-\infty}^{\infty} f(\mathbf{r}')\delta(\mathbf{r} - \mathbf{r}')d^3r',$$

we have

$$f(\mathbf{r}) = \iiint_{-\infty}^{\infty} f(\mathbf{r}') \frac{1}{(2\pi)^3} \iiint_{-\infty}^{\infty} e^{i\mathbf{k}\cdot(\mathbf{r}-\mathbf{r}')} d^3k\, d^3r',$$

which can be written as

$$f(\mathbf{r}) = \frac{1}{(2\pi)^{3/2}} \iiint_{-\infty}^{\infty} \left[\frac{1}{(2\pi)^{3/2}} \iiint_{-\infty}^{\infty} f(\mathbf{r}')e^{-i\mathbf{k}\cdot\mathbf{r}'} d^3r' \right] e^{i\mathbf{k}\cdot\mathbf{r}}\, d^3k.$$

Thus, in three dimensions, we can define a Fourier transform pair

$$\widehat{f}(\mathbf{k}) = \frac{1}{(2\pi)^{3/2}} \iiint_{-\infty}^{\infty} f(\mathbf{r})e^{-i\mathbf{k}\cdot\mathbf{r}}\, d^3r = \mathcal{F}\{f(\mathbf{r})\},$$

$$f(\mathbf{r}) = \frac{1}{(2\pi)^{3/2}} \iiint_{-\infty}^{\infty} \widehat{f}(\mathbf{k})\, e^{i\mathbf{k}\cdot\mathbf{r}}\, d^3k = \mathcal{F}^{-1}\left\{\widehat{f}(\mathbf{k})\right\}.$$

Again, how to split $1/(2\pi)^3$ between the Fourier transform and its inverse is somewhat arbitrary. Here we split them equally to conform with most of the quantum mechanics text books.

In quantum mechanics, the momentum $\mathbf{p}$ is given by $\mathbf{p} = \hbar\mathbf{k}$. The Fourier transform pair in terms of $\mathbf{r}$ and $\mathbf{p}$ is therefore given by

$$\widehat{f}(\mathbf{p}) = \frac{1}{(2\pi\hbar)^{3/2}} \iiint_{-\infty}^{\infty} f(\mathbf{r})e^{-i\mathbf{p}\cdot\mathbf{r}/\hbar} d^3r,$$

$$f(\mathbf{r}) = \frac{1}{(2\pi\hbar)^{3/2}} \iiint_{-\infty}^{\infty} \widehat{f}(\mathbf{p})e^{i\mathbf{p}\cdot\mathbf{r}/\hbar} d^3p.$$

If $f(\mathbf{r})$ is the Schrödinger wave function, then its Fourier transform $\widehat{f}(\mathbf{p})$ is the momentum wave function. In describing a dynamic system, either space or momentum wave functions may be used, depending on which is more convenient for the particular problem.

If in three-dimensional space, a function possesses spherical symmetry, that is, $f(\mathbf{r}) = f(r)$, then its Fourier transform is reduced to a one-dimensional integral. In this case, let the wave vector $\mathbf{k}$ be along the z-axis of the coordinate space, so

$$\mathbf{k} \cdot \mathbf{r} = kr\cos\theta$$

and
$$d^3r = r^2 \sin\theta \, d\theta \, dr \, d\varphi.$$

The Fourier transform of $f(r)$ becomes

$$\mathcal{F}\{f(\mathbf{r})\} = \frac{1}{(2\pi)^{3/2}} \int_0^{2\pi} d\varphi \int_0^\infty f(r) \left[\int_0^\pi e^{-ikr\cos\theta} \sin\theta \, d\theta \right] r^2 \, dr$$

$$= \frac{1}{(2\pi)^{3/2}} 2\pi \int_0^\infty f(r) \left[\frac{1}{ikr} e^{-ikr\cos\theta} \right]_0^\pi r^2 \, dr$$

$$= \frac{1}{(2\pi)^{3/2}} 2\pi \int_0^\infty f(r) \frac{2\sin kr}{kr} r^2 \, dr = \sqrt{\frac{2}{\pi}}\frac{1}{k} \int_0^\infty f(r) r \sin kr \, dr.$$

Example 2.4.1. Find the Fourier transform of

$$f(r) = \frac{z^3}{\pi} e^{-2zr},$$

where z is a constant.

Solution 2.4.1.

$$\mathcal{F}\{f(\mathbf{r})\} = \sqrt{\frac{2}{\pi}}\frac{1}{k} \int_0^\infty \frac{z^3}{\pi} e^{-2zr} r \sin kr \, dr.$$

One way to evaluate this integral is to recall the Laplace transform of $\sin kr$

$$\int_0^\infty e^{-sr} \sin kr \, dr = \frac{k}{s^2 + k^2},$$

$$\frac{d}{ds} \int_0^\infty e^{-sr} \sin kr \, dr = \int_0^\infty (-r) e^{-sr} \sin kr \, dr,$$

$$\frac{d}{ds} \frac{k}{s^2 + k^2} = \frac{-2sk}{(s^2 + k^2)^2}.$$

So

$$\int_0^\infty r e^{-sr} \sin kr \, dr = \frac{2sk}{(s^2 + k^2)^2}.$$

With $s = 2z$, we have

$$\int_0^\infty e^{-2zr} r \sin kr \, dr = \frac{4zk}{(4z^2 + k^2)^2}.$$

It follows that:

$$\mathcal{F}\{f(\mathbf{r})\} = \sqrt{\frac{2}{\pi}}\frac{1}{k}\frac{z^3}{\pi} \frac{4zk}{(4z^2 + k^2)^2} = \left(\frac{2}{\pi}\right)^{3/2} \frac{2z^4}{(4z^2 + k^2)^2}.$$

2.5 Some Important Transform Pairs

There are some prototype Fourier transform pairs that we should be familiar with. Not only they frequently occur in engineering and physics, they also form the base upon which transforms of other functions can derived.

2.5.1 Rectangular Pulse Function

The rectangular function is defined as

$$\Pi_a(t) = \begin{cases} 1 & -a \leq t \leq a, \\ 0 & \text{otherwise.} \end{cases}$$

This function is sometimes called box function or top-hat function. It can be expressed as

$$\Pi_a(t) = u(t+a) - u(t-a),$$

where $u(t)$ is the Heaviside step function,

$$u(t) = \begin{cases} 1 & t > 0, \\ 0 & t < 0. \end{cases}$$

The Fourier transform of this function is given by

$$\mathcal{F}\{\Pi_a'(t)\} = \int_{-\infty}^{\infty} \Pi_a(t) e^{-i\omega t} dt = \int_{-a}^{a} e^{-i\omega t} dt$$

$$= \frac{e^{-i\omega t}}{-i\omega}\Big|_{-a}^{a} = \frac{e^{-i\omega a} - e^{i\omega a}}{-i\omega} = \frac{2\sin\omega a}{\omega} = \widehat{f}(\omega).$$

In terms of "sinc function," defined as $\sin c\,(x) = \dfrac{\sin x}{x}$, we have

$$\mathcal{F}\{\Pi_a(t)\} = 2a \sin c\,(a\omega).$$

This Fourier transform pair is shown in Fig. 2.4.

2.5.2 Gaussian Function

The Gaussian function is defined as

$$f(t) = e^{-\alpha t^2}.$$

Its Fourier transform is given by

$$\mathcal{F}\left\{e^{-\alpha t^2}\right\} = \int_{-\infty}^{\infty} e^{-\alpha t^2} e^{-i\omega t}\, dt = \int_{-\infty}^{\infty} e^{-\alpha t^2 - i\omega t}\, dt = \widehat{f}(\omega).$$

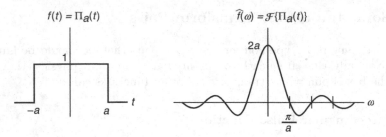

Fig. 2.4. Fourier transform pair of a rectangular function. Note that $\widehat{f}(0) = 2a$, and the zeros of $\widehat{f}(\omega)$ are at $\omega = \pi/a,\; 2\pi/a,\; 3\pi/a, \cdots$

Completing the square of the exponential

$$\alpha t^2 + i\omega t = \left(\sqrt{\alpha}t + \frac{i\omega}{2\sqrt{\alpha}}\right)^2 + \frac{\omega^2}{4\alpha},$$

we have

$$\int_{-\infty}^{\infty} \exp\left\{-\left[\left(\sqrt{\alpha}t + \frac{i\omega}{2\sqrt{\alpha}}\right)^2 + \frac{\omega^2}{4\alpha}\right]\right\} dt$$

$$= \exp\left(-\frac{\omega^2}{4\alpha}\right) \int_{-\infty}^{\infty} \exp\left\{-\left(\sqrt{\alpha}t + \frac{i\omega}{2\sqrt{\alpha}}\right)^2\right\} dt.$$

Let

$$u = \sqrt{\alpha}t + \frac{i\omega}{2\sqrt{\alpha}}, \qquad du = \sqrt{\alpha}\, dt$$

then we can write the Fourier transform as

$$\widehat{f}(\omega) = \exp\left(-\frac{\omega^2}{4\alpha}\right) \frac{1}{\sqrt{\alpha}} \int_{-\infty}^{\infty} e^{-u^2}\, du.$$

Since

$$\int_{-\infty}^{\infty} e^{-u^2}\, du = \sqrt{\pi},$$

thus

$$\widehat{f}(\omega) = \sqrt{\frac{\pi}{\alpha}} \exp\left(-\frac{\omega^2}{4\alpha}\right).$$

It is interesting to note that $\widehat{f}(\omega)$ is also of a Gaussian function with a peak at the origin, monotonically decreasing as $k \to \pm\infty$. If $f(t)$ is sharply peaked (large α), then $\widehat{f}(\omega)$ is flattened, and vice versa. This is a general feature in the theory of Fourier transforms. In quantum-mechanical applications it is related to the Heisenberg uncertainty principle. The pair of Gaussian transforms is shown in Fig. 2.5.

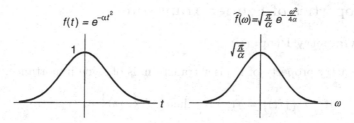

Fig. 2.5. The Fourier transform of a Gaussian function is another Gaussian function

2.5.3 Exponentially Decaying Function

The Fourier transform of the exponentially decaying function

$$f(t) = e^{-a|t|}, \quad a > 0$$

is given by

$$\mathcal{F}\left\{e^{-a|t|}\right\} = \int_{-\infty}^{\infty} e^{-a|t|} e^{-i\omega t} \, dt$$

$$= \int_{-\infty}^{0} e^{at} e^{-i\omega t} \, dt + \int_{0}^{\infty} e^{-at} e^{-i\omega t} \, dt$$

$$= \left. \frac{e^{(a-i\omega)t}}{a - i\omega} \right|_{-\infty}^{0} + \left. \frac{e^{-(a+i\omega)t}}{-(a + i\omega)} \right|_{0}^{\infty}$$

$$= \frac{1}{a - i\omega} + \frac{1}{a + i\omega} = \frac{2a}{a^2 + \omega^2} = \widehat{f}(\omega).$$

This is a bell-shaped curve, similar in appearance to a Gaussian curve and is known as a Lorentz profile. This pair of transforms is shown in Fig. 2.6.

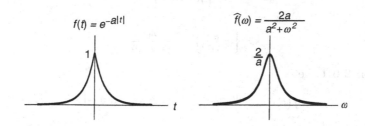

Fig. 2.6. The Fourier transform of an exponentail decaying function is Lorentz profile

2.6 Properties of Fourier Transform

2.6.1 Symmetry Property

The symmetry property of Fourier transform is of some importance.

$$\text{If} \quad \mathcal{F}\{f(t)\} = \widehat{f}(\omega), \quad \text{then} \quad \mathcal{F}\left\{\widehat{f}(t)\right\} = 2\pi f(-\omega).$$

Proof. Since

$$\widehat{f}(\omega) = \int_{-\infty}^{\infty} f(t) e^{-i\omega t}\, dt,$$

by definition

$$f(t) = \frac{1}{2\pi} \int_{-\infty}^{\infty} \widehat{f}(\omega)\, e^{i\omega t} d\omega.$$

Interchanging t and ω, we have

$$f(\omega) = \frac{1}{2\pi} \int_{-\infty}^{\infty} \widehat{f}(t) e^{i\omega t}\, dt.$$

Clearly,

$$f(-\omega) = \frac{1}{2\pi} \int_{-\infty}^{\infty} \widehat{f}(t) e^{-i\omega t}\, dt.$$

Therefore

$$\mathcal{F}\left\{\widehat{f}(t)\right\} = \int_{-\infty}^{\infty} \widehat{f}(t) e^{-i\omega t}\, dt = 2\pi f(-\omega).$$

Using this simple relation, we can avoid many complicated mathematical manipulations.

Example 2.6.1. Find

$$\mathcal{F}\left\{\frac{1}{a^2 + t^2}\right\}$$

from

$$\mathcal{F}\left\{e^{-a|t|}\right\} = \frac{2a}{a^2 + \omega^2}.$$

Solution 2.6.1. Let

$$f(t) = e^{-a|t|}, \quad \text{so} \quad f(-\omega) = e^{-a|\omega|}$$

and

$$\mathcal{F}\{f(t)\} = \widehat{f}(\omega) = \frac{2a}{a^2 + \omega^2}.$$

Thus

$$\widehat{f}(t) = \frac{2a}{a^2 + t^2},$$

$$\mathcal{F}\left\{\widehat{f}(t)\right\} = \mathcal{F}\left\{\frac{2a}{a^2+t^2}\right\} = 2\pi f(-\omega).$$

Therefore

$$\mathcal{F}\left\{\frac{1}{a^2+t^2}\right\} = \frac{\pi}{a}\,e^{-a|\omega|}.$$

This result can also be found by complex contour integration.

2.6.2 Linearity, Shifting, Scaling

Linearity of the Transform and its Inverse. If $\mathcal{F}\{f(t)\} = \widehat{f}(\omega)$ and $\mathcal{F}\{g(t)\} = \widehat{g}(\omega)$, then

$$\mathcal{F}\{af(t)+bg(t)\} = \int_{-\infty}^{\infty}\left[af(t)+bg(t)\right]e^{-i\omega t}\,dt$$

$$= a\int_{-\infty}^{\infty}f(t)e^{-i\omega t}\,dt + b\int_{-\infty}^{\infty}g(t)e^{-i\omega t}\,dt$$

$$= a\mathcal{F}\{f(t)\} + b\mathcal{F}\{f(t)\} = a\widehat{f}(\omega) + b\widehat{g}(\omega).$$

Similarly,

$$\mathcal{F}^{-1}\left\{a\widehat{f}(\omega)+b\widehat{g}(\omega)\right\} = a\mathcal{F}^{-1}\left\{\widehat{f}(\omega)\right\} + b\mathcal{F}^{-1}\left\{\widehat{g}(\omega)\right\}$$

$$= af(t)+bg(t).$$

These simple relations are of considerable importance because it reflects the applicability of the Fourier transform to the analysis of linear systems.

Time Shifting. If time is shifted by a in the Fourier transform

$$\mathcal{F}\{f(t-a)\} = \int_{-\infty}^{\infty}f\left(t-a\right)e^{-i\omega t}\,dt,$$

then by substituting $t-a=x,\quad dt=dx,\quad t=x+a$, we have

$$\mathcal{F}\{f(t-a)\} = \int_{-\infty}^{\infty}f\left(x\right)e^{-i\omega(x+a)}\,dx$$

$$= e^{-i\omega a}\int_{-\infty}^{\infty}f(x)e^{-i\omega x}\,dx = e^{-i\omega a}\,\widehat{f}(\omega).$$

Note that a time delay will only change the phase of the Fourier transform and not its magnitude. For example,

$$\sin\omega_0 t = \cos\left(\omega_0 t - \frac{\pi}{2}\right) = \cos\omega_0\left(t - \frac{\pi}{2}\frac{1}{\omega_0}\right).$$

Thus, if $f(t) = \cos \omega_0 t$, then $\sin \omega_0 t = f(t - a)$ with $a = \dfrac{\pi}{2} \dfrac{1}{\omega_0}$. Therefore

$$
\begin{aligned}
\mathcal{F}\{A \sin \omega_0 t\} &= e^{-i\omega \frac{\pi}{2} \frac{1}{\omega_0}} \mathcal{F}\{A \cos \omega_0 t\} \\
&= e^{-i\omega \frac{\pi}{2} \frac{1}{\omega_0}} [A\pi\delta(\omega - \omega_0) + A\pi\delta(\omega + \omega_0)] \\
&= e^{-i\frac{\pi}{2}} A\pi\delta(\omega - \omega_0) + e^{i\frac{\pi}{2}} A\pi\delta(\omega + \omega_0) \\
&= -iA\pi\delta(\omega - \omega_0) + iA\pi\delta(\omega + \omega_0),
\end{aligned}
$$

as shown in (2.16).

Frequency Shifting. If the frequency in $\widehat{f}(\omega)$ is shifted by a constant a, its inverse is multiplied by a factor of e^{iat}. Since

$$
\mathcal{F}^{-1}\left\{\widehat{f}(\omega - a)\right\} = \frac{1}{2\pi} \int_{-\infty}^{\infty} \widehat{f}(\omega - a) e^{i\omega t} \, d\omega,
$$

substituting $\varpi = \omega - a$, we have

$$
\mathcal{F}^{-1}\left\{\widehat{f}(\omega - a)\right\} = \frac{1}{2\pi} \int_{-\infty}^{\infty} \widehat{f}(\varpi) e^{i(\varpi + a)t} \, d\varpi = e^{iat} f(t)
$$

or

$$
\widehat{f}(\omega - a) = \mathcal{F}\left\{e^{iat} f(t)\right\}.
$$

To illustrate the effect of frequency shifting, let us consider the case that $f(t)$ is multiplied by $\cos \omega_0 t$. Since $\cos \omega_0 t = \frac{1}{2}\left(e^{i\omega_0 t} + e^{-i\omega_0 t}\right)$, so

$$
f(t) \cos \omega_0 t = \frac{1}{2} e^{i\omega_0 t} f(t) + \frac{1}{2} e^{-i\omega_0 t} f(t)
$$

and

$$
\begin{aligned}
\mathcal{F}\{f(t) \cos \omega_0 t\} &= \frac{1}{2} \mathcal{F}\left\{e^{i\omega_0 t} f(t)\right\} + \frac{1}{2} \mathcal{F}\left\{e^{-i\omega_0 t} f(t)\right\} \\
&= \frac{1}{2} \widehat{f}(\omega - \omega_0) + \frac{1}{2} \widehat{f}(\omega + \omega_0).
\end{aligned}
$$

This process is known as modulation. In other words, when $f(t)$ is modulated by $\cos \omega_0 t$, its frequency is symmetrically shifted up and down by ω_0.

Time Scaling. If $\mathcal{F}\{f(t)\} = \widehat{f}(\omega)$, then the Fourier transform of $f(at)$ can be determined by substituting $t' = at$ in the Fourier integral

$$
\begin{aligned}
\mathcal{F}\{f(at)\} &= \int_{-\infty}^{\infty} f(at) e^{-i\omega t} \, dt \\
&= \int_{-\infty}^{\infty} f(t') e^{-i\omega t'/a} \frac{1}{a} \, dt' = \frac{1}{a} \widehat{f}\left(\frac{\omega}{a}\right).
\end{aligned}
$$

This is correct for $a > 0$. However, if a is negative, then $t' = at = -|a|\,t$. As a consequence, when the integration variable is changed from t to t', the integration limits should also be interchanged. That is,

$$\mathcal{F}\{f(at)\} = \int_{-\infty}^{\infty} f(at)\,e^{-i\omega t}\,dt = \int_{\infty}^{-\infty} f(t')\,e^{-i\omega t'/a}\frac{1}{-|a|}\,dt'$$

$$= \frac{1}{|a|}\int_{-\infty}^{\infty} f(t')\,e^{-i\omega t'/a}\,dt' = \frac{1}{|a|}\widehat{f}\left(\frac{\omega}{a}\right).$$

Therefore, in general

$$\mathcal{F}\{f(at)\} = \frac{1}{|a|}\widehat{f}\left(\frac{\omega}{a}\right).$$

This means that as the time scale expands, the frequency scale not only contracts, its amplitude will also increase. It increases in such a way as to keep the area constant.

Frequency Scaling. This is just the reverse of time scaling. If $\mathcal{F}^{-1}\left\{\widehat{f}(\omega)\right\} = f(t)$, then

$$\mathcal{F}^{-1}\left\{\widehat{f}(a\omega)\right\} = \frac{1}{2\pi}\int_{-\infty}^{\infty} \widehat{f}(a\omega)\,e^{i\omega t}\,d\omega$$

$$= \frac{1}{2\pi}\int_{-\infty}^{\infty} \widehat{f}(\omega')\,e^{i\omega' t/a}\frac{1}{|a|}\,d\omega' = \frac{1}{|a|}f\left(\frac{t}{a}\right).$$

This means that as the frequency scale expands, the time scale will contract and the amplitude of the time function will also increase.

2.6.3 Transform of Derivatives

If the transform of nth derivative $f^n(t)$ exists, then $f^n(t)$ must be integrable over $(-\infty, \infty)$. That means $f^n(t) \to 0$, as $t \to \pm\infty$. With this assumption, the Fourier transforms of derivatives of $f(t)$ can be expressed in terms of the transform of $f(t)$. This can be shown as follows:

$$\mathcal{F}\{f'(t)\} = \int_{-\infty}^{\infty} f'(t)e^{-i\omega t}\,dt = \int_{-\infty}^{\infty} \frac{df(t)}{dt}e^{-i\omega t}\,dt$$

$$= f(t)e^{-i\omega t}\Big|_{-\infty}^{\infty} + i\omega \int_{-\infty}^{\infty} f(t)e^{-i\omega t}\,dt.$$

The integrated term is equal to zero at both limits. Thus

$$\mathcal{F}\{f'(t)\} = i\omega \int_{-\infty}^{\infty} f(t)e^{-i\omega t}\,dt = i\omega\mathcal{F}\{f(t)\} = i\omega\widehat{f}(\omega).$$

It follows that:

$$\mathcal{F}\{f''(t)\} = i\omega\mathcal{F}\{f'(t)\} = (i\omega)^2\,\mathcal{F}\{f(t)\} = (i\omega)^2\,\widehat{f}(\omega).$$

Therefore

$$\mathcal{F}\{f^n(t)\} = (i\omega)^n\,\mathcal{F}\{f(t)\} = (i\omega)^n\,\widehat{f}(\omega).$$

Thus a differentiation in the time domain becomes a simple multiplication in the frequency domain.

2.6.4 Transform of Integral

The Fourier transform of the following integral:

$$I(t) = \int_{-\infty}^{t} f(x)\,dx$$

can be found by using the relation for Fourier transform of derivatives. Since

$$\frac{d}{dt}I(t) = f(t),$$

it follows that:

$$\mathcal{F}\{f(t)\} = \mathcal{F}\left\{\frac{dI(t)}{dt}\right\} = i\omega\mathcal{F}\{I(t)\} = i\omega\mathcal{F}\left\{\int_{-\infty}^{t} f(x)\,dx\right\}.$$

Therefore

$$\mathcal{F}\left\{\int_{-\infty}^{t} f(x)dx\right\} = \frac{1}{i\omega}\mathcal{F}\{f(t)\}.$$

Thus an integration in the time domain becomes a division in the frequency domain.

2.6.5 Parseval's Theorem

The Parseval's theorem in Fourier series is equally valid in Fourier transform. The integral of the square of a function is related to the integral of the square of its transform in the following way:

$$\int_{-\infty}^{\infty} |f(t)|^2\,dt = \frac{1}{2\pi}\int_{-\infty}^{\infty}\left|\widehat{f}(\omega)\right|^2\,d\omega.$$

Since

$$f(t) = \frac{1}{2\pi}\int_{-\infty}^{\infty}\widehat{f}(\omega)\,e^{i\omega t}\,d\omega,$$

its complex conjugate is

$$f^*(t) = \left[\frac{1}{2\pi}\int_{-\infty}^{\infty}\widehat{f}(\omega)e^{i\omega t}\,d\omega\right]^* = \frac{1}{2\pi}\int_{-\infty}^{\infty}\widehat{f}^*(\omega)e^{-i\omega t}\,d\omega.$$

Thus

$$\int_{-\infty}^{\infty} |f(t)|^2 \, dt = \int_{-\infty}^{\infty} f(t) f^*(t) dt = \int_{-\infty}^{\infty} f(t) \left[\frac{1}{2\pi} \int_{-\infty}^{\infty} \widehat{f}^*(\omega) e^{-i\omega t} \, d\omega \right] dt.$$

Interchanging the ω and t integration,

$$\int_{-\infty}^{\infty} |f(t)|^2 \, dt = \frac{1}{2\pi} \int_{-\infty}^{\infty} \widehat{f}^*(\omega) \left[\int_{-\infty}^{\infty} f(t) e^{-i\omega t} \, dt \right] d\omega$$

$$= \frac{1}{2\pi} \int_{-\infty}^{\infty} \widehat{f}^*(\omega) \widehat{f}(\omega) d\omega = \frac{1}{2\pi} \int_{-\infty}^{\infty} \left| \widehat{f}(\omega) \right|^2 d\omega.$$

Written in terms of frequency ν, instead of angular frequency ω $(\omega = 2\pi\nu)$, this theorem is expressed as

$$\int_{-\infty}^{\infty} |f(t)|^2 \, dt = \int_{-\infty}^{\infty} \left| \widehat{f}(\nu) \right|^2 d\nu.$$

In physics, the total energy associated with wave form $f(t)$ (electromagnetic radiation, water waves, etc.) is proportional to $\int_{-\infty}^{\infty} |f(t)|^2 \, dt$. By Parseval's theorem, this energy is also given by $\int_{-\infty}^{\infty} \left| \widehat{f}(\nu) \right|^2 d\nu$. Therefore $\left| \widehat{f}(\nu) \right|^2$ is the energy content per unit frequency interval, and is known as "power density." For this reason, Parseval's theorem is also known as power theorem.

Example 2.6.2. Find the value of

$$I = \int_{-\infty}^{\infty} \frac{\sin^2 x}{x^2} dx$$

from the Parseval's theorem and the Fourier transform of

$$\Pi_1(t) = \begin{cases} 1 & |t| < 1, \\ 0 & |t| > 1. \end{cases}$$

Solution 2.6.2. Let $f(t) = \Pi_1(t)$, so

$$\mathcal{F}\{f(t)\} = \widehat{f}(\omega) = \int_{-\infty}^{\infty} \Pi_1(t) e^{-i\omega t} \, dt = \int_{-1}^{1} e^{-i\omega t} \, dt$$

$$= -\frac{1}{i\omega} e^{-i\omega t} \Big|_{-1}^{1} = \frac{1}{i\omega} \left(e^{i\omega} - e^{-i\omega} \right) = \frac{2\sin\omega}{\omega}$$

and

$$\int_{-\infty}^{\infty} |f(t)|^2 \, dt = \int_{-1}^{1} dt = 2.$$

On the other hand

$$\int_{-\infty}^{\infty} \left| \widehat{f}(\omega) \right|^2 d\omega = \int_{-\infty}^{\infty} \left| \frac{2\sin\omega}{\omega} \right|^2 d\omega = 4 \int_{-\infty}^{\infty} \frac{\sin^2\omega}{\omega^2} d\omega.$$

Therefore from Parseval's theorem

$$\int_{-\infty}^{\infty} |f(t)|^2 \, dt = \frac{1}{2\pi} \int_{-\infty}^{\infty} \left| \hat{f}(\omega) \right|^2 d\omega,$$

we have

$$2 = \frac{2}{\pi} \int_{-\infty}^{\infty} \frac{\sin^2 \omega}{\omega^2} d\omega.$$

It follows that:

$$\int_{-\infty}^{\infty} \frac{\sin^2 \omega}{\omega^2} d\omega = \pi.$$

Since $\frac{\sin^2 \omega}{\omega^2}$ is an even function, so

$$\int_{0}^{\infty} \frac{\sin^2 x}{x^2} dx = \frac{1}{2} \int_{-\infty}^{\infty} \frac{\sin^2 x}{x^2} dx = \frac{\pi}{2}.$$

2.7 Convolution

2.7.1 Mathematical Operation of Convolution

Convolution is an important and useful concept. The convolution $c(t)$ of two functions $f(t)$ and $g(t)$ is usually written as $f(t) * g(t)$ and is defined as

$$c(t) = \int_{-\infty}^{\infty} f(\tau) g(t - \tau) d\tau = f(t) * g(t).$$

The mathematical operation of convolution consists of the following steps:

1. Take the mirror image of $g(\tau)$ about the coordinate axis to create $g(-\tau)$ from $g(\tau)$.
2. Shift $g(-\tau)$ by an amount t to get $g(t - \tau)$. If t is positive, the shift is to the right, if it is negative, to the left.
3. Multiply the shifted function $g(t - \tau)$ by $f(\tau)$.
4. The area under the product of $f(\tau)$ and $g(t-\tau)$ is the value of convolution at t.

   Let us illustrate these steps with a simple example shown in Fig. 2.7. Suppose that $f(\tau)$ is given in (a) and $g(\tau)$ in (b). The mirror image of $g(\tau)$ is $g(-\tau)$, which is shown in (c). In (d), $g(t - \tau)$ is shown as $g(-\tau)$ shifted by an amount t.

It is clear, if $t < 0$, there is no overlap between $f(\tau)$ and $g(t - \tau)$. That means that at any value of τ, either $f(\tau)$ or $g(t - \tau)$, or both are zero. Since $f(\tau)g(t - \tau) = 0$ for $t < 0$, therefore

$$c(t) = 0, \quad \text{if} \quad t < 0.$$

Between $t = 0$ and $t = 1$, the convolution integral is simply equal to abt,

$$c(t) = abt, \quad 0 < t < 1.$$

There is full overlap at $t = 1$, so

$$c(t) = ab \quad \text{at} \quad t = 1.$$

Between $t = 1$ and $t = 2$, the overlap is steadily decreasing. The convolution integral is equal to

$$c(t) = ab[1 - (t - 1)] = ab(2 - t), \quad \text{if} \quad 1 < t < 2.$$

For $t > 2$, there will be no overlap and the convolution integral is equal to zero. Thus the convolution of $f(t)$ and $g(t)$ is given by the triangle shown in (e).

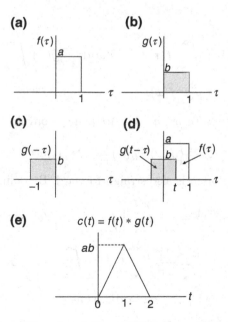

Fig. 2.7. Convolution. The convolution of $f(t)$ shown in (a) and $g(t)$ shown in (b) is given in (e).

2.7.2 Convolution Theorems

Time Convolution Theorem. The time convolution theorem

$$\mathcal{F}\{f(t) * g(t)\} = \widehat{f}(\omega)\,\widehat{g}(\omega)$$

can be proved as follows.
   By definition

$$\mathcal{F}\{f(t) * g(t)\} = \int_{-\infty}^{\infty}\left[\int_{-\infty}^{\infty} f(\tau)\,g(t-\tau)\,\mathrm{d}\tau\right]\mathrm{e}^{-\mathrm{i}\omega t}\,\mathrm{d}t.$$

Interchanging the τ and t integration, we have

$$\mathcal{F}\{f(t) * g(t)\} = \int_{-\infty}^{\infty} f(\tau)\left[\int_{-\infty}^{\infty} g(t-\tau)\,\mathrm{e}^{-\mathrm{i}\omega t}\,\mathrm{d}t\right]\mathrm{d}\tau$$

Let $t - \tau = x$, $t = x + \tau$, $\mathrm{d}t = \mathrm{d}x$, then

$$\int_{-\infty}^{\infty} g(t-\tau)\,\mathrm{e}^{-\mathrm{i}\omega t}\,\mathrm{d}t = \int_{-\infty}^{\infty} g(x)\mathrm{e}^{-\mathrm{i}\omega(x+\tau)}\mathrm{d}x$$

$$= \mathrm{e}^{-\mathrm{i}\omega\tau}\int_{-\infty}^{\infty} g(x)\mathrm{e}^{-\mathrm{i}\omega x}\,\mathrm{d}x = \mathrm{e}^{-\mathrm{i}\omega\tau}\widehat{g}(\omega).$$

Therefore

$$\mathcal{F}\{f(t) * g(t)\} = \int_{-\infty}^{\infty} f(\tau)\mathrm{e}^{-\mathrm{i}\omega\tau}\widehat{g}(\omega)\mathrm{d}\tau = \widehat{g}(\omega)\int_{-\infty}^{\infty} f(\tau)\mathrm{e}^{-\mathrm{i}\omega\tau}\mathrm{d}\tau$$

$$= \widehat{g}(\omega)\widehat{f}(\omega).$$

Frequency Convolution Theorem. The frequency convolution theorem can be written as

$$\mathcal{F}^{-1}\left\{\widehat{f}(\omega) * \widehat{g}(\omega)\right\} = 2\pi f(t)g(t).$$

The proof of this theorem is also straightforward. By definition

$$\mathcal{F}^{-1}\left\{\widehat{f}(\omega) * \widehat{g}(\omega)\right\} = \frac{1}{2\pi}\int_{-\infty}^{\infty}\left[\int_{-\infty}^{\infty} \widehat{f}(\varpi)\,\widehat{g}(\omega-\varpi)\,\mathrm{d}\varpi\right]\mathrm{e}^{\mathrm{i}\omega t}\,\mathrm{d}\omega$$

$$= \frac{1}{2\pi}\int_{-\infty}^{\infty} \widehat{f}(\varpi)\left[\int_{-\infty}^{\infty} \widehat{g}(\omega-\varpi)\,\mathrm{e}^{\mathrm{i}\omega t}\,\mathrm{d}\omega\right]\mathrm{d}\varpi.$$

Let $\omega - \varpi = \Omega$, $\omega = \Omega + \varpi$, $\mathrm{d}\omega = \mathrm{d}\Omega$, thus

$$\mathcal{F}^{-1}\left\{\widehat{f}(\omega) * \widehat{g}(\omega)\right\} = \frac{1}{2\pi}\int_{-\infty}^{\infty} \widehat{f}(\varpi)\,\mathrm{e}^{\mathrm{i}\varpi t}\mathrm{d}\varpi\int_{-\infty}^{\infty} \widehat{g}(\Omega)\mathrm{e}^{\mathrm{i}\Omega t}\,\mathrm{d}\Omega$$

$$= 2\pi f(t)g(t).$$

Clearly this theorem can also be written as

$$\mathcal{F}\{f(t)g(t)\} = \frac{1}{2\pi}\widehat{f}(\omega) * \widehat{g}(\omega).$$

Example 2.7.1. (a) Use

$$\mathcal{F}\{\cos\omega_0 t\} = \pi\delta\left(\omega + \omega_0\right) + \pi\delta\left(\omega - \omega_0\right),$$

$$\mathcal{F}\{\Pi_a(t)\} = \frac{2\sin a\omega}{\omega},$$

and the convolution theorem to find the Fourier transform of the finite wave train $f(t)$

$$f(t) = \begin{cases} \cos\omega_0 t & |t| < a, \\ 0 & |t| > a. \end{cases}$$

(b) Use direct integration to verify the result.

Solution 2.7.1. (a) Since

$$\Pi_a(t) = \begin{cases} 1 & |t| < a, \\ 0 & |t| > a, \end{cases}$$

so we can write $f(t)$ as

$$f(t) = \cos\omega_0 t \cdot \Pi_a(t).$$

According to the convolution theorem

$$\mathcal{F}\{f(t)\} = \frac{1}{2\pi}\mathcal{F}\{\cos\omega_0 t\} * \mathcal{F}\{\Pi_a(t)\}$$

$$= \frac{1}{2\pi}[\pi\delta(\omega + \omega_0) + \pi\delta(\omega - \omega_0)] * \frac{2\sin a\omega}{\omega}$$

$$= \int_{-\infty}^{\infty} [\delta(\omega' + \omega_0) + \delta(\omega' - \omega_0)]\frac{\sin a(\omega - \omega')}{(\omega - \omega')}d\omega'$$

$$= \frac{\sin a(\omega + \omega_0)}{\omega + \omega_0} + \frac{\sin a(\omega - \omega_0)}{\omega - \omega_0}.$$

(b) By definition

$$\mathcal{F}\{f(t)\} = \int_{-\infty}^{\infty} f(t)e^{-i\omega t}\,dt = \int_{-a}^{a} \cos\omega_0 t\, e^{-i\omega t}\,dt.$$

Since

$$\cos\omega_0 t = \frac{1}{2}\left(e^{i\omega_0 t} + e^{-i\omega_0 t}\right),$$

so

$$\mathcal{F}\{f(t)\} = \frac{1}{2}\int_{-a}^{a}(e^{i(\omega_0-\omega)t} + e^{-i(\omega_0+\omega)t})dt$$

$$= \frac{1}{2}\left[\frac{e^{i(\omega_0-\omega)t}}{i(\omega_0-\omega)}\bigg|_{-a}^{a} - \frac{e^{-i(\omega_0+\omega)t}}{i(\omega_0+\omega)}\bigg|_{-a}^{a}\right]$$

$$= \frac{\sin a\,(\omega-\omega_0)}{\omega-\omega_0} + \frac{\sin a(\omega+\omega_0)}{\omega+\omega_0}.$$

This pair of Fourier transform is shown in Fig. 2.8.

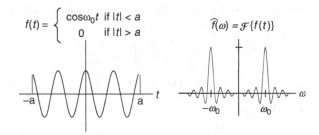

$$f(t) = \begin{cases} \cos\omega_0 t & \text{if } |t| < a \\ 0 & \text{if } |t| > a \end{cases} \qquad \widehat{f}(\omega) = \mathcal{F}\{f(t)\}$$

Fig. 2.8. The Fourier transform pair of a finite cosine wave

Example 2.7.2. Find the Fourier transform of the triangle function

$$f(t) = \begin{cases} t + 2a & -2a < t < 0 \\ -t + 2a & 0 < t < 2a \\ 0 & \text{otherwise} \end{cases}.$$

Solution 2.7.2. Following the procedure shown in Fig. 2.7, one can easily show that the triangle function is the convolution of two identical rectangle pulse function

$$f(t) = \Pi_a(t) * \Pi_a(t).$$

According to the time convolution theorem

$$\mathcal{F}\{f(t)\} = \mathcal{F}\{\Pi_a(t) * \Pi_a(t)\} = \mathcal{F}\{\Pi_a(t)\}\,\mathcal{F}\{\Pi_a(t)\}.$$

Since

$$\mathcal{F}\{\Pi_a(t)\} = \frac{2\sin a\omega}{\omega},$$

therefore

$$\mathcal{F}\{f(t)\} = \frac{2\sin a\omega}{\omega} \cdot \frac{2\sin a\omega}{\omega} = \frac{4\sin^2 a\omega}{\omega^2}.$$

This pair of transforms is shown in Fig. 2.9.

We can obtain the same result by calculating the transform directly, but that would be much more tedious.

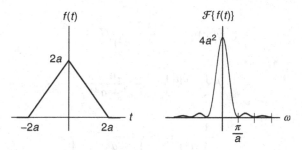

Fig. 2.9. Fourier transform of a triangular function

2.8 Fourier Transform and Differential Equations

A characteristic property of Fourier transform, like other integral transforms, is that it can be used to reduce the number of independent variables in a differential equation by one. For example, if we apply the transform to an ordinary differential equation (which has only one independent variable), then we just get an algebraic equation for the transformed function. A one-dimensional wave equation is a partial differential equation with two independent variables. It can be transformed into an ordinary differential equation in the transformed function. Usually it is easier to solve the resultant equation for the transformed function than it is to solve the original equation, since the equation for the transformed function has one less independent variable. After the transformed function is determined, we can get the solution of the original equation by an inverse transform. We will illustrate this method with the following two examples.

Example 2.8.1. Solve the following differential equation:

$$y''(t) - a^2 y(t) = f(t)$$

where a is a constant and $f(t)$ is given function. The only imposed conditions are that all functions must vanish as $t \to \pm\infty$. This ensures that their Fourier transforms exist.

Solution 2.8.1. Apply the Fourier transform to the equation, and let

$$\widehat{y}(\omega) = \mathcal{F}\left\{y(t)\right\}, \quad \widehat{f}(\omega) = \mathcal{F}\left\{f(t)\right\}.$$

Since

$$\mathcal{F}\left\{y''(t)\right\} = (i\omega)^2 \mathcal{F}\{y(t)\} = -\omega^2 \widehat{y}(\omega),$$

the differential equation becomes

$$-\left(\omega^2 + a^2\right) \widehat{y}\left(\omega\right) = \widehat{f}(\omega).$$

Thus

$$\widehat{y}(\omega) = -\frac{1}{(\omega^2 + a^2)}\widehat{f}(\omega)$$

Recall

$$\mathcal{F}\left\{e^{-a|t|}\right\} = \frac{2a}{(\omega^2 + a^2)},$$

therefore

$$-\frac{1}{(\omega^2 + a^2)} = \mathcal{F}\left\{-\frac{1}{2a}e^{-a|t|}\right\}.$$

In other words, if we define

$$\widehat{g}(\omega) = -\frac{1}{(\omega^2 + a^2)}, \quad \text{then} \quad g(t) = -\frac{1}{2a}e^{-a|t|}.$$

According to the convolution theorem,

$$\widehat{g}(\omega)\widehat{f}(\omega) = \mathcal{F}\{g(t) * f(t)\}.$$

Since

$$\widehat{y}(\omega) = -\frac{1}{(\omega^2 + a^2)}\widehat{f}(\omega) = \widehat{g}(\omega)\widehat{f}(\omega) = \mathcal{F}\{g(t) * f(t)\},$$

it follows that:

$$y(t) = \mathcal{F}^{-1}\{\widehat{y}(\omega)\} = \mathcal{F}^{-1}\mathcal{F}\{g(t) * f(t)\} = g(t) * f(t).$$

Therefore

$$y(t) = -\frac{1}{2a}\int_{-\infty}^{\infty} e^{-a|t-\tau|}f(\tau)\mathrm{d}\tau.$$

This is the particular solution of the equation. With a given $f(t)$, this equation can be evaluated.

Example 2.8.2. Use the Fourier transform to solve the one-dimensional classical wave equation

$$\frac{\partial^2 y(x,t)}{\partial x^2} = \frac{1}{v^2}\frac{\partial^2 y(x,t)}{\partial t^2} \tag{2.17}$$

with an initial condition

$$y(x,0) = f(x), \tag{2.18}$$

where v^2 is a constant.

Solution 2.8.2. Let us Fourier analyze $y(x,t)$ with respect to x. First express $y(x,t)$ in terms of the Fourier integral

$$y(x,t) = \frac{1}{2\pi} \int_{-\infty}^{\infty} \widehat{y}(k,t) e^{ikx}\, dk, \tag{2.19}$$

so the Fourier transform is

$$\widehat{y}(k,t) = \int_{-\infty}^{\infty} y(x,t) e^{-ikx}\, dx. \tag{2.20}$$

It follows form (2.19) and (2.18) that:

$$y(x,0) = \frac{1}{2\pi} \int_{-\infty}^{\infty} \widehat{y}(k,0) e^{ikx}\, dk = f(x). \tag{2.21}$$

Since the Fourier integral of $f(x)$ is

$$f(x) = \frac{1}{2\pi} \int_{-\infty}^{\infty} \widehat{f}(k) e^{ikx}\, dk, \tag{2.22}$$

clearly

$$\widehat{y}(k,0) = \widehat{f}(k). \tag{2.23}$$

Taking the Fourier transform of the original equation, we have

$$\int_{-\infty}^{\infty} \frac{\partial^2 y(x,t)}{\partial x^2} e^{-ikx}\, dx = \frac{1}{v^2} \int_{-\infty}^{\infty} \frac{\partial^2 y(x,t)}{\partial t^2} e^{-ikx}\, dx,$$

which can be written as

$$\int_{-\infty}^{\infty} \frac{\partial^2 y(x,t)}{\partial x^2} e^{-ikx}\, dx = \frac{1}{v^2} \frac{\partial^2}{\partial t^2} \int_{-\infty}^{\infty} y(x,t) e^{-ikx}\, dx.$$

The first term is just the Fourier transform of the second derivative of $y(x,t)$ with respect to x

$$\int_{-\infty}^{\infty} \frac{\partial^2 y(x,t)}{\partial x^2} e^{-ikx}\, dx = (ik)^2 \widehat{y}(k,t),$$

therefore the equation becomes

$$-k^2 \widehat{y}(k,t) = \frac{1}{v^2} \frac{\partial^2}{\partial t^2} \widehat{y}(k,t).$$

Clearly the general solution of this equation is

$$\widehat{y}(k,t) = c_1(k) e^{ikvt} + c_2(k) e^{-ikvt}.$$

where $c_1(k)$ and $c_2(k)$ are constants with respect to t. At $t = 0$, according to (2.23)

$$\widehat{y}(k, 0) = c_1(k) + c_2(k) = \widehat{f}(k).$$

This equation can be satisfied by the following symmetrical and antisymmetrical forms:

$$c_1(k) = \frac{1}{2}\left[\widehat{f}(k) + \widehat{g}(k)\right],$$

$$c_2(k) = \frac{1}{2}\left[\widehat{f}(k) - \widehat{g}(k)\right],$$

where $\widehat{g}(k)$ is a yet undefined function. Thus

$$\widehat{y}(k, t) = \frac{1}{2}\widehat{f}(k)\left(e^{ikvt} + e^{-ikvt}\right) + \frac{1}{2}\widehat{g}(k)\left(e^{ikvt} - e^{-ikvt}\right).$$

Substituting it into (2.19), we have

$$
\begin{aligned}
y(x, t) \quad &= \quad \frac{1}{2\pi}\int_{-\infty}^{\infty}\frac{1}{2}\widehat{f}(k)\left[e^{ik(x+vt)} + e^{ik(x-vt)}\right]dk \\
&+ \frac{1}{2\pi}\int_{-\infty}^{\infty}\frac{1}{2}\widehat{g}(k)\left[e^{ik(x+vt)} - e^{ik(x-vt)}\right]dk.
\end{aligned}
$$

Comparing the integral

$$I_1 = \frac{1}{2\pi}\int_{-\infty}^{\infty}\widehat{f}(k)e^{ik(x+vt)}dk$$

with (2.22), we see that the integral is the same except the argument x is changed to $x + vt$. Therefore

$$I_1 = f(x + vt).$$

It follows that:

$$y(x, t) = \frac{1}{2}\left[f(x + vt) + f(x - vt)\right] + \frac{1}{2}\left[g(x + vt) - g(x - vt)\right]$$

where $g(x)$ is the Fourier inverse transform of $\widehat{g}(k)$. The function $g(x)$ is determined by additional initial, or boundary conditions.

In Chap. 5, we will have a more detailed discussion on this type of problems.

2.9 The Uncertainty of Waves

Fourier transform enables us to break a complicated, even nonperiodic wave down into simple waves. The way of doing it is to assume that the wave is a periodic function with an infinite period. Since it is not possible to observe the wave over an infinite amount of time, we have to do the analysis based on our observation over a finite period of time. Consequently we can never be 100% certain of the characteristics of a given wave.

For example, a constant function $f(t)$ has no oscillation, therefore the frequency is zero. Thus the Fourier transform $\widehat{f}$ is a delta function at $\omega = 0$, as shown in Fig. 2.2. However, this is true only if the function $f(t)$ is a constant from $-\infty$ to $+\infty$. But under no circumstances can we be sure of that. What we can say is that during certain time interval Δt, the function is a constant. This is represented by a rectangular pulse function shown in Fig. 2.4. Outside this time interval, we have no information, therefore the function is given a value of zero. The Fourier transform of this function is $2\sin a\omega/\omega$. As we see in Fig. 2.4, now there is a spread of frequency around $\omega = 0$. In other words, there is an uncertainty of wave's frequency. We can tell how uncertain is the frequency by measuring the width $\Delta\omega$ of the central peak. In this example, $\Delta t = 2a$, $\Delta\omega = 2\pi/a$. It is interesting to note that $\Delta t \, \Delta\omega = 4\pi$, which is a constant. Since it is a constant, it can never be zero, no matter how large or small Δt may be. Therefore there is always some degree of uncertainty.

According to quantum mechanics, photons and electrons can also be thought of as waves. As waves, they are also subject to the uncertainty that applies to all waves. Therefore in the subatomic world, phenomena can only be described within a range of precision that allows for the uncertainty of waves. This is known as the Uncertainty Principle, first formulated by Werner Heisenberg.

In quantum mechanics, if $f(t)$ is normalized wave function, that is

$$\int_{-\infty}^{\infty} |f(t)|^2 \, dt = 1,$$

then the expectation value $\langle t^n \rangle$ is defined as

$$\langle t^n \rangle = \int_{-\infty}^{\infty} |f(t)|^2 \, t^n \, dt.$$

The uncertainty Δt is given by the "root mean square" deviation, that is

$$\Delta t = \left\langle t^2 - \langle t \rangle^2 \right\rangle^{1/2}.$$

If $\widehat{f}(\omega)$ is the Fourier transform of $f(t)$, then according to Parseval's theorem

$$\int_{-\infty}^{\infty} \left| \widehat{f}(\omega) \right|^2 \, d\omega = 2\pi \int_{-\infty}^{\infty} |f(t)|^2 \, dt = 2\pi.$$

Therefore the expectation value of $\langle \omega^n \rangle$ is given by

$$\langle \omega^n \rangle = \frac{1}{2\pi} \int_{-\infty}^{\infty} \left| \hat{f}(\omega) \right|^2 \omega^n \, d\omega.$$

The uncertainty $\Delta\omega$ is similarly defined as

$$\Delta\omega = \left\langle \omega^2 - \langle \omega \rangle^2 \right\rangle^{1/2}.$$

If $f(t)$ is given by a normalized Gaussian function

$$f(t) = \left(\frac{2a}{\pi} \right)^{1/4} \exp\left(-at^2 \right),$$

then clearly $\langle t \rangle = 0$, since the integrand of $\int_{-\infty}^{\infty} |f(t)|^2 \, t \, dt$ is an odd function, and $\Delta t = \langle t^2 \rangle^{1/2}$. By definition

$$\langle t^2 \rangle = \left(\frac{2a}{\pi} \right)^{1/2} \int_{-\infty}^{\infty} \exp\left(-2at^2 \right) t^2 \, dt.$$

With integration by parts, it can be easily shown that

$$\int_{-\infty}^{\infty} \exp(-2at^2) t^2 \, dt = -\frac{1}{4a} t \exp(-2at^2) \Big|_{-\infty}^{\infty} + \frac{1}{4a} \int_{-\infty}^{\infty} \exp(-2at^2) dt$$

$$= \frac{1}{4a} \left(\frac{1}{2a} \right)^{1/2} \int_{-\infty}^{\infty} \exp(-u^2) du = \frac{1}{4a} \left(\frac{\pi}{2a} \right)^{1/2}.$$

Thus

$$\Delta t = \langle t^2 \rangle^{1/2} = \left[\left(\frac{2a}{\pi} \right)^{1/2} \frac{1}{4a} \left(\frac{\pi}{2a} \right)^{1/2} \right]^{1/2} = \left(\frac{1}{4a} \right)^{1/2}.$$

Now

$$\hat{f}(\omega) = \mathcal{F}\{f(t)\} = \left(\frac{2a}{\pi} \right)^{1/4} \left(\frac{\pi}{a} \right)^{1/2} \exp\left(-\frac{\omega^2}{4a} \right).$$

So $\langle \omega \rangle = 0$, and

$$\langle \omega^2 \rangle = \frac{1}{2\pi} \int \left| \hat{f}(\omega) \right|^2 \omega^2 \, d\omega = \frac{1}{2\pi} \left(\frac{2a}{\pi} \right)^{1/2} \left(\frac{\pi}{a} \right) \int_{-\infty}^{\infty} \exp\left(-\frac{\omega^2}{2a} \right) \omega^2 \, d\omega$$

$$= \frac{1}{2\pi} \left(\frac{2a}{\pi} \right)^{1/2} \left(\frac{\pi}{a} \right) a(2a\pi)^{1/2} = a$$

Thus

$$\Delta\omega = \langle \omega^2 \rangle^{1/2} = (a)^{1/2}.$$

Therefore

$$\Delta t \cdot \Delta \omega = \left(\frac{1}{4a}\right)^{1/2} (a)^{1/2} = \frac{1}{2}.$$

As we have discussed, if we change the name of the variable t (representing time) to x (representing distance), the angular frequency ω is changed to the wave number k. This relation is then written as

$$\Delta x \cdot \Delta k = \frac{1}{2}.$$

The two most fundamental relations in quantum mechanics are

$$E = \hbar\omega \quad \text{and} \quad p = \hbar k,$$

where E is the energy, p the momentum, and $\hbar$ is the Planck constant, $h/2\pi$. It follows that the uncertainty in energy is $\Delta E = \hbar\, \Delta\omega$, and the uncertainty in momentum is $\Delta p = \hbar\, \Delta k$. Therefore, with a Gaussian wave, we have

$$\Delta t \cdot \Delta E = \frac{\hbar}{2}, \qquad \Delta x \cdot \Delta p = \frac{\hbar}{2}.$$

Since no other form of wave function can reduce the product of uncertainties below this value, these relations are usually presented as

$$\Delta t \cdot \Delta E \geq \frac{\hbar}{2}, \qquad \Delta x \cdot \Delta p \geq \frac{\hbar}{2},$$

which are the formal statements of uncertainty principle in quantum mechanics.

Exercises

1. Use an odd function to show that

$$\int_0^\infty \frac{1 - \cos \pi\omega}{\omega} \sin \omega t \, d\omega = \begin{cases} \dfrac{\pi}{2} & 0 < t < \pi \\ 0 & t > \pi \end{cases}.$$

2. Use an even function to show that

$$\int_0^\infty \frac{\cos \omega t}{1 + \omega^2} d\omega = \frac{\pi}{2} e^{-t}.$$

3. Show that

$$\int_0^\infty \frac{\cos \omega t + \omega \sin \omega t}{1 + \omega^2} d\omega = \begin{cases} 0 & t < 0, \\ \dfrac{\pi}{2} & t = 0, \\ \pi e^{-t} & t > 0. \end{cases}$$

4. Show that

$$\int_0^\infty \frac{\sin \pi\omega \sin \omega t}{1 - \omega^2} d\omega = \begin{cases} \dfrac{\pi \sin t}{2} & 0 \le t \le \pi. \\ 0 & t > \pi \end{cases}$$

5. Find the Fourier integral of

$$f(t) = \begin{cases} 1 & 0 < t < a, \\ 0 & t > a. \end{cases}$$

Ans. $f(t) = \dfrac{2}{\pi} \displaystyle\int_0^\infty \dfrac{\sin a\omega \cos \omega t}{\omega} d\omega.$

6. Find the Fourier integral of

$$f(t) = \begin{cases} t & 0 < t < a, \\ 0 & t > a. \end{cases}$$

Ans. $f(t) = \dfrac{2}{\pi} \displaystyle\int_0^\infty \left(\dfrac{a \sin a\omega}{\omega} + \dfrac{\cos a\omega - 1}{\omega^2} \right) \cos \omega t \, d\omega.$

7. Find the Fourier integral of

$$f(t) = e^{-t} + e^{-2t}, \quad t > 0.$$

Ans. $f(t) = \dfrac{6}{\pi} \displaystyle\int_0^\infty \dfrac{2 + \omega^2}{\omega^4 + 5\omega^2 + 4} \cos \omega t \, d\omega.$

8. Find the Fourier integral of

$$f(t) = \begin{cases} t^2 & 0 < t < a, \\ 0 & t > a. \end{cases}$$

Ans. $f(t) = \dfrac{2}{\pi} \displaystyle\int_0^\infty \left[\left(a^2 - \dfrac{2}{\omega^2} \right) \sin a\omega + \dfrac{2a}{\omega} \cos a\omega \right] \dfrac{\cos \omega t}{\omega} d\omega.$

9. Find Fourier cosine and sine transform of

$$f(t) = \begin{cases} 1 & 0 < t < 1, \\ 0 & t > 1. \end{cases}$$

Ans. $\widehat{f_s} = \dfrac{2}{\pi} \dfrac{1 - \cos \omega}{\omega}, \qquad \widehat{f_c} = \dfrac{2}{\pi} \dfrac{\sin \omega}{\omega}.$

10. Find Fourier transform of

$$f(t) = \begin{cases} e^{-t} & 0 < t, \\ 0 & t < 0. \end{cases}$$

Ans. $\dfrac{1}{(1 + i\omega)}.$

11. Find Fourier transform of

$$f(t) = \begin{cases} 1-t & |t| < 1, \\ 0 & 1 < |t|. \end{cases}$$

Ans. $\left(\dfrac{2e^{i\omega}}{i\omega} + \dfrac{e^{i\omega} - e^{-i\omega}}{\omega^2}\right).$

12. Find Fourier transform of

$$f(t) = \begin{cases} e^t & |t| < 1, \\ 0 & 1 < |t|. \end{cases}$$

Ans. $\dfrac{e^{1-i\omega} - e^{-1+i\omega}}{1 - i\omega}.$

13. Show that if $f(t)$ is an even function, then the Fourier transform reduces to the Fourier cosine transform, and if $f(t)$ is an odd function it reduces to Fourier sine transform.

Note that the multiplicative constants α and β may not come out the same as we have defined. But remember that as long as $\alpha \times \beta$ is equal to $2/\pi$, they are equivalent.

14. If $\widehat{f}(\omega) = \mathcal{F}\{f(t)\}$, show that

$$\mathcal{F}\{(-it)^n f(t)\} = \dfrac{d^n}{d\omega^n} \widehat{f}(\omega).$$

Hint: First show that $\frac{d\widehat{f}}{d\omega} = -i\mathcal{F}\{tf(t)\}$.

15. Show that

$$\mathcal{F}\{\tfrac{1}{t}f(t)\} = -i \int_{-\infty}^{\omega} \widehat{f}(\omega')d\omega'$$

16. (a) Find the normalization constant A of the Gaussian function $\exp(-at^2)$ such that

$$\int_{-\infty}^{\infty} \left|A\exp(-at^2)\right|^2 dt = 1.$$

(b) Find the Fourier transform $\widehat{f}(\omega)$ of the normalized Gaussian function and verify the Parseval's theorem with explicit integration that

$$\int_{-\infty}^{\infty} \left|\widehat{f}(\omega)\right|^2 d\omega = 2\pi.$$

Ans. $A = (2a/\pi)^{1/4}.$

17. Use Fourier transform of $\exp(-|t|)$ and the Parseval's theorem to show
that

$$\int_{-\infty}^{\infty} \frac{d\omega}{(1+\omega^2)^2} = \frac{\pi}{2}.$$

18. (a) Find the Fourier transform of

$$f(t) = \begin{cases} 1 - \left|\dfrac{t}{2}\right| & -2 < t < 2 \\ 0 & \text{otherwise} \end{cases}$$

(b) Use the result of (a) and the Parseval's theorem to evaluate the integral

$$I = \int_{-\infty}^{\infty} \left(\frac{\sin t}{t}\right)^4 dt.$$

Ans. $I = 2\pi/3$.

19. The function $f(r)$ has a Fourier transform

$$\widehat{f}(\mathbf{k}) = \frac{1}{(2\pi)^{3/2}} \int f(r) e^{i\mathbf{k}\cdot\mathbf{r}} \, d^3r = \frac{1}{(2\pi)^{3/2}} \frac{1}{k^2}.$$

Determine $f(r)$.

Ans. $f(r) = \dfrac{1}{4\pi r}$.

20. Find the Fourier transform of

$$f(t) = te^{-4t^2}.$$

Ans. $\widehat{f}(\omega) = -i\frac{\sqrt{\pi}}{16}\omega e^{-\omega^2/16}$.

21. Find the inverse Fourier transform of

$$\widehat{f}(\omega) = e^{-2|\omega|}.$$

Ans. $f(t) = \dfrac{2}{\pi}\dfrac{1}{t^2+4}$.

22. Evaluate

$$\mathcal{F}^{-1}\left\{\frac{1}{\omega^2+4\omega+13}\right\}.$$

Hint: $\omega^2 + 4\omega + 13 = (\omega+2)^2 + 9$.

Ans. $f(t) = \frac{1}{6}e^{-i2t}e^{-3|t|}$.

Part II

Sturm–Liouville Theory and Special Functions

3

Orthogonal Functions and Sturm–Liouville Problems

In Fourier series we have seen that a function can be expressed in terms of an infinite series of sines and cosines. This is possible mainly because these trigonometrical functions form a complete orthogonal set.

The concept of an orthogonal set of functions is a natural generalization of the concept of an orthogonal set of vectors. In fact, a function can be considered as a generalized vector in an infinite dimensional vector space and sines and cosines as basis vectors of this space. This make us ask where does such basis come from. Are there other bases as well? In this chapter we discover that such bases arise as the eigenfunctions of self-adjoint (Hermitian) linear differential operators, just as Hermitian $n \times n$ matrices provide us with sets of eigenvectors that are orthogonal bases for n-dimensional space.

Many important physical problems are described by differential equations which can be put into a form known as Sturm–Liouville equation. We will show that under certain boundary conditions of the solution of the equation, the Sturm–Liouville operators are self-adjoint. Therefore many basis sets of orthogonal functions can be generated by Sturm–Liouville equations. Viewed from a broader Sturm–Liouville theory, Fourier series is only a special case.

Some Sturm–Liouville equations are of great importance, we give names to them. Solutions of these equations are known as special functions. In this chapter we will discuss the origin and properties of some special functions that are frequently encountered in mathematical physics. A more detailed discussion of the most important ones will be given in Chap. 4.

3.1 Functions as Vectors in Infinite Dimensional Vector Space

3.1.1 Vector Space

When we construct our number system, first we find that additions and multiplications of positive integers satisfy certain rules concerning the order in which the computation can proceed. Then we use these rules to define a wider class of numbers.

Here we are going to do the same thing with vectors. Based on the properties of ordinary three-dimensional vectors, we abstract a set of rules that these vectors satisfy. Then we use this set of rules as the definition of a vector space. Any set of objects that satisfies these rules is said to form a linear vector space.

As a consequence of the definition of ordinary vectors, it can be easily shown that they satisfy the following set of rules:

– Vector addition is commutative and associative

$$\mathbf{a} + \mathbf{b} = \mathbf{b} + \mathbf{a},$$
$$(\mathbf{a} + \mathbf{b}) + \mathbf{c} = \mathbf{a} + (\mathbf{b} + \mathbf{c}).$$

– Multiplication by a scalar is distributive and associative

$$\alpha(\mathbf{a} + \mathbf{b}) = \alpha\mathbf{a} + \alpha\mathbf{b},$$
$$(\alpha + \beta)\mathbf{a} = \alpha\mathbf{a} + \beta\mathbf{a},$$
$$\alpha(\beta\mathbf{a}) = (\alpha\beta)\mathbf{a},$$

where α and β are arbitrary scalars.

– There exists a null vector $\mathbf{0}$, such that

$$\mathbf{a} + \mathbf{0} = \mathbf{a}.$$

– All vectors $\mathbf{a}$ have a corresponding negative vector $-\mathbf{a}$, such that

$$\mathbf{a} + (-\mathbf{a}) = \mathbf{0}.$$

– Multiplication by unit scalar leaves any vector unchanged,

$$1\mathbf{a} = \mathbf{a}.$$

– Multiplication by zero gives a null vector,

$$0\mathbf{a} = \mathbf{0}.$$

Now let us consider all well behaved functions $f(x)$, $g(x)$, $h(x), \ldots$ defined in the interval $a \leq x \leq b$. Clearly, they form a linear vector space, since it can be readily verified that

$$f(x) + g(x) = g(x) + f(x),$$
$$[f(x) + g(x)] + h(x) = f(x) + [g(x) + h(x)].$$

$$\alpha\,[f(x) + g(x)] = \alpha f(x) + \alpha g(x),$$

$$(\alpha + \beta)f(x) = \alpha f(x) + \beta f(x),$$

$$\alpha(\beta f(x)) = (\alpha\beta)f(x).$$

$$f(x) + 0 = f(x).$$

$$f(x) + (-f(x)) = 0.$$

$$1 \times f(x) = f(x).$$

$$0 \times f(x) = 0.$$

Therefore a collection of all functions of x defined in a certain interval of x constitutes a vector space.

Dimension of a Vector Space. A three-dimensional ordinary vector $\mathbf{v}$ is described by its three components $(v_1,\ v_2,\ v_3)$. It can be regarded a function with three distinct values $[v(1),\ v(2),\ v(3)]$. A n-dimensional vector is defined by n-tuples $[v(1),\ v(2),\ldots,v(n)]$, as we have seen in the matrix theory. Now the function $f(x)$ is a vector, what is its dimension?

Let us imagine approximating the function $f(x)$ between $a \leq x \leq b$ in a piecewise constant manner. Divide the x interval $(a \leq x \leq b)$ into n equal parts. Approximate the function by a sequence of values $(f_1,\ f_2,\ldots,f_n)$, where f_i is the value of $f(x)$ at the left endpoint of the ith subinterval, except f_n which is the value of $f(b)$. For example, if we approximate $f(x) = 1 + x$ in $0 \leq x \leq 1$ by dividing the interval into two equal parts, then $f(x)$ is approximated by $[f(0),\ f(0.5),\ f(1)]$, or $(1,\ 1.5,\ 2.0)$. Of course this is a very poor approximation. A better approximation would be to divide the interval in ten equal parts and approximate the function with 11 tuples of numbers $(1,\ 1.1,\ 1.2,\ldots,2)$. Since the function is actually defined by all possible values of x between 0 and 1, which consists of infinite number of values of x from 0 to 1, the function is described by n-tuples of numbers with $n \to \infty$. In this sense, we say that the function is a vector in an infinite dimensional vector space.

3.1.2 Inner Product and Orthogonality

So far we have not mentioned dot product of vectors. Dot product is also called inner product or scalar product. Often it is written as $\mathbf{u} \cdot \mathbf{v}$, or as $\langle \mathbf{u}|\ \mathbf{v}\rangle$, or $\langle \mathbf{u}, \mathbf{v}\rangle$.

$$\mathbf{u} \cdot \mathbf{v} = \langle \mathbf{u}|\ \mathbf{v}\rangle = \langle \mathbf{u}, \mathbf{v}\rangle.$$

A vector space does not need to have a dot product. But a function space without an inner product defined is too large a vector space to be useful in physical applications.

If we choose to introduce an inner product for the function space, how is it to be defined? Again we elevate the properties of dot product of familiar vectors to axioms and require the inner product of any vector space to satisfy these axioms.

From the definition of the dot product of two three-dimensional vectors **u** and **v** :

$$\mathbf{u} \cdot \mathbf{v} = u_1 v_1 + u_2 v_2 + u_3 v_3 = \sum_{j=1}^{3} u_j v_j,$$

it can be easily deduced that dot product is

- commutative

$$\mathbf{u} \cdot \mathbf{v} = \mathbf{v} \cdot \mathbf{u},$$

- and linear

$$(\alpha \mathbf{u} + \beta \mathbf{v}) \cdot \mathbf{w} = \alpha (\mathbf{u} \cdot \mathbf{w}) + \beta (\mathbf{v} \cdot \mathbf{w}).$$

The norm (or length) of vector is defined as

$$\|\mathbf{u}\| = (\mathbf{u} \cdot \mathbf{u})^{1/2} = \left(\sum_{j=1}^{3} u_j u_j \right)^{1/2}.$$

- Therefore the norm is non-negative

$$\mathbf{u} \cdot \mathbf{u} > 0 \quad \text{for} \quad \text{all} \quad \mathbf{u} \neq \mathbf{0}.$$

In complex space, the components of a vector can assume complex values. As we have seen in matrix theory, the inner product in complex space is defined as

$$\mathbf{u} \cdot \mathbf{v} = u_1^* v_1 + u_2^* v_2 + u_3^* v_3 = \sum_{j=1}^{3} u_j^* v_j,$$

where u^* is the complex conjugate of u. Therefore in complex space,

- The commutative rule $\mathbf{u} \cdot \mathbf{v} = \mathbf{v} \cdot \mathbf{u}$ is replaced by

$$\mathbf{u} \cdot \mathbf{v} = (\mathbf{v} \cdot \mathbf{u})^*, \tag{3.1}$$

This follows from the fact that

$$\mathbf{u} \cdot \mathbf{v} = \sum_{j=1}^{3} u_j^* v_j = \sum_{j=1}^{3} \left(u_j v_j^* \right)^* = \left(\sum_{j=1}^{3} v_j^* u_j \right)^* = (\mathbf{v} \cdot \mathbf{u})^*.$$

Thus, if α is a complex number, then

$$(\alpha \mathbf{u} \cdot \mathbf{v}) = \alpha^* (\mathbf{u} \cdot \mathbf{v}), \tag{3.2}$$

$$(\mathbf{u} \cdot \alpha \mathbf{v}) = \alpha (\mathbf{u} \cdot \mathbf{v}). \tag{3.3}$$

Now if we use these properties as axioms to define a wider class of inner products, then we can see that for two n-dimensional vectors $\mathbf{u}$ and $\mathbf{v}$ in complex space, the expression

$$\mathbf{u} \cdot \mathbf{v} = u_1^* v_1 w_1 + u_2^* v_2 w_2 + \cdots + u_n^* v_n w_n = \sum_{j=1}^{n} u_j^* v_j w_j \tag{3.4}$$

is also a legitimate inner product as long as w_j is a fixed real positive constant for each j.

Let us use two-dimensional real space for illustration. Suppose that $\mathbf{u} = (1, 2)$ and $\mathbf{v} = (3, -4)$, with $w_1 = 2$, $w_2 = 3$, then

$$\mathbf{u} \cdot \mathbf{v} = (1)(3)(2) + (2)(-4)(3) = -18$$

$$\mathbf{v} \cdot \mathbf{u} = (3)(1)(2) + (-4)(2)(3) = -18,$$

in agreement with the axiom $\mathbf{u} \cdot \mathbf{v} = \mathbf{v} \cdot \mathbf{u}$.

On the other hand, if $w_1 = 2$, $w_2 = -3$, then

$$\mathbf{u} \cdot \mathbf{u} = (1)(1)(2) + (2)(2)(-3) = -10,$$

in violation of the axiom $\mathbf{u} \cdot \mathbf{u} > 0$ for $\mathbf{u} \neq \mathbf{0}$.

It can be readily verified that with real positive w_j, (3.4) satisfies all the axioms of inner product. The w_js are known as "weights" because they attach more or less weight to the different components of the vector. Of course, w_j can all be equal to one. In many applications, this is indeed the case.

To define an inner product in a function space in the interval $a \leq x \leq b$, let us divide the interval into $n-1$ equal parts and imagine that the functions $f(x)$ and $g(x)$ are approximated in a piecewise constant manner as discussed before:

$$f(x) = (f_1, \ f_2, \ldots, f_n),$$

$$g(x) = (g_1, \ g_2, \ldots, g_n).$$

We can adopt the inner product as

$$\langle f \,|\, g \rangle = \sum_{j=1}^{n} f_j^* g_j \Delta x_j,$$

where Δx_j is the width of the subinterval. Regarding Δx_j as the weights, this definition is in accordance with (3.4). If we let $n \to \infty$, this sum becomes an integral

$$\langle f \,|\, g \rangle = \int_a^b f^*(x) g(x) \mathrm{d}x.$$

The weight could also be $w(x)\mathrm{d}x$, as long as $w(x)$ is a real positive function. In that case, the inner product is defined to be

$$\langle f \,|\, g \rangle = \int_a^b f^*(x) g(x) w(x) \mathrm{d}x.$$

This is the general definition of an inner product of an infinite dimensional vector space of functions. It can be readily shown that this definition satisfies all the axioms of an inner product. As mentioned before, in many problems the weight function $w(x)$ is equal to one for all x. It is to be emphasized that our heuristic approach is neither a derivation nor a proof, it only provides the motivation for this definition.

Two functions are said to be orthogonal in the interval between a and b if

$$\langle f\,|g\rangle = \int_a^b f^*(x)g(x)w(x)\mathrm{d}x = 0.$$

The norm of a function is defined as

$$\|f\| = \langle f\,|f\rangle^{1/2} = \left[\int_a^b f^*(x)f(x)w(x)\mathrm{d}x\right]^{1/2} = \left[\int_a^b |f(x)|^2 w(x)\mathrm{d}x\right]^{1/2}.$$

The function is said to be normalized if

$$\|f\| = 1.$$

An infinite dimensional vector space of functions, for which an inner product is defined is called a Hilbert space. In quantum mechanics, all legitimate wavefunctions live in Hilbert space.

3.1.3 Orthogonal Functions

Orthonormal Set. A collection of functions $\{\psi_n(x)\}$, where $n = 1, 2, \ldots$ is called an orthogonal set if $\langle \psi_n\,|\psi_m\rangle = 0$ whenever $n \neq m$.

Dividing each function by its norm

$$\phi_n(x) = \frac{1}{\|\psi_n\|}\psi_n(x),$$

we have an orthonormal set $\{\phi_n(x)\}$, which satisfies the relation

$$\langle \phi_n\,|\phi_m\rangle = \begin{cases} 0 & n \neq m \\ 1 & n = m \end{cases}.$$

It is to be noted that the functions in the set and their inner products are to be defined in the same interval of x.

For example, with a unit weight function $w(x) = 1$, the set of functions $\{\sin\frac{n\pi x}{L}\}$ $(n = 1, 2, \ldots)$ is orthogonal on the interval $0 \leq x \leq L$, since

$$\int_0^L \sin\frac{n\pi x}{L}\sin\frac{m\pi x}{L}\mathrm{d}x = \begin{cases} 0 & n \neq m \\ \frac{L}{2} & n = m \end{cases}.$$

Furthermore, $\{\phi_n(x)\}$ where

$$\phi_n(x) = \sqrt{\frac{2}{L}} \sin \frac{n\pi x}{L},$$

is an orthonormal set in the interval of $t(0, L)$.

Gram–Schmidt Orthogonalization. Out of a linearly independent (but not orthogonal) set of functions $\{u_n(x)\}$, an orthonormal set $\{\phi_n\}$ over an arbitrary interval and with respect to an arbitrary weight function can be constructed by the Gram–Schmidt orthogonalization method. The procedure is similar to that we have used in the construction of a set of orthogonal eigenvectors of a Hermitian matrix.

From a given linearly independent set $\{u_n\}$, an orthogonal set $\{\psi_n\}$ can be constructed. We start with $n = 0$. Let

$$\psi_0(x) = u_0(x)$$

and normalized it to unity and denote the result as ϕ_0

$$\phi_0(x) = \frac{1}{\left[\int |\psi_0(x)|^2 w(x)dx\right]^{1/2}} \psi_0(x).$$

Clearly,

$$\int |\phi_0(x)|^2 w\, dx = \frac{1}{\left[\int |\psi_0(x)|^2 w(x)dx\right]} \int |\psi_0(x)|^2 w(x)\, dx = 1.$$

For $n = 1$, let

$$\psi_1(x) = u_1(x) + a_{10}\phi_0(x).$$

we require $\psi_1(x)$ to be orthogonal to $\phi_0(x)$,

$$\int \phi_0^*(x)\psi_1(x)w(x)\, dx$$

$$= \int \phi_0^*(x)u_1(x)w(x)\, dx + a_{10} \int |\phi_0(x)|^2 w(x)\, dx = 0.$$

Since ϕ_0 is normalized to unity, we have

$$a_{10} = - \int \phi_0^*(x)u_1(x)w(x)dx.$$

With a_{10} so determined, $\psi_1(x)$ is a known function, which can be normalized. Let

$$\phi_1(x) = \frac{1}{\left[\int |\psi_1(x)|^2 w(x)dx\right]^{1/2}} \psi_1(x).$$

For $n = 2$, let
$$\psi_2(x) = u_2(x) + a_{21}\phi_1(x) + a_{20}\phi_0\,(x).$$

The requirement that $\psi_2(x)$ be orthogonal to $\phi_1(x)$ and to $\phi_0(x)$ leads to

$$a_{21} = -\int \phi_1^*(x)u_2(x)w(x)\,dx,$$

$$a_{20} = -\int \phi_0^*(x)u_2(x)w(x)\,dx.$$

Thus $\psi_2(x)$ is determined. Clearly this process can be continued. We take ψ_i as the ith function of $\{\psi_n\}$ and set it to equal u_i plus an unknown linear combination of the previously determined ϕ_j, $j = 0, 1, \ldots i - 1$. The requirement that ψ_i be orthogonal to each of the previous ϕ_j yields just enough constraints to determine each of the unknown coefficients. Then the fully determined ψ_i can be normalized to unity and the steps are repeated for ψ_{i+1}. In terms of the inner products, the procedure can be expressed as:

$$\psi_0 = u_0 \qquad\qquad \phi_0 = \psi_0\,\langle\psi_0\,|\psi_0\rangle^{-1/2}$$

$$\psi_1 = u_1 - \phi_0\,\langle\phi_0\,|u_1\rangle \qquad\qquad \phi_1 = \psi_1\,\langle\psi_1\,|\psi_1\rangle^{-1/2}$$

$$\psi_2 = u_2 - \phi_1\,\langle\phi_1\,|u_2\rangle - \phi_0\,\langle\phi_0\,|u_2\rangle \qquad \phi_2 = \psi_2\,\langle\psi_2\,|\psi_2\rangle^{-1/2}$$

$$\psi_i = u_i - \phi_{i-1}\,\langle\phi_{i-1}\,|u_i\rangle - \cdots \qquad \phi_i = \psi_i\,\langle\psi_i\,|\psi_i\rangle^{-1/2}.$$

Clearly $\{\psi_n\}$ is an orthogonal set and $\{\phi_n\}$ is an orthonormal set.

Example 3.1.1. Legendre Polynomials. Construct an orthonormal set from the linear independent functions $u_n(x) = x^n$, $n = 0, 1, 2, \ldots$ in the interval of $-1 \le x \le 1$ with a weight function $w(x) = 1$.

Solution 3.1.1. According to the Gram–Schmidt process, the first unnormalized function of the orthogonal set $\{\psi_n\}$ is simply u_0,

$$\psi_0 = u_0 = 1.$$

The first normalized function of the orthonormal set $\{\phi_n\}$ is

$$\phi_0 = \psi_0\,\langle\psi_0\,|\psi_0\rangle^{-1/2} = \psi_0\left[\int_{-1}^{1} dx\right]^{-1/2} = \frac{1}{\sqrt{2}}.$$

The next function in the orthogonal set is

$$\psi_1 = u_1 - \phi_0\,\langle\phi_0\,|u_1\rangle.$$

Since

$$\langle \phi_0 \, | u_1 \rangle = \int_{-1}^{1} \frac{1}{\sqrt{2}} x \, \mathrm{d}x = 0,$$

so

$$\psi_1 = x$$

and

$$\phi_1 = \psi_1 \langle \psi_1 \, | \psi_1 \rangle^{-1/2} = x \left[\int_{-1}^{1} x^2 \, \mathrm{d}x \right]^{-1/2} = \sqrt{\frac{3}{2}} x.$$

Continue the process

$$\psi_2 = u_2 - \phi_1 \langle \phi_1 \, | u_2 \rangle - \phi_0 \langle \phi_0 \, | u_2 \rangle.$$

Since

$$\langle \phi_1 \, | u_2 \rangle = \int_{-1}^{1} \sqrt{\frac{3}{2}} x^3 \, \mathrm{d}x = 0, \qquad \langle \phi_0 \, | u_2 \rangle = \int_{-1}^{1} \sqrt{\frac{1}{2}} x^2 \, \mathrm{d}x = \frac{\sqrt{2}}{3},$$

so

$$\psi_2 = x^2 - 0 - \frac{1}{\sqrt{2}} \frac{\sqrt{2}}{3} = x^2 - \frac{1}{3},$$

and

$$\phi_2 = \psi_2 \langle \psi_2 \, | \psi_2 \rangle^{-1/2} = \left(x^2 - \frac{1}{3} \right) \left[\int_{-1}^{1} \left(x^2 - \frac{1}{3} \right)^2 \, \mathrm{d}x \right]^{-1/2}$$

$$= \left(x^2 - \frac{1}{3} \right) \sqrt{\frac{45}{8}} = \sqrt{\frac{5}{2}} \left(\frac{3}{2} x^2 - \frac{1}{2} \right).$$

The next normalized function is

$$\phi_3 = \sqrt{\frac{7}{2}} \left(\frac{5}{2} x^3 - \frac{3}{2} x \right).$$

It is straight-forward, although tedious, to show that

$$\phi_n = \sqrt{\frac{2n+1}{2}} P_n(x),$$

where $P_n(x)$ is a polynomial of order n, and

$$P_n(1) = 1,$$

$$\int_{-1}^{1} P_n(x) P_m(x) \, \mathrm{d}x = \frac{2}{2n+1} \delta_{nm}.$$

These polynomials are known as Legendre polynomials. They are one of the most useful and most frequently encountered special functions in mathematical physics. Fortunately, as we shall see later, there are much easier methods to derive them.

In this example, we have used the Gram–Schmidt procedure to rearrange the set of linear independent functions $\{x^n\}$ into an orthonormal set for the given interval $-1 \le x \le 1$ and given weight function $w(x) = 1$. With other choices of intervals and weight functions, we will get other sets of orthogonal polynomials. For example, with the same set of functions $\{x^n\}$ and the same weight function $w(x) = 1$, if the interval is chosen to be $[0, 1]$, instead of $[-1, 1]$, the Gram–Schmidt process will lead to another set of orthogonal polynomials known as shifted Legendre polynomial $\{P_n^s(x)\}$. With $P_n^s(x)$ normalized in such a way that $P_n^s(1) = 1$,

$$P_n^s(x) = P_n\left(2\left(x - \frac{1}{2}\right)\right).$$

The first few shifted Legendre polynomials are

$$P_0^s(x) = 1, \quad P_1^s(x) = 2x - 1, \quad P_2^s(x) = 6x^2 - 6x + 1.$$

As another example, with the weight function chosen as $w(x) = e^{-x}$ in the interval of $0 \le x < \infty$, the orthonormal set constructed from $\{x^n\}$ is known as the Laguerre polynomial $\{L_n(x)\}$. The first three Laguerre polynomials are

$$L_0(x) = 1, \quad L_1(x) = 1 - x, \quad L_2(x) = \frac{1}{2}(2 - 4x + x^2).$$

It can be readily verified that

$$\int_0^\infty L_n(x)L_m(x)e^{-x}\mathrm{d}x = \delta_{nm}.$$

Sometimes Laguerre polynomials are defined with a normalization

$$\int_0^\infty L_n(x)L_m(x)e^{-x}\mathrm{d}x = \delta_{nm}(n!)^2.$$

In that case, the first three Laguerre polynomials are

$$L_0(x) = 1, \quad L_1(x) = 1 - x, \quad L_2(x) = 2 - 4x + x^2.$$

Obviously infinitely many orthogonal sets of functions can be generated from $\{x^n\}$ by the Gram–Schmidt process. With a given weight function and a specified interval, the Gram–Schmidt process is unique up to a multiplication constant, positive or negative. This process is rather cumbersome. Fortunately, almost all interesting orthogonal polynomials constructed by this method are solutions of particular differential equations. Therefore they can be discussed from the perspective of differential equations.

3.2 Generalized Fourier Series

By analogy with finite dimensional vector space, we can consider an ortho-normal set of functions $\{\phi_n(x)\}$ $(n = 0, 1, 2, \ldots)$ on the interval $a \leq x \leq b$ as basis vectors in an infinite dimensional vector space of functions, in which

$$\langle \phi_m \,|\phi_n\rangle = \int_a^b \phi_m^*(x)\phi_n(x)w(x)\,\mathrm{d}x = \delta_{nm}.$$

If any arbitrary piecewise continuous bounded function $f(x)$ in the same interval can be represented as the linear sum of these functions

$$f(x) = c_0\phi_0(x) + c_1\phi_1(x) + \cdots = \sum_{n=0}^{\infty} c_n\phi_n(x), \tag{3.5}$$

then $\{\phi_n(x)\}$ is said to be complete. If this equation is valid, taking the inner product with $\phi_m(x)$, we have

$$\langle \phi_m \,|f\rangle = \sum_{n=0}^{\infty} c_n \langle \phi_m \,|\phi_n\rangle = \sum_{n=0}^{\infty} c_n\delta_{nm} = c_m.$$

The coefficients c_n

$$c_n = \langle \phi_n \,|f\rangle = \int_a^b \phi_m^*(x)f(x)w(x)\,\mathrm{d}x \tag{3.6}$$

are called Fourier coefficients and the series (3.5) with these coefficients

$$f(x) = \sum_{n=0}^{\infty} \langle \phi_n \,|f\rangle \,\phi_n(x)$$

is called the generalized Fourier series. Clearly if a different set of basis $\{\varphi_n\}$ is chosen, then the function can be expressed in terms of the new basis with a different set of coefficients.

The nature of the representation of $f(x)$ by a generalized Fourier series is that the series representation converges to the mean. Let us use real functions to illustrate. Select M equally spaced points in the interval $a \leq x \leq b$ at $x_1 = a$, $x_2 = a + \Delta x$, $x_3 = a + 2\,\Delta x, \ldots$ where $\Delta x = (b - a)/(M - 1)$. Then approximate the function at any one of these M points by the finite series

$$f(x_i) = \sum_{n=0}^{N} A_n\phi_n(x_i).$$

In order to make this approximation as good as possible in the least square sense, we have to minimize the mean square error. This means we have to differentiate the mean square error D,

$$D = \sum_{i=1}^{M} \left[f(x_i) - \sum_{n=0}^{N} A_n \phi_n(x_i) \right]^2 w(x_i) \, \Delta x$$

with respect to each of coefficient A_n and set it zero. Let A_k be one of the A_ns. The differentiation with respect to A_k

$$\frac{\partial D}{\partial A_k} = 0$$

leads to

$$\sum_{i=1}^{M} 2 \left[f(x_i) - \sum_{n=0}^{N} A_n \phi_n(x_i) \right] [-\phi_k(x_i)] \, w(x_i) \, \Delta x = 0,$$

or

$$\sum_{i=1}^{M} \phi_k(x_i) f(x_i) w(x_i) \Delta x - \sum_{n=0}^{N} A_n \sum_{i=1}^{M} \phi_k(x_i) \phi_n(x_i) w(x_i) \, \Delta x = 0.$$

Now if we take the limit as $M \to \infty$ and $\Delta x \to 0$, we see this approaching the limit

$$\int_a^b \phi_k(x) f(x) w(x) \, dx - \sum_{n=0}^{N} A_n \int_a^b \phi_k(x) \phi_n(x) w(x) \, dx = 0.$$

With real functions, the orthogonality condition is

$$\int_a^b \phi_k(x) \phi_n(x) w(x) \, dx = \delta_{nk}.$$

Therefore

$$A_k = \int_a^b \phi_k(x) f(x) w(x) \, dx$$

which is exactly the same as the Fourier coefficient. In this approximation, the mean square error is minimized. For the generalized Fourier series, in which $\{\phi_n\}$ is a complete set and $N \to \infty$, the integral of the error squared goes to zero.

Of crucial importance is that the basis set must be complete. The set $\{\phi_n\}$ is complete in the function space if there is no nonzero function that is orthogonal to each of the function ϕ_n. For example, $\left\{ \frac{1}{\sqrt{\pi}} \sin nx \right\}$ $(n = 1, 2, \ldots)$ is an orthonormal set on the interval $-\pi \le x \le \pi$. But it is not complete since any even function in that interval is orthogonal to any of ϕ_n in the set.

It is not always that easy to use the definition to test if a set is complete. Fortunately, complete sets of orthogonal functions are provided by the eigenfunctions of certain type of differential operators known as Hermitian (or self-adjoint) operators.

3.3 Hermitian Operators

3.3.1 Adjoint and Self-adjoint (Hermitian) Operators

If the functions $f(x)$ and $g(x)$ in the vector space of functions, satisfy certain boundary conditions, the adjoint of a linear differential operator L, denoted by L^+, is defined by the relation

$$\langle Lf\,|g\rangle = \langle f\,|L^+ g\rangle.$$

For example, in an infinite dimensional vector space consisting of all square-integrable functions with the inner product defined as

$$\langle f\,|f\rangle = \int_{-\infty}^{\infty} |f|^2\, dx < \infty,$$

all functions must satisfy the boundary conditions

$$f(x) \to 0, \text{ as } x \to \pm\infty.$$

If the differential operator L in this space, in which $w(x)=1$, is d/dx; ($L =$ d/dx), then the inner product $\langle Lf\,|g\rangle$ is given by

$$\langle Lf\,|g\rangle = \left\langle \frac{d}{dx}f\,\middle|\,g\right\rangle = \int_{-\infty}^{\infty}\left(\frac{d}{dx}f\right)^* g\, dx = \int_{-\infty}^{\infty}\frac{d}{dx}f^* g\, dx.$$

With integration by parts,

$$\int_{-\infty}^{\infty}\frac{d}{dx}f^* g\, dx = f^*(x)g(x)\Big|_{-\infty}^{\infty} - \int_{-\infty}^{\infty} f^*\frac{d}{dx}g\, dx = \left\langle f\,\middle|\,-\frac{d}{dx}g\right\rangle = \langle f\,|L^+ g\rangle,$$

since the integrated part is equal to zero because of the boundary conditions $f(\pm\infty) \to 0$. Thus, the adjoint of the operator $L =$ d/dx is $L^+ = -$d/dx in this space.

Example 3.3.1. In the space of square integrable functions $f(x)$ on the interval $-\infty < x < \infty$, find the adjoint of the operators (a) $L = d^2/dx^2$, and (b) $L = \frac{1}{i}\frac{d}{dx}$.

Solution 3.3.1. (a) $L = \dfrac{d^2}{dx^2}$,

$$\langle Lf\,|g\rangle = \left\langle \frac{d^2}{dx^2}f\,\middle|\,g\right\rangle = \left\langle \frac{d}{dx}f\,\middle|\,-\frac{d}{dx}g\right\rangle = \left\langle f\,\middle|\,\frac{d^2}{dx^2}g\right\rangle = \langle f\,|L^+ g\rangle.$$

Therefore the adjoint of d^2/dx^2 is $L^+ = d^2/dx^2$.

(b) $L = \frac{1}{i}\frac{d}{dx}$,

$$\langle Lf\,|g\rangle = \left\langle \frac{1}{i}\frac{d}{dx}f \Big| g \right\rangle = \frac{1}{-i}\left\langle \frac{d}{dx}f \Big| g \right\rangle = \frac{1}{-i}\left\langle f \Big| -\frac{d}{dx}g \right\rangle = \left\langle f \Big| \frac{1}{i}\frac{d}{dx}g \right\rangle,$$

where we have used (3.2) and (3.3). Therefore the adjoint of $L = \frac{1}{i}\frac{d}{dx}$ is

$$L^+ = \frac{1}{i}\frac{d}{dx}.$$

An operator is said to be self-adjoint (or Hermitian) if $L^+ = L$. Thus, in the above example, the operators $\frac{d^2}{dx^2}$ and $\frac{1}{i}\frac{d}{dx}$ are Hermitian, but d/dx is not Hermitian since $L^+ = -d/dx$ which is not the same as $L = d/dx$.

In this example, the weight function $w(x)$ is taken to be unity. In general, $w(x)$ can be any real and positive function. Furthermore, the space can be defined in any interval. If x is specified to be on the interval $a \le x \le b$, the general expressions of inner products take the following forms.

$$\langle Lf\,|g\rangle = \int_{-\infty}^{\infty} (Lf(x))^* g(x)w(x)dx,$$

$$\langle f\,|Lg\rangle = \int_{-\infty}^{\infty} f^*(x)Lg(x)w(x)dx.$$

Since $w(x)$ is real, and

$$\int_{-\infty}^{\infty} (Lf(x))^* g(x)w(x)\,dx = \left(\int_{-\infty}^{\infty} g^*(x)Lf(x)w(x)\,dx \right)^*,$$

a self-adjoint operator L can also be expressed as

$$\int_{-\infty}^{\infty} f^*(x)Lg(x)w(x)\,dx = \left(\int_{-\infty}^{\infty} g^*(x)Lf(x)w(x)dx \right)^*.$$

Symbolically, this also follows from the fact that $\langle Lf\,|g\rangle = \langle f\,|Lg\rangle$ and $\langle Lf\,|g\rangle = \langle g\,|Lf\rangle^*$, so

$$\langle f\,|Lg\rangle = \langle g\,|Lf\rangle^*. \tag{3.7}$$

In a finite dimensional space, the eigenvalues of a Hermitian matrix are real and the eigenvectors form an orthogonal basis. In an infinite dimensional space, the Hermitian differential operator plays the same role as the Hermitian matrix in the finite dimensional space. Corresponding to the matrix eigenvalue problem, we have the eigenvalue problem of differential operator

$$L\phi(x) = \lambda\phi(x),$$

where λ is a constant. For a given choice of λ, a function which satisfies the equation and the imposed boundary conditions is called an eigenfunction corresponding to λ. The constant λ is then called an eigenvalue. There is no guarantee the eigenfunction $\phi(x)$ will exist for any arbitrary choice of the parameter λ. The requirement that there be an eigenfunction often restricts the acceptable values of λ to a discrete set. We shall see in Sect. 3.3.2 that the eigenvalues of a Hermitian operator are real and the eigenfunctions form a complete orthogonal basis set.

Furthermore, the elements a_{ij} of a Hermitian matrix are characterized by the relation

$$a_{ij} = a_{ji}^*. \tag{3.8}$$

In analogy, we often define a "matrix element" L_{ij} of a Hermitian operator

$$L_{ij} = \langle \phi_i \,|L\phi_j \rangle.$$

By (3.7),

$$\langle \phi_i \,|L\phi_j \rangle = \langle \phi_j \,|L\phi_i \rangle^*.$$

Therefore

$$L_{ij} = L_{ji}^*. \tag{3.9}$$

The similarity between (3.8) and (3.9) is obvious.

In quantum mechanics, the expectation value of an observable (a physical quantity that can be observed), such as energy and momentum, is the average value of many measurements of that quantity. The outcome of a measurement is of course a real number. Furthermore, the observable is represented by an operator O and the expectation value is given by $\langle \Psi \,|O\,\Psi \rangle$ where Ψ is the wave function describing the state of the system. Thus $\langle \Psi \,|O\,\Psi \rangle$ must be real, that is

$$\langle \Psi \,|O\,\Psi \rangle^* = \langle \Psi \,|O\,\Psi \rangle.$$

Since

$$\langle \Psi \,|O\,\Psi \rangle^* = \langle O\Psi \,|\Psi \rangle,$$

it follows

$$\langle O\Psi \,|\Psi \rangle = \langle \Psi \,|O\,\Psi \rangle.$$

Therefore any operator representing an observable must be Hermitian.

3.3.2 Properties of Hermitian Operators

The Eigenvalues of a Hermitian Operator are Real. Let λ be an eigenvalue of the operator L and ϕ be the corresponding eigenfunction

$$L\phi = \lambda\phi.$$

So

$$\langle L\phi \,|\phi \rangle = \langle \lambda\phi \,|\phi \rangle = \lambda^* \langle \phi \,|\phi \rangle.$$

Since L is Hermitian, it follows that

$$\langle L\phi \,|\phi\rangle = \langle \phi \,|L\phi\rangle = \langle \phi \,|\lambda\phi\rangle = \lambda \langle \phi \,|\phi\rangle .$$

Thus

$$\lambda^* \langle \phi \,|\phi\rangle = \lambda \langle \phi \,|\phi\rangle .$$

Therefore

$$\lambda^* = \lambda,$$

the eigenvalue of a Hermitian operator must be real.

It is interesting to note that the Hermitian operator can be imaginary. Even if the operator is real, the eigenfunction can be complex. But in all cases, the eigenvalues must be real.

Because the eigenvalues are real, the eigenfunctions of a real Hermitian operator can always be made real by taking a suitable linear combinations. Since by definition

$$L\phi_i = \lambda_i \phi_i,$$

the complex conjugate is given by

$$L\phi_i^* = \lambda_i^* \phi_i^* = \lambda_i \phi_i^*,$$

where we have used the fact $\lambda^* = \lambda$. Thus both ϕ_i and ϕ_i^* are eigenfunctions corresponding to the same eigenvalue. Because of the linearity of L, any linear combination of ϕ_i and ϕ_i^* must also be an eigenfunction. Now both $\phi_i + \phi_i^*$ and $i(\phi_i - \phi_i^*)$ are real, so we can take them as eigenfunctions for the eigenvalue λ_i. So for a real operator, we can assume both eigenvalues and eigenfunctions are real.

The Eigenfunctions of a Hermitian Operator are Orthogonal. Let ϕ_i and ϕ_j be eigenfunctions corresponding to two different eigenvalues λ_i and λ_j,

$$L\phi_i = \lambda_i \phi_i,$$
$$L\phi_j = \lambda_j \phi_j.$$

It follows that

$$\langle L\phi_i \,|\phi_j\rangle = \langle \lambda_i\phi_i \,|\phi_j\rangle = \lambda_i^* \langle \phi_i \,|\phi_j\rangle = \lambda_i \langle \phi_i \,|\phi_j\rangle ,$$

the last equality follows from the fact that the eigenvalues are real. Since L is Hermitian,

$$\langle L\phi_i \,|\phi_j\rangle = \langle \phi_i \,|L\phi_j\rangle = \langle \phi_i \,|\lambda_j\phi_j\rangle = \lambda_j \langle \phi_i \,|\phi_j\rangle .$$

Thus

$$\lambda_i \langle \phi_i \,|\phi_j\rangle = \lambda_j \langle \phi_i \,|\phi_j\rangle ,$$
$$(\lambda_i - \lambda_j) \langle \phi_i \,|\phi_j\rangle = 0.$$

Since $\lambda_i \neq \lambda_j$, we must have

$$\langle \phi_i \,|\phi_j \rangle = 0.$$

Therefore ϕ_i and ϕ_j are orthogonal.

Degeneracy. If n linear independent eigenfunctions correspond to the same eigenvalue, the eigenvalue is said to be n-fold degenerate. If this is the case, we cannot use the above argument to show that these eigenfunctions are orthogonal and they may not be. However, if they are not orthogonal, we can use the Gram–Schmidt process to construct n-orthogonal functions out of the n linearly independent eigenfunctions. These newly constructed functions will satisfy the same equation and be orthogonal to each other and to other eigenfunctions belonging to different eigenvalues.

The Eigenfunctions of an Hermitian Operator form a Complete Set. Recall that a Hermitian matrix can always be diagonalized. The eigenvector of a diagonalized matrix is a column vector with only one nonzero element. For example

$$\begin{pmatrix} \lambda_1 & 0 \\ 0 & \lambda_2 \end{pmatrix} \begin{pmatrix} 1 \\ 0 \end{pmatrix} = \lambda_1 \begin{pmatrix} 1 \\ 0 \end{pmatrix}, \quad \begin{pmatrix} \lambda_1 & 0 \\ 0 & \lambda_2 \end{pmatrix} \begin{pmatrix} 0 \\ 1 \end{pmatrix} = \lambda_2 \begin{pmatrix} 0 \\ 1 \end{pmatrix}.$$

Any vector in this two-dimensional space can be expressed in terms of these two eigenvectors

$$\begin{pmatrix} c_1 \\ c_2 \end{pmatrix} = c_1 \begin{pmatrix} 1 \\ 0 \end{pmatrix} + c_2 \begin{pmatrix} 0 \\ 1 \end{pmatrix}.$$

We say that these two eigenvectors form a complete orthogonal basis. Clearly, the eigenvectors of a $n \times n$ Hermitian matrix will form a complete orthogonal basis for the n-dimensional space.

One would expect that in an infinite dimensional vector space of functions, the eigenfunctions of a Hermitian operator will form a complete set of orthogonal basis. This is indeed the case. A proof of this fact can be found in "Methods of Mathematical Physics", Chap. 6, by Courant and Hilbert, Interscience Publishers (1953), Reprinted by Wiley (1989).

Thus, in the interval where the linear operator L is Hermitian, any piecewise continuous function $f(x)$ can be expressed in a generalized Fourier series of eigenfunctions of L, that is, if the set of eigenfunctions $\{\phi_n\}$ $(n = 0, 1, 2, \ldots)$ is normalized, then

$$f(x) = \sum_{n=0}^{\infty} \langle f \,|\phi_n \rangle \, \phi_n,$$

where $L\phi_n = \lambda_n \phi_n$.

It is to be emphasized that in the space where L is Hermitian, the functions in this space have to satisfy certain boundary conditions. It is these boundary conditions that determine the eigenfunctions. Let us illustrate this point with the following example.

Example 3.3.2. (a) Let the weight function be equal to unity $w(x) = 1$, find the required boundary conditions for the differential operator $L = \mathrm{d}^2/\mathrm{d}x^2$ to be Hermitian over the interval $a \leq x \leq b$. (b) Show that if the solutions of $Ly = \lambda y$ in the interval $0 \leq x \leq 2\pi$ satisfy the boundary conditions $y(0) = y(2\pi)$, $y'(0) = y'(2\pi)$, (where y' means the derivative of y with respect to x), then the operator L in this interval is Hermitian. (c) Find the complete set of eigenfunctions of L.

Solution 3.3.2. (a) Let $y_i(x)$ and $y_j(x)$ be two functions in this space. Integrating the inner product $\langle y_i \,|Ly_j\rangle$ by parts gives

$$\langle y_i \,|Ly_j\rangle = \int_a^b y_i^* \frac{\mathrm{d}^2 y_j}{\mathrm{d}x^2}\,\mathrm{d}x = \left[y_i^* \frac{\mathrm{d}y_j}{\mathrm{d}x} \right]_a^b - \int_a^b \frac{\mathrm{d}y_i^*}{\mathrm{d}x} \frac{\mathrm{d}y_j}{\mathrm{d}x}\,\mathrm{d}x.$$

Integrating the second term on the right-hand side by parts again yields

$$\int_a^b \frac{\mathrm{d}y_i^*}{\mathrm{d}x} \frac{\mathrm{d}y_j}{\mathrm{d}x}\,\mathrm{d}x = \left[\frac{\mathrm{d}y_i^*}{\mathrm{d}x} y_j \right]_a^b - \int_a^b \frac{\mathrm{d}^2 y_i^*}{\mathrm{d}x^2} y_j\,\mathrm{d}x.$$

Thus

$$\langle y_i \,|Ly_j\rangle = \left[y_i^* \frac{\mathrm{d}y_j}{\mathrm{d}x} \right]_a^b - \left[\frac{\mathrm{d}y_i^*}{\mathrm{d}x} y_j \right]_a^b + \langle Ly_i \,|y_j\rangle .$$

Therefore L is Hermitian provided

$$\left[y_i^* \frac{\mathrm{d}y_j}{\mathrm{d}x} \right]_a^b - \left[\frac{\mathrm{d}y_i^*}{\mathrm{d}x} y_j \right]_a^b = 0.$$

(b) Because of the boundary conditions $y(0) = y(2\pi)$, $y'(0) = y'(2\pi)$,

$$\left[y_i^* \frac{\mathrm{d}y_j}{\mathrm{d}x} \right]_0^{2\pi} = y_i^*(2\pi)y_j'(2\pi) - y_i^*(0)y_j'(0) = 0,$$

$$\left[\frac{\mathrm{d}y_i^*}{\mathrm{d}x} y_j \right]_0^{2\pi} = y_i^{*\prime}(2\pi)y_j(2\pi) - y_i^{*\prime}(0)y_j(0) = 0.$$

Therefore L is Hermitian in this interval, since

$$\langle y_i \,|Ly_j\rangle = \left[y_i^* \frac{\mathrm{d}y_j}{\mathrm{d}x} \right]_a^b - \left[\frac{\mathrm{d}y_i^*}{\mathrm{d}x} y_j \right]_a^b + \langle Ly_i \,|y_j\rangle = \langle y_i \,|L^+y_j\rangle .$$

(c) To find the eigenfunctions of L, we must solve the differential equation

$$\frac{\mathrm{d}^2 y(x)}{\mathrm{d}x^2} = \lambda y(x),$$

subject to the boundary conditions

$$y(0) = y(2\pi), \qquad y'(0) = y'(2\pi).$$

The solution of the differential equation is

$$y(x) = A \cos \sqrt{\lambda}x + B \sin \sqrt{\lambda}x,$$

where A and B are two arbitrary constants. So

$$y'(x) = -\sqrt{\lambda}A \sin \sqrt{\lambda}x + \sqrt{\lambda}B \cos \sqrt{\lambda}x,$$

and

$$y(0) = A, \qquad y(2\pi) = A \cos \sqrt{\lambda}2\pi + B \sin \sqrt{\lambda}2\pi,$$
$$y'(0) = \sqrt{\lambda}B, \qquad y'(2\pi) = -\sqrt{\lambda}A \sin \sqrt{\lambda}2\pi + \sqrt{\lambda}B \cos \sqrt{\lambda}2\pi.$$

Because of the boundary conditions $y(0) = y(2\pi)$, $y'(0) = y'(2\pi)$,

$$A = A \cos \sqrt{\lambda}2\pi + B \sin \sqrt{\lambda}2\pi,$$
$$\sqrt{\lambda}B = -\sqrt{\lambda}A \sin \sqrt{\lambda}2\pi + \sqrt{\lambda}B \cos \sqrt{\lambda}2\pi,$$

or

$$A(1 - \cos \sqrt{\lambda}2\pi) - B \sin \sqrt{\lambda}2\pi = 0$$
$$A \sin \sqrt{\lambda}2\pi + B(1 - \cos \sqrt{\lambda}2\pi) = 0.$$

A and B will have nontrivial solutions if and only if

$$\begin{vmatrix} 1 - \cos \sqrt{\lambda}2\pi & -\sin \sqrt{\lambda}2\pi \\ \sin \sqrt{\lambda}2\pi & 1 - \cos \sqrt{\lambda}2\pi \end{vmatrix} = 0.$$

It follows that

$$1 - 2\cos \sqrt{\lambda}2\pi + \cos^2 \sqrt{\lambda}2\pi + \sin^2 \sqrt{\lambda}2\pi = 0,$$

or

$$2 - 2\cos \sqrt{\lambda}2\pi = 0.$$

Thus

$$\cos \sqrt{\lambda}2\pi = 1$$

and

$$\sqrt{\lambda} = n, \qquad n = 0, 1, 2, \ldots.$$

Hence, for each integer n, the solution is

$$y_n(x) = A_n \cos nx + B_n \sin nx.$$

In other words, for this periodic boundary conditions, the eigenfunctions of this Hermitian operator $\mathrm{d}^2/\mathrm{d}x^2$ are $\cos nx$ and $\sin nx$. This means that the collection of $\{\cos nx,\ \sin nx\}$ $(n = 0, 1, 2, \ldots)$ is a complete basis set for this space. Therefore, any piecewise continuous periodic function with period of 2π can be expanded in terms of these eigenfunctions. This expansion is, of course, just the regular Fourier series.

A systematic account of the relations between the boundary conditions and the eigenfunctions of the second-order differential equations is provided by the Sturm–Liouville theory.

3.4 Sturm–Liouville Theory

In the last example, we have seen that the eigenfunctions of the differential operator $\mathrm{d}^2/\mathrm{d}x^2$ with some boundary conditions form a complete set of orthogonal basis. A far more general eigenvalue problem of second-order differential operators is the Sturm–Liouville problem.

3.4.1 Sturm–Liouville Equations

A linear second-order differential equation

$$A(x)\frac{\mathrm{d}^2}{\mathrm{d}x^2}y + B(x)\frac{\mathrm{d}}{\mathrm{d}x}y + C(x)y + \lambda D(x)y = 0, \tag{3.10}$$

where λ is a parameter to be determined by the boundary conditions, can be put in the form of

$$\frac{\mathrm{d}^2}{\mathrm{d}x^2}y + b(x)\frac{\mathrm{d}}{\mathrm{d}x}y + c(x)y + \lambda\,\mathrm{d}(x)y = 0 \tag{3.11}$$

by dividing every term by $A(x)$, provided $A(x) \neq 0$. Let us define an integrating factor $p(x)$,

$$p(x) = e^{\int^x b(x')\mathrm{d}x'}.$$

Multiplying (3.11) by $p(x)$, we have

$$p(x)\frac{\mathrm{d}^2}{\mathrm{d}x^2}y + p(x)b(x)\frac{\mathrm{d}}{\mathrm{d}x}y + p(x)c(x)y + \lambda p(x)d(x)y = 0. \tag{3.12}$$

Since

$$\frac{\mathrm{d}p(x)}{\mathrm{d}x} = \frac{\mathrm{d}}{\mathrm{d}x}e^{\int^x b(x')\mathrm{d}x'} = e^{\int^x b(x')\mathrm{d}x'}\frac{\mathrm{d}}{\mathrm{d}x}\int^x b(x')\mathrm{d}x = p(x)b(x),$$

so

$$\frac{\mathrm{d}}{\mathrm{d}x}\left[p(x)\frac{\mathrm{d}}{\mathrm{d}x}y\right] = p(x)\frac{\mathrm{d}^2}{\mathrm{d}x^2}y + \frac{\mathrm{d}p(x)}{\mathrm{d}x}\frac{\mathrm{d}}{\mathrm{d}x}y = p(x)\frac{\mathrm{d}^2}{\mathrm{d}x^2}y + p(x)b(x)\frac{\mathrm{d}}{\mathrm{d}x}y.$$

Thus, (3.12) can be written as

$$\frac{\mathrm{d}}{\mathrm{d}x}\left[p(x)\frac{\mathrm{d}}{\mathrm{d}x}y\right] + q(x)y + \lambda w(x)y = 0, \tag{3.13}$$

where $q(x) = p(x)c(x)$ and $w(x) = p(x)d(x)$. Since the factor $p(x)$ is every-where nonzero, the solutions of (3.10)–(3.13) are identical, so these equations are equivalent.

Under the general conditions that p, q, w are real and continuous, and both $p(x)$ and $w(x)$ are positive on certain interval, equations in the form of (3.13) are known as Sturm–Liouville equations, named after French mathematicians Sturm (1803–1855) and Liouville (1809–1882), who first developed an extensive theory of these equations.

These equations can be put in the usual eigenvalue problem form

$$Ly = \lambda y$$

by defining a Sturm–Liouville operator

$$L = -\frac{1}{w(x)}\left[\frac{\mathrm{d}}{\mathrm{d}x}\left(p(x)\frac{\mathrm{d}}{\mathrm{d}x}\right) + q(x)\right]. \tag{3.14}$$

Sturm–Liouville theory is very important in engineering and physics, because under a variety of boundary conditions on the solution $y(x)$, linear operators that can be written in this form are Hermitian. Therefore the eigenfunctions of the Sturm–Liouville equations form complete sets of orthogonal bases for the function space in which the weight function is $w(x)$. The set of cosine and sine functions of Fourier series is just one example within a broader Sturm–Liouville theory.

We note that the definitions of the Sturm–Liouville operator vary; some authors use

$$L = \frac{\mathrm{d}}{\mathrm{d}x}\left(p\frac{\mathrm{d}}{\mathrm{d}x}\right) + q(x)$$

and write the eigenvalue equation as

$$Ly = -\lambda wy.$$

As long as it is consistent, the difference is just a matter of convention. We will use (3.14) as the definition of the Sturm–Liouville operator.

3.4.2 Boundary Conditions of Sturm–Liouville Problems

Sturm–Liouville Operators as Hermitian Operators. Let L be the Sturm–Liouville operator in (3.14), and $f(x)$ and $g(x)$ be two functions having continuous second derivatives on the interval $a \leq x \leq b$, then

$$\langle Lf \,|g\rangle = \int_a^b \left\{ -\frac{1}{w} \left[\frac{\mathrm{d}}{\mathrm{d}x} \left(p\frac{\mathrm{d}}{\mathrm{d}x} \right) + q \right] f \right\}^* gw \, \mathrm{d}x.$$

Since p, q, w are real, the integral can be written as

$$\langle Lf \,|g\rangle = - \int_a^b \frac{\mathrm{d}}{\mathrm{d}x} \left(p\frac{\mathrm{d}}{\mathrm{d}x} f^* \right) g \, \mathrm{d}x - \int_a^b qf^*g \, \mathrm{d}x.$$

With integration by parts,

$$\int_a^b \frac{\mathrm{d}}{\mathrm{d}x} \left(p\frac{\mathrm{d}f^*}{\mathrm{d}x} \right) g \, \mathrm{d}x = \left. p\frac{\mathrm{d}f^*}{\mathrm{d}x} g \right|_a^b - \int_a^b p\frac{\mathrm{d}f^*}{\mathrm{d}x}\frac{\mathrm{d}g}{\mathrm{d}x} \mathrm{d}x,$$

and

$$\int_a^b p\frac{\mathrm{d}f^*}{\mathrm{d}x}\frac{\mathrm{d}g}{\mathrm{d}x}\mathrm{d}x = \int_a^b \frac{\mathrm{d}f^*}{\mathrm{d}x}p\frac{\mathrm{d}g}{\mathrm{d}x}\mathrm{d}x = \left. f^*p\frac{\mathrm{d}g}{\mathrm{d}x} \right|_a^b - \int_a^b f^* \frac{\mathrm{d}}{\mathrm{d}x} \left(p\frac{\mathrm{d}g}{\mathrm{d}x} \right) \mathrm{d}x.$$

It follows that

$$\langle Lf \,|g\rangle = - \left. p\frac{\mathrm{d}f^*}{\mathrm{d}x} g \right|_a^b + \left. f^*p\frac{\mathrm{d}g}{\mathrm{d}x} \right|_a^b - \int_a^b f^* \frac{\mathrm{d}}{\mathrm{d}x} \left(p\frac{\mathrm{d}g}{\mathrm{d}x} \right) \mathrm{d}x - \int_a^b qf^*g \, \mathrm{d}x,$$

or

$$\langle Lf \,|g\rangle = \left[p \left(f^*\frac{\mathrm{d}g}{\mathrm{d}x} - \frac{\mathrm{d}f^*}{\mathrm{d}x} g \right) \right]_a^b + \int_a^b f^* \left\{ -\frac{1}{w} \left[\frac{\mathrm{d}}{\mathrm{d}x} \left(p\frac{\mathrm{d}}{\mathrm{d}x} \right) + q \right] g \right\} w \, \mathrm{d}x$$

$$= \left[p \left(f^*\frac{\mathrm{d}g}{\mathrm{d}x} - \frac{\mathrm{d}f^*}{\mathrm{d}x} g \right) \right]_a^b + \langle f \,|Lg\rangle.$$

It is clear that if

$$\left[p \left(f^*\frac{\mathrm{d}g}{\mathrm{d}x} - \frac{\mathrm{d}f^*}{\mathrm{d}x} g \right) \right]_a^b = 0, \tag{3.15}$$

then

$$\langle Lf \,|g\rangle = \langle f \,|Lg\rangle.$$

In other words, if the function space consists of functions that satisfy (3.15), then the Sturm–Liouville operator L is Hermitian in that space.

Sturm–Liouville Problems. It is customary to refer to the Sturm–Liouville equation and the boundary conditions together as the Sturm–Liouville problem. Since the operator is Hermitian, the eigenfunctions of the Sturm–Liouville problem are orthogonal to each other with respect to the weight function $w(x)$ and they are complete. Therefore they can be used as basis for the generalized Fourier series, which is also called eigenfunction expansion.

If any two solutions $y_n(x)$ and $y_m(x)$ of the linear homogeneous second-order differential equation

$$[p(x)y'(x)]' + q(x)y(x) + \lambda w y(x) = 0, \quad a \leq x \leq b$$

satisfy the boundary condition (3.15), then the equation together with its boundary conditions is called a Sturm–Liouville problem. Since the operator is real, the eigenfunctions can also be taken as real. Therefore the boundary condition (3.15) can be conveniently written as

$$p(b) \begin{vmatrix} y_n(b) & y_n'(b) \\ y_m(b) & y_m'(b) \end{vmatrix} - p(a) \begin{vmatrix} y_n(a) & y_n'(a) \\ y_m(a) & y_m'(a) \end{vmatrix} = 0. \tag{3.16}$$

Depending on how the boundary conditions are met, Sturm–Liouville problems are divided into the following subgroups.

3.4.3 Regular Sturm–Liouville Problems

In this case, $p(a) \neq 0$ and $p(b) \neq 0$. The Sturm–Liouville problem consists of the equation

$$Ly(x) = \lambda y(x)$$

with L given by (3.14), and the boundary conditions

$$\alpha_1 y(a) + \alpha_2 y'(a) = 0,$$
$$\beta_1 y(b) + \beta_2 y'(b) = 0,$$

where the constants α_1 and α_2 cannot both be zero, and β_1 and β_2 also cannot both be zero.

Let us show that these boundary conditions satisfy (3.16). If $y_n(x)$ and $y_m(x)$ are two different solutions of the problem, both have to satisfy the boundary conditions. The first boundary condition requires

$$\alpha_1 y_n(a) + \alpha_2 y_n'(a) = 0,$$
$$\alpha_1 y_m(a) + \alpha_2 y_m'(a) = 0.$$

This is a system of two simultaneous equations in α_1 and α_2. Since α_1 and α_2 cannot both be zero, the determinant of the coefficients must be zero,

$$\begin{vmatrix} y_n(a) & y_n'(a) \\ y_m(a) & y_m'(a) \end{vmatrix} = 0.$$

Similarly, the second boundary condition requires

$$\begin{vmatrix} y_n(b) & y_n'(b) \\ y_m(b) & y_m'(b) \end{vmatrix} = 0.$$

Clearly,

$$p(b)\begin{vmatrix} y_n(b) & y_n'(b) \\ y_m(b) & y_m'(b) \end{vmatrix} - p(a)\begin{vmatrix} y_n(a) & y_n'(a) \\ y_m(a) & y_m'(a) \end{vmatrix} = 0.$$

Therefore the boundary condition of (3.16) is satisfied.

Example 3.4.1. (a) Show that for $0 \le x \le 1$,

$$y'' + \lambda y = 0$$
$$y(0) = 0, \quad y(1) = 0,$$

constitute a regular Sturm–Liouville problem.
  (b) Find the eigenvalues and eigenfunctions of the problem.

Solution 3.4.1. (a) With $p(x) = 1$, $q(x) = 0$, $w(x) = 1$, the Sturm–Liouville equation

$$(py')' + qy + \lambda wy = 0$$

becomes

$$y'' + \lambda y = 0.$$

Furthermore, with $a = 0$, $b = 1$, $\alpha_1 = 1$, $\alpha_2 = 0$, $\beta_1 = 1$, $\beta_2 = 0$, the boundary conditions

$$\alpha_1 y(a) + \alpha_2 y'(a) = 0,$$

$$\beta_1 y(b) + \beta_2 y'(b) = 0,$$

become

$$y(0) = 0, \quad y(1) = 0.$$

Therefore the given equation and the boundary conditions constitute a regular Sturm–Liouville problem.

  (b) To find the eigenvalues, let us look at the possibilities of $\lambda = 0, \lambda < 0, \lambda > 0$.
  If $\lambda = 0$, the solution of the equation is given by

$$y(x) = c_1 x + c_2.$$

Applying the boundary conditions, we have

$$y(0) = c_2 = 0, \quad y(1) = c_1 + c_2 = 0,$$

so $c_1 = 0$ and $c_2 = 0$. This is a trivial solution. Therefore $\lambda = 0$ is not an eigenvalue.

If $\lambda < 0$, let $\lambda = -\mu^2$ with real μ, so the solution of the equation is

$$y(x) = c_1 e^{\mu x} + c_2 e^{-\mu x}.$$

The condition $y(0) = 0$ makes $c_2 = -c_1$. The condition $y(1) = 0$ requires

$$y(1) = c_1(e^{u} - e^{-\mu}) = 0.$$

Since $\mu \neq 0$, so $c_1 = 0$. Again this gives the trivial solution.

This leaves the only possibility that $\lambda > 0$. Let $\lambda = \mu^2$ with real μ, so the solution of the equation becomes

$$y(x) = c_1 \cos \mu x + c_2 \sin \mu x.$$

Applying the boundary condition $y(0) = 0$ leads to

$$y(0) = c_1 = 0.$$

Therefore we are left with

$$y(x) = c_2 \sin \mu x.$$

The boundary condition $y(1) = 0$ requires

$$c_2 \sin \mu = 0.$$

For the nontrivial solution, we must have

$$\sin \mu = 0.$$

This will occur if μ is an integer multiple of π,

$$\mu = n\pi, \quad n = 1,\, 2,\ldots.$$

Thus the eigenvalues are

$$\lambda_n = \mu^2 = (n\pi)^2, \quad n = 1,\, 2,\ldots,$$

and the corresponding eigenfunctions are

$$y_n(x) = \sin n\pi x.$$

Of course, we can solve this problem without knowing that it is a Sturm–Liouville problem. The advantage of knowing that $\{\sin n\pi x\}\, (n = 1, 2, \ldots)$ are eigenfunctions of a Sturm–Liouville problem is that immediately we know that they are orthogonal to each other. More importantly, we know that they form a complete set in the interval $0 \leq x \leq 1$.

Example 3.4.2. (a) Put the following problem into the Sturm–Liouville form,

$$y'' - 2y' + \lambda y = 0, \qquad 0 \leq x \leq \pi$$
$$y(0) = 0, \quad y(\pi) = 0.$$

(b) Find the eigenvalues and eigenfunctions of the problem.
(c) Find the eigenfunction expansion of a given function $f(x)$ on the interval $0 \leq x \leq \pi$.

Solution 3.4.2. (a) Let us first find the integrating factor p,

$$p(x) = e^{\int^x (-2) dx'} = e^{-2x}.$$

Multiplying the differential equation by $p(x)$, we have

$$e^{-2x} y'' - 2e^{-2x} y' + \lambda e^{-2x} y = 0,$$

which can be written as

$$(e^{-2x} y')' + \lambda\, e^{-2x} y = 0.$$

This is a Sturm–Liouville equation with $p(x) = e^{-2x}$, $q(x) = 0$, and $w(x) = e^{-2x}$.

(b) Since the original differential equation is an equation with constant coefficients, we seek the solution in the form of $y(x) = e^{mx}$. With this trial solution, the equation becomes

$$(m^2 - 2m + \lambda)e^{mx} = 0.$$

The roots of the characteristic equation $m^2 - 2m + \lambda = 0$ are

$$m = 1 \pm \sqrt{1 - \lambda},$$

therefore

$$y(x) = e^x \left(c_1 e^{\sqrt{1-\lambda}\,x} + c_2 e^{-\sqrt{1-\lambda}\,x} \right)$$

for $\lambda \neq 1$.

For $\lambda = 1$, the characteristic equation has a double root at $m = 1$, and the solution becomes

$$y_2(x) = c_3 x + c_4.$$

The boundary conditions $y_2(0) = 0$ and $y_2(\pi) = 0$ require that $c_3 = c_4 = 0$. Therefore there is no nontrivial solution in this case, so $\lambda = 1$ is not an eigenvalue.

For $\lambda \neq 1$, the boundary condition $y(0) = 0$ requires

$$y(0) = c_1 + c_2 = 0.$$

Therefore the solution becomes

$$y(x) = c_1 e^x \left(e^{\sqrt{1-\lambda}x} - e^{-\sqrt{1-\lambda}x} \right).$$

If $\lambda < 1$, the other boundary condition $y(\pi) = 0$ requires

$$y(\pi) = c_1 e^\pi \left(e^{\sqrt{1-\lambda}\pi} - e^{-\sqrt{1-\lambda}\pi} \right) = 0.$$

This is possible only for the trivial solution of $c_1 = 0$. Therefore there is no eigenvalue less than 1.

For $\lambda > 1$, the solution can be written in the form of

$$y(x) = c_1 e^x \left(e^{i\sqrt{\lambda-1}x} - e^{-i\sqrt{\lambda-1}x} \right)$$

$$= 2ic_1 e^x \sin \sqrt{\lambda - 1}x.$$

The boundary condition $y(\pi) = 0$ is satisfied if

$$\sin \sqrt{\lambda - 1}\pi = 0.$$

This can occur if

$$\sqrt{\lambda - 1} = n, \quad n = 1, 2, \ldots.$$

Therefore the eigenvalues are

$$\lambda_n = n^2 + 1, \quad n = 1, 2, \ldots,$$

and the eigenfunction associated with each eigenvalue λ_n is

$$\phi_n(x) = e^x \sin nx.$$

Any arbitrary constant can be multiplied to $\phi_n(x)$ to give a solution for the problem with $\lambda = \lambda_n$.

(c) For a given function $f(x)$ on the interval $0 \le x \le \pi$, the eigenfunction expansion is

$$f(x) = \sum_{n=1}^{\infty} c_n \phi_n(x).$$

Since $\{\phi_n\}$ $(n = 1, 2, \ldots)$ is a set of eigenfunctions of the Sturm–Liouville problem, it is an orthogonal set with respect to the weight function $w(x) = e^{-2x}$,

$$\langle \phi_n | \phi_m \rangle = \int_0^\pi (e^x \sin nx)(e^x \sin mx)e^{-2x}dx = 0, \quad for \ n \ne m.$$

For $n = m$,

$$\langle \phi_n | \phi_n \rangle = \int_0^\pi (e^x \sin nx)(e^x \sin nx)e^{-2x}dx = \int_0^\pi \sin^2 nx \ dx = \frac{\pi}{2}.$$

Therefore

$$\langle \phi_n \,|\phi_m \rangle = \frac{\pi}{2}\delta_{nm}.$$

Taking the inner product of both sides of the eigenfunction expansion with ϕ_m, we have

$$\langle f \,|\phi_m \rangle = \sum_{n=1}^{\infty} c_n \langle \phi_n \,|\phi_m \rangle = \sum_{n=1}^{\infty} c_n \frac{\pi}{2}\delta_{nm} = \frac{\pi}{2}c_m.$$

Therefore

$$c_n = \frac{2}{\pi} \langle f \,|\phi_n \rangle,$$

where

$$\langle f \,|\phi_n \rangle = \int_0^{\pi} f(x)e^x \sin nx\; e^{-2x}\mathrm{d}x = \int_0^{\pi} f(x)e^{-x} \sin nx\; \mathrm{d}x.$$

It follows that

$$f(x) = \sum_{n=1}^{\infty} \frac{2}{\pi} \langle f \,|\phi_n \rangle \, \phi_n$$

$$= \sum_{n=1}^{\infty} \frac{2}{\pi} \left(\int_0^{\pi} f(x)e^{-x} \sin nx\; \mathrm{d}x \right) e^x \sin nx.$$

Example 3.4.3. (a) Find the eigenvalues and eigenfunctions of the following Sturm–Liouville problem:

$$y'' + \lambda y = 0,$$

$$y(0) = 0, \quad y(1) - y'(1) = 0.$$

(b) Show that the eigenfunctions are orthogonal by explicit integration,

$$\int_0^1 y_n(x)y_m(x)\mathrm{d}x = 0, \quad n \neq m.$$

(c) Find the orthonormal set of the eigenfunctions.

Solution 3.4.3. (a) It can be easily shown that for $\lambda < 0$, no solution can satisfy the equation and the boundary conditions. For $\lambda = 0$, it is actually an eigenvalue with an associated eigenfunction $y_0(x) = x$, since it satisfies both the equation and the boundary conditions

$$\frac{\mathrm{d}^2}{\mathrm{d}x^2}x = 0, \quad y_0(0) = 0, \quad y_0(1) - y_0'(1) = 1 - 1 = 0.$$

Most of the eigenvalues come from the branch where $\lambda = \alpha^2 > 0$. In that case, the solution of

$$\frac{\mathrm{d}^2}{\mathrm{d}x^2}y(x) + \alpha^2 y(x) = 0$$

is given by

$$y(x) = A\cos\alpha x + B\sin\alpha x.$$

The boundary condition $y(0) = A = 0$ leaves us with

$$y(x) = B\sin\alpha x.$$

The other boundary condition $y(1) - y'(1) = 0$ requires that

$$\sin\alpha - \alpha\cos\alpha = 0. \tag{3.17}$$

Therefore α has to be the positive roots of

$$\tan\alpha = \alpha.$$

These roots are labeled as α_n in Fig. 3.1. The roots of the equation $\tan x = \mu x$ are frequently needed in many applications, and they are listed in Tables 4.19 and 4.20 in "Handbook of Mathematical Functions" by M. Abramowitz and I.A. Stegun, Dover Publications, 1970. For example, in our case $\mu = 1$, $\alpha_1 = 4.49341$, $\alpha_2 = 7.72525$, $\alpha_3 = 10.90412$, $\alpha_4 = 14.06619\dots$. Thus the eigenvalues of this Sturm–Liouville problem are $\lambda_0 = 0, \lambda_n = \alpha_n^2 (n = 1, 2, \dots)$, the corresponding eigenfunctions are

$$y_0(x) = x, \ y_n(x) = \sin\alpha_n x \ (n = 1, 2, \dots).$$

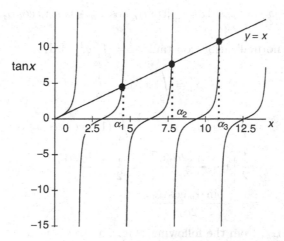

Fig. 3.1. Roots of $\tan x = x$, α_n is the nth root. $\alpha_1 = 4.49341, \alpha_2 = 7.72525$, $\alpha_3 = 10.90412, \dots$ as listed in Table 4.19 of "Handbook of Mathematical Functions", by M. Abramowitz and I.A. Stegun, Dover Publications, 1970

(b) According to the Sturm–Liouville theory, these eigenfunctions are orthogonal to each other. It is instructive to show this explicitly. First,

$$\int_0^1 x \sin \alpha_n x \, dx = \left[-\frac{x}{\alpha_n} \cos \alpha_n x + \frac{1}{\alpha_n^2} \sin \alpha_n x \right]_0^1$$

$$= \frac{1}{\alpha_n^2} \left[-\alpha_n \cos \alpha_n + \sin \alpha_n \right] = 0,$$

since α_n satisfies (3.17). Next

$$\int_0^1 \sin \alpha_n x \sin \alpha_m x \, dx = \frac{1}{2} \int_0^1 \left[\cos(\alpha_n - \alpha_m)x - \cos(\alpha_n + \alpha_m)x \right] \, dx$$

$$= \frac{1}{2} \left[\frac{\sin(\alpha_n - \alpha_m)}{\alpha_n - \alpha_m} - \frac{\sin(\alpha_n + \alpha_m)}{\alpha_n + \alpha_m} \right].$$

Now

$$\alpha_n - \alpha_m = \tan \alpha_n - \tan \alpha_m = \frac{\sin \alpha_n}{\cos \alpha_n} - \frac{\sin \alpha_m}{\cos \alpha_m}$$

$$= \frac{\sin \alpha_n \cos \alpha_m - \cos \alpha_n \sin \alpha_m}{\cos \alpha_n \cos \alpha_m} = \frac{\sin(\alpha_n - \alpha_m)}{\cos \alpha_n \cos \alpha_m},$$

thus

$$\frac{\sin(\alpha_n - \alpha_m)}{\alpha_n - \alpha_m} = \cos \alpha_n \cos \alpha_m.$$

Similarly

$$\frac{\sin(\alpha_n + \alpha_m)}{\alpha_n + \alpha_m} = \cos \alpha_n \cos \alpha_m.$$

It follows that

$$\int_0^1 \sin \alpha_n x \sin \alpha_m x \, dx = \frac{1}{2} \left[\cos \alpha_n \cos \alpha_m - \cos \alpha_n \cos \alpha_m \right] = 0.$$

(c) To find the normalization constant $\beta_n^2 = \int_0^1 y_n^2(x) \, dx$:

$$\beta_0^2 = \int_0^1 x^2 dx = \frac{1}{3}.$$

$$\beta_n^2 = \int_0^1 \sin^2 \alpha_n x \, dx = \frac{1}{2} \int_0^1 (1 - \cos 2\alpha_n x) \, dx$$

$$= \frac{1}{2} \left[x - \frac{\sin 2\alpha_n x}{2\alpha_n} \right]_0^1 = \frac{1}{2} - \frac{\sin 2\alpha_n}{4\alpha_n}$$

$$= \frac{1}{2} - \frac{\sin \alpha_n \cos \alpha_n}{2\alpha_n}.$$

Since $\tan \alpha_n = \alpha_n$, from the following diagram, we see that

$$\sin \alpha_n = \frac{\alpha_n}{\sqrt{1 + \alpha_n^2}}, \qquad \cos \alpha_n = \frac{1}{\sqrt{1 + \alpha_n^2}}.$$

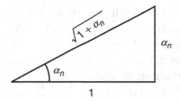

Thus

$$\beta_n^2 = \frac{1}{2}\left(1 - \frac{1}{\alpha_n}\frac{\alpha_n}{\sqrt{1+\alpha_n^2}}\frac{1}{\sqrt{1+\alpha_n^2}}\right) = \frac{\alpha_n^2}{2(1+\alpha_n^2)}$$

Therefore, the orthonormal set of the eigenfunctions is as follows:

$$\left\{\sqrt{3}x,\ \frac{\sqrt{2(1+\alpha_n^2)}}{\alpha_n}\sin\alpha_n x\right\}\ (n=1,\,2,\,3,\ldots).$$

3.4.4 Periodic Sturm–Liouville Problems

On the interval $a \le x \le b$, if $p(a) = p(b)$, then the periodic boundary conditions

$$y(a) = y(b), \qquad y'(a) = y'(b)$$

also satisfy the condition (3.16). This is very easy to show. Let $y_n(x)$ and $y_m(x)$ be two functions that satisfy these boundary conditions, that is

$$y_n(a) = y_n(b), \qquad y_n'(a) = y_n'(b),$$
$$y_m(a) = y_m(b), \qquad y_m'(a) = y_m'(b).$$

Clearly

$$p(b)\begin{vmatrix} y_n(b) & y_n'(b) \\ y_m(b) & y_m'(b) \end{vmatrix} - p(a)\begin{vmatrix} y_n(a) & y_n'(a) \\ y_m(a) & y_m'(a) \end{vmatrix} = 0,$$

since the two terms are equal.

Therefore, a Sturm–Liouville equation plus these periodic boundary conditions also constitute a Sturm–Liouville problem. Note that the difference between the regular and periodic Sturm–Liouville problems is that the boundary conditions in the regular Sturm–Liouville problem are separated, with one condition applying at $x = a$ and the other at $x = b$, whereas the boundary conditions in the periodic Sturm–Liouville problem relate the values at $x = a$ to the values at $x = b$. In addition, in the periodic Sturm–Liouville problem, $p(a)$ must equal to $p(b)$.

For example,

$$y'' + \lambda y = 0, \qquad a \le x \le b$$

is a Sturm–Liouville equation with $p = 1$, $q = 0$, and $w = 1$. Since $p(a) = p(b) = 1$, a periodic boundary condition will make this a Sturm–Liouville problem. As we have seen, if $y(0) = y(2\pi)$, $y'(0) = y'(2\pi)$, the eigenfunctions are $\{\cos nx,\ \sin nx\}$ $(n = 0, 1, 2, \ldots)$, which is the basis of the ordinary Fourier series for any periodic function of period 2π.

Note that, within the interval of $0 \leq x \leq 2\pi$, any piece-wise continuous function $f(x)$, not necessarily periodic, can be expanded into a Fourier series of cosines and sines. However, outside the interval, since the trigonometric functions are periodic, $f(x)$ will also be periodic with period 2π.

If the period is not 2π, we can either make change of scale in the Fourier series, or change the boundary in the Sturm–Liouville problem. The result will be the same.

3.4.5 Singular Sturm–Liouville Problems

In this case, $p(x)$ (and possibly $w(x)$) vanishes at one or both endpoints. We call it singular, because Sturm–Liouville equation

$$(py')' + qy + \lambda wy = 0$$

can be written as

$$py'' + p'y' + qy + \lambda wy = 0,$$

or

$$y'' + \frac{1}{p}p'y' + \frac{1}{p}qy + \lambda\frac{1}{p}wy = 0.$$

If $p(a) = 0$, then clearly at $x = a$, this equation is singular.

If both $p(a)$ and $p(b)$ are zero, $p(a) = 0$ and $p(b) = 0$, the boundary condition (3.16) is automatically satisfied. This may suggest that there is no restriction on the eigenvalue λ. However, for an arbitrary λ, the equation may have no meaningful solution. The requirement that the solution and its derivative must remain bounded even at the singular points often restricts the acceptable values of λ to a discrete set. In other words, the boundary conditions in this case are replaced by the requirement that $y(x)$ must be bounded at $x = a$ and $x = b$.

If $p(a) = 0$ and $p(b) \neq 0$, then the boundary condition (3.16) becomes

$$\begin{vmatrix} y_n(b) & y_n'(b) \\ y_m(b) & y_m'(b) \end{vmatrix} = 0.$$

This condition will be met, if all solutions of the equation satisfy the boundary condition

$$\beta_1 y(b) + \beta_2 y'(b) = 0,$$

where constants β_1 and β_2 are not both zero. In addition, solutions must be bounded at $x = a$.

Similarly, if $p(b) = 0$ and $p(a) \neq 0$, then $y(x)$ must be bounded at $x = b$, and

$$\alpha_1 y(a) + \alpha_2 y'(a) = 0,$$

where α_1 and α_2 are not both equal to zero.

Many physically important and named differential equations are singular Sturm–Liouville problems. The following are a few examples.

Legendre Equation. The Legendre differential equation

$$(1 - x^2)y'' - 2xy' + \lambda y = 0, \qquad (-1 \leq x \leq 1)$$

is one of the most important equations in mathematical physics. The details of the solutions of this equation will be studied in Chap. 4. Here we only want to note that it is a singular Sturm–Liouville problem because this equation can be obviously written as

$$\left[(1 - x^2)y'\right]' + \lambda y = 0,$$

which is in the form of Sturm–Liouville equation with $p(x) = 1 - x^2$, $q = 0, w = 1$. Since $p(x)$ vanishes at both ends, $p(1) = p(-1) = 0$, it is a singular Sturm–Liouville problem. As we will see in Chap. 4, in order to have a bounded solution on $-1 \leq x \leq 1$, λ has to assume one of the following values

$$\lambda_n = n(n + 1), \qquad n = 0, \ 1, \ 2, \ \ldots.$$

Corresponding to each λ_n, the eigenfunction is the Legendre function $P_n(x)$, which is a polynomial of order n. We have met these functions when we constructed an orthogonal set out of $\{x^n\}$ in the interval $-1 \leq x \leq 1$, with a unit weight function. The properties of this function will be discussed again in Chap. 4. Since $P_n(x)$ are eigenfunctions of a Sturm–Liouville problem, they are orthogonal to each other in the interval $-1 \leq x \leq 1$ with respect to a unit weight function $w(x) = 1$. Furthermore, the set $\{P_n(x)\}$ $(n = 0, 1, 2, \ldots)$ is complete. Therefore, any piece-wise continuous function $f(x)$ in the interval $-1 \leq x \leq 1$ can be expressed as

$$f(x) = \sum_{n=0}^{\infty} c_n P_n(x),$$

where

$$c_n = \frac{\langle f \, | P_n \rangle}{\langle P_n \, | P_n \rangle} = \frac{\int_{-1}^{1} f(x) P_n(x) \mathrm{d}x}{\int_{-1}^{1} P_n^2(x) \mathrm{d}x}.$$

This series is known a Fourier–Legendre series, which is very important in solving partial differential equations with spherical symmetry, as we shall see.

Bessel Equation. The problem consists of the differential equation

$$x^2 y''(x) + xy'(x) - \nu^2 y + \lambda^2 x^2 y(x) = 0, \qquad 0 \leq x \leq L \qquad (3.18)$$

and the boundary condition

$$y(L) = 0.$$

It is a singular Sturm–Liouville problem. In the equation, ν^2 is a given constant and λ^2 is a parameter that can be chosen to fit the boundary condition. To convert this equation into the standard Sturm–Liouville form, let us first divide the equation by x^2,

$$y''(x) + \frac{1}{x}y'(x) - \frac{1}{x^2}\nu^2 y + \lambda^2 y(x) = 0, \tag{3.19}$$

and then find the integrating factor

$$p(x) = e^{\int^x \frac{1}{x'}dx'} = e^{\ln x} = x.$$

Multiplying (3.19) by this integrating factor, we have

$$xy''(x) + y'(x) - \frac{1}{x}\nu^2 y(x) + \lambda^2 x y(x) = 0, \tag{3.20}$$

which can be written as

$$[xy']' - \frac{1}{x}\nu^2 y + \lambda^2 x y = 0.$$

This is a Sturm–Liouville equation with $p(x) = x$, $q(x) = -\nu^2/x$, $w(x) = x$. Of course, (3.20) can be obtained directly from (3.18) by dividing (3.18) by x. However, a step by step approach will enable us to handle more complicated equations, as we shall soon see.

Since $p(0) = 0$, there is a singular point at $x = 0$. So we only need the other boundary condition $y(L) = 0$ at $x = L$ to make it a Sturm–Liouville problem.

Equation (3.18) is closely related to the well known Bessel equation. To see this connection, let us make a change of variable, $t = \lambda x$,

$$\frac{dy}{dx} = \frac{dy}{dt}\frac{dt}{dx} = \lambda\frac{dy}{dt},$$

$$\frac{d^2y}{dx^2} = \frac{d}{dx}\left(\frac{dy}{dx}\right) = \lambda\frac{d}{dt}\left(\lambda\frac{dy}{dt}\right) = \lambda^2\frac{d^2y}{dt^2}.$$

Thus

$$x\frac{dy}{dx} = \frac{t}{\lambda}\lambda\frac{dy}{dt} = t\frac{dy}{dt},$$

$$x^2\frac{d^2y}{dx^2} = \left(\frac{t}{\lambda}\right)^2\lambda^2\frac{d^2y}{dt^2} = t^2\frac{d^2y}{dt^2}.$$

It follows that (3.18) can be written as

$$t^2 \frac{d^2y}{dt^2} + t\frac{dy}{dt} - \nu^2 y + t^2 y = 0.$$

This is the Bessel equation which is very important in both pure mathematics and applied sciences. A great deal of information about this equation is known. We shall discuss some of its properties in Chap. 4.

There are two linearly independent solutions of this equation. One is known as the Bessel function $J_\nu(t)$, and the other, the Neumann function $N_\nu(t)$. The Bessel function is everywhere bounded, but the Neumann function goes to infinity as $t \to 0$.

Since $t = \lambda x$, the solution $y(x)$ of (3.18) must be

$$y(x) = AJ_\nu(\lambda x) + BN_\nu(\lambda x).$$

Since the solution must be bounded at $x = 0$, therefore the constant B must be zero. Now the values of the Bessel functions $J_\nu(t)$ can be calculated, as we shall see in Chap. 4. As an example, we show in Fig. 3.2 the Bessel function of zeroth order $J_0(t)$ as a function t. Note that at certain values of t, it becomes zero. These values are known as the zeros of the Bessel functions, they are tabulated for many values of ν. For example, the first zero of $J_0(t)$ occur at $t = 2.405$, the second zero at $t = 5.520,\dots$. These values are listed as $z_{01} = 2.405$, $z_{02} = 5.520,\dots$.

The boundary condition $y(L) = 0$ requires that

$$J_v(\lambda L) = 0.$$

This means that λ can only assume certain discrete values such that

$$\lambda_1 L = z_{\nu 1}, \quad \lambda_2 L = z_{\nu 2}, \quad \lambda_3 L = z_{\nu 3},\dots .$$

That is,

$$\lambda_n = \frac{z_{\nu n}}{L}.$$

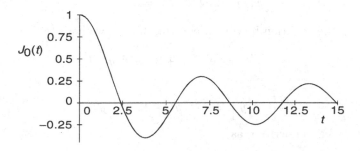

Fig. 3.2. Bessel function of zeroth order $J_0(t)$

It follows that the eigenfunctions of our Sturm–Liouville problem are

$$y_n(x) = J_\nu(\lambda_n x).$$

Now $J_\nu(\lambda_n x)$ and $J_\nu(\lambda_m x)$ are two different eigenfunctions corresponding two different eigenvalues λ_n and λ_m. The eigenfunctions are orthogonal to each other with respect to the weight function $w(x) = x$. Furthermore, $\{J_\nu(\lambda_n x)\}$ $(n = 1, 2, 3, \ldots.)$ is a complete set in the interval $0 \le x \le L$. Therefore any piece-wise continuous function $f(x)$ in this interval can be expanded in terms of these eigenfunctions,

$$f(x) = \sum_{n=1}^{\infty} c_n J_v(\lambda_n x),$$

where

$$c_n = \frac{\langle f(x) \,|\, J_\nu(\lambda_n x)\rangle}{\langle J_\nu(\lambda_n x) \,|\, J_\nu(\lambda_n x)\rangle} = \frac{\int_0^L f(x) J_\nu(\lambda_n x) x \,\mathrm{d}x}{\int_0^L \left[J_\nu(\lambda_n x)\right]^2 x \,\mathrm{d}x}.$$

This expansion is known as Fourier–Bessel series. It is needed in solving partial differential equations with cylindrical symmetry.

Example 3.4.4. Hermite Equation. Show that the following differential equation

$$y'' - 2xy' + 2\alpha y = 0, \qquad -\infty < x < \infty$$

forms a singular Sturm–Liouville problem. If $H_n(x)$ and $H_m(x)$ are two solutions of this problem, show that

$$\int_{-\infty}^{\infty} H_n(x) H_m(x) \mathrm{e}^{-x^2} \mathrm{d}x = 0 \quad \text{for } n \ne m.$$

Solution 3.4.4. To put the equation into the Sturm–Liouville form, let us first calculate the integrating factor

$$p(x) = \mathrm{e}^{-\int^x 2x' \mathrm{d}x'} = \mathrm{e}^{-x^2}.$$

Multiplying the equation by this integrating factor, we have

$$\mathrm{e}^{-x^2} y'' - 2x\, \mathrm{e}^{-x^2} y' + 2\alpha\, \mathrm{e}^{-x^2} y = 0.$$

Since

$$\left[\mathrm{e}^{-x^2} y'\right]' = \mathrm{e}^{-x^2} y'' - 2x\, \mathrm{e}^{-x^2} y',$$

the equation can be written as

$$\left[\mathrm{e}^{-x^2} y'\right]' + 2\alpha\, \mathrm{e}^{-x^2} y = 0.$$

This is in the form of a Sturm–Liouville equation with $p(x) = e^{-x^2}, q = 0$, $w(x) = e^{-x^2}$. Since $p(\infty) = p(-\infty) = 0$, this is a singular Sturm–Liouville problem. Therefore, if $H_n(x)$ and $H_m(x)$ are two solutions of this problem, then they must be orthogonal with respect to the weight function $w(x) = e^{-x^2}$, that is

$$\int_{-\infty}^{\infty} H_n(x) H_m(x) e^{-x^2} \, dx = 0 \quad \text{for } n \neq m.$$

Example 3.4.5. Laguerre Equation. Show that the following differential equation

$$xy'' + (1 - x)y' + ny = 0, \quad 0 < x < \infty$$

forms a singular Sturm–Liouville problem. If $L_n(x)$ and $L_m(x)$ are two solutions of this problem, show that

$$\int_0^{\infty} L_n(x) L_m(x) e^{-x} \, dx = 0 \quad \text{for } n \neq m.$$

Solution 3.4.5. To put the equation into the Sturm–Liouville form, let us first divide the equation by x

$$y'' + \frac{1 - x}{x} y' + n \frac{1}{x} y = 0$$

and then calculate the integrating factor

$$p(x) = e^{\int^x \frac{1 - x'}{x'} dx'} = e^{\ln x - x} = x \, e^{-x}.$$

Multiplying the last equation by this integrating factor, we have

$$x \, e^{-x} y'' + (1 - x) e^{-x} y' + n \, e^{-x} y = 0.$$

Since

$$\left[x \, e^{-x} y' \right]' = x \, e^{-x} y'' + (1 - x) e^{-x} y',$$

the equation can be written as

$$\left[x \, e^{-x} y' \right]' + n \, e^{-x} y = 0.$$

This is in the form of a Sturm–Liouville equation with $p(x) = x \, e^{-x}, q = 0$, $w(x) = e^{-x}$. Since $p(0) = p(\infty) = 0$, this is a singular Sturm–Liouville problem. Therefore, if $L_n(x)$ and $L_m(x)$ are two solutions of this problem, then they must be orthogonal with respect to the weight function $w(x) = e^{-x}$, that is

$$\int_{-\infty}^{\infty} L_n(x) L_m(x) e^{-x} \, dx = 0 \quad \text{for } n \neq m.$$

Example 3.4.6. Chebyshev Equation. Show that the following differential equation

$$(1 - x^2)y'' - xy' + n^2 y = 0, \quad -1 < x < 1$$

forms a singular Sturm–Liouville problem. If $T_n(x)$ and $T_m(x)$ are two solutions of this problem, show that

$$\int_0^\infty T_n(x) T_m(x) \frac{1}{\sqrt{1 - x^2}} dx = 0 \quad \text{for } n \neq m.$$

Solution 3.4.6. To put the equation into the Sturm–Liouville form, let us first divide the equation by $(1 - x^2)$

$$y'' - \frac{x}{1 - x^2} y' + n^2 \frac{1}{1 - x^2} y = 0$$

and then calculate the integrating factor

$$p(x) = e^{-\int^x \frac{x'}{1 - x'^2} dx'}.$$

To evaluate the integral, let $u = 1 - x^2, du = -2x \, dx$, so

$$\int^x \frac{x'}{1 - x'^2} dx' = -\frac{1}{2} \int \frac{du}{u} = -\frac{1}{2} \ln u = -\frac{1}{2} \ln(1 - x^2).$$

Thus,

$$p(x) = e^{-\int^x \frac{x'}{1 - x'^2} dx'} = e^{\frac{1}{2} \ln(1 - x^2)} = \left[e^{\ln(1 - x^2)} \right]^{1/2} = (1 - x^2)^{1/2}.$$

Multiplying the last equation by this integrating factor, we have

$$(1 - x^2)^{1/2} y'' - (1 - x^2)^{-1/2} xy' + n^2 (1 - x^2)^{-1/2} y = 0.$$

Since

$$\left[(1 - x^2)^{1/2} y' \right]' = (1 - x^2)^{1/2} y'' - (1 - x^2)^{-1/2} xy',$$

the equation can be written as

$$\left[(1 - x^2)^{1/2} y' \right]' + n^2 (1 - x^2)^{-1/2} y = 0.$$

This is in the form of a Sturm–Liouville equation with $p(x) = (1 - x^2)^{1/2}$, $q = 0$, $w(x) = (1 - x^2)^{-1/2}$. Since $p(-1) = p(1) = 0$, this is a singular Sturm–Liouville problem. Therefore, if $T_n(x)$ and $T_m(x)$ are two solutions of this problem, then they must be orthogonal with respect to the weight function $w(x) = (1 - x^2)^{-1/2}$, that is

$$\int_{-\infty}^\infty T_n(x) T_m(x) \frac{1}{\sqrt{1 - x^2}} dx = 0 \quad \text{for } n \neq m.$$

3.5 Green's Function

3.5.1 Green's Function and Inhomogeneous Differential Equation

So far we have shown that if the solutions of the Sturm–Liouville equation satisfy certain boundary conditions, then they become a set of orthogonal eigenfunctions $y_n(x)$, with associated eigenvalues λ_n.

Now suppose that we want to solve the following inhomogeneous differential equation in the interval $a \leq x \leq b$,

$$\frac{d}{dx}\left[p(x)\frac{d}{dx}y\right] + q(x)y + kw(x)y = f(x), \qquad (3.21)$$

where $f(x)$ is a given function. The boundary conditions to be satisfied by the solution $y(x)$ are the same as that satisfied by eigenfunctions $y_n(x)$ of the Sturm–Liouville problem

$$\frac{d}{dx}\left[p(x)\frac{d}{dx}y_n\right] + q(x)y_n + \lambda_n w(x)y_n = 0.$$

Note that $k \neq \lambda_n$. In fact, k can even be zero.

It is more convenient to work with the normalized eigenfunctions. If $y_n(x)$ is not yet normalized, we can define

$$\phi_n(x) = \frac{1}{\langle y_n \,|\, y_n \rangle^{1/2}} y_n(x),$$

so that

$$\langle \phi_m \,|\, \phi_n \rangle = \int_a^b \phi_m(x)\phi_n(x)w(x)dx = \delta_{nm}.$$

Since $\{\phi_n\}$ $(n = 1, 2, \ldots)$ is a complete orthonormal set, the solution $y(x)$ of (3.21) can be expanded in terms of ϕ_n,

$$y(x) = \sum_{n=1}^{\infty} c_n \phi_n(x).$$

Putting it into (3.21), we have

$$\sum_{n=1}^{\infty} c_n \left\{ \frac{d}{dx}\left[p(x)\frac{d}{dx}\right] + q(x) \right\} \phi_n(x) + kw(x) \sum_{n=1}^{\infty} c_n \phi_n(x) = f(x).$$

Since

$$\left\{ \frac{d}{dx}\left[p(x)\frac{d}{dx}\right] + q(x) \right\} \phi_n(x) = -\lambda_n w(x)\phi_n(x),$$

so

$$\sum_{n=1}^{\infty} c_n(-\lambda_n + k)w(x)\phi_n(x) = f(x).$$

Multiplying both sides by $\phi_m(x)$ and integrating,

$$\sum_{n=1}^{\infty} c_n(-\lambda_n + k)\int_a^b w(x)\phi_n(x)\phi_m(x)\mathrm{d}x = \int_a^b f(x)\phi_m(x)\mathrm{d}x.$$

Because of the orthogonality condition, we have

$$c_m(-\lambda_m + k) = \int_a^b f(x)\phi_m(x)\mathrm{d}x,$$

or

$$c_n = \frac{1}{k - \lambda_n}\int_a^b f(x)\phi_n(x)\mathrm{d}x.$$

Hence the solution $y(x)$ is given by

$$y(x) = \sum_{n=1}^{\infty} c_n\phi_n(x) = \sum_{n=1}^{\infty}\left[\frac{1}{k - \lambda_n}\int_a^b f(x')\phi_n(x')\mathrm{d}x'\right]\phi_n(x).$$

Since $f(x)$ is a given function, presumably this series can be computed. However, we want to put it in a somewhat different form, and introduce a conceptually important function, known as the Green's function. Assuming the summation and the integration can be interchanged, we can write the last expression as:

$$y(x) = \int_a^b f(x')\sum_{n=1}^{\infty}\frac{\phi_n(x')\phi_n(x)}{k - \lambda_n}\mathrm{d}x'.$$

Now if we define the Green's function as:

$$G(x', x) = \sum_{n=1}^{\infty}\frac{\phi_n(x')\phi_n(x)}{k - \lambda_n}, \tag{3.22}$$

then the solution $y(x)$ can be written as:

$$y(x) = \int_a^b f(x')G(x', x)\mathrm{d}x'.$$

3.5.2 Green's Function and Delta Function

To appreciate the meaning of the Green's function, we will first show that $G(x', x)$ is the solution of (3.21), except with $f(x)$ replaced by the delta function $\delta(x' - x)$. That is, we will show that

$$\frac{d}{dx}\left[p(x)\frac{d}{dx}G(x',x)\right] + q(x)G(x',x) + kw(x)G(x',x) = \delta(x'-x), \quad (3.23)$$

where the delta function $\delta(x'-x)$ is defined by the relation

$$F(x) = \int_a^b F(x')\delta(x'-x)dx', \quad a < x < b.$$

With $G(x',x)$ given by (3.22),

$$\frac{d}{dx}\left[p(x)\frac{d}{dx}G(x',x)\right] + q(x)G(x',x) + kw(x)G(x',x)$$

$$= \left\{\frac{d}{dx}\left[p(x)\frac{d}{dx}\right] + q(x)\right\}\sum_{n=1}^{\infty}\frac{\phi_n(x')\phi_n(x)}{k-\lambda_n} + kw(x)\sum_{n=1}^{\infty}\frac{\phi_n(x')\phi_n(x)}{k-\lambda_n}$$

$$= \sum_{n=1}^{\infty}\frac{-\lambda_n w(x)\phi_n(x')\phi_n(x)}{k-\lambda_n} + kw(x)\sum_{n=1}^{\infty}\frac{\phi_n(x')\phi_n(x)}{k-\lambda_n} = w(x)\sum_{n-1}^{\infty}\phi_n(x')\phi_n(x),$$

which can be shown as the eigenfunction expansion of the delta function. Let

$$\delta(x'-x) = \sum_{n=1}^{\infty} a_n\phi_n(x).$$

The inner product of both sides with one of the eigenfunctions shows that

$$a_n = \langle \delta(x'-x)\,|\phi_n(x)\rangle.$$

Therefore

$$\delta(x'-x) = \sum_{n=1}^{\infty} a_n\phi_n(x) = \sum_{n=1}^{\infty}\langle \delta(x'-x)\,|\phi_n(x)\rangle\,\phi_n(x)$$

$$= \sum_{n=1}^{\infty}\left[\int_a^b \delta(x'-x)\phi_n(x)w(x)dx\right]\phi_n(x) = w(x')\sum_{n=1}^{\infty}\phi_n(x')\phi_n(x).$$

Furthermore, since $\delta(x'-x)$ is nonzero only for $x = x'$,

$$\delta(x'-x) = \delta(x-x') = w(x)\sum_{n=1}^{\infty}\phi_n(x)\phi_n(x'). \quad (3.24)$$

Equation (3.23) is thus established.

Now the Green's function can be interpreted as follows. The linear differential equation, such as (3.21), can be used to describe a linear physical system. The function $f(x)$ in the right-hand side of the equation represents the "force," or forcing function applied to the system. In other words, $f(x)$

is the input to the system. The solution $y(x)$ of the equation represents the response of the system.

The Green's function $G(x', x)$ describes the response of the physical system to a unit delta function, which represents the impulse of a point source at x' with a unit strength.

We can model any input $f(x)$ as the sum of a set of point inputs. This is expressed as

$$f(x) = \int f(x')\delta(x' - x)\mathrm{d}x'.$$

The value of $f(x')$ is simply the strength of the delta function at x'. Since $G(x', x)$ is the response of a unit delta function, if the strength of the delta function is $f(x')$ times larger, the response will also be larger by that amount. That is, the response will be $f(x')\,G(x', x)$. Since the system is linear, we can find the response of the system to the input $f(x)$ by adding up the responses of the point inputs. That is

$$y(x) = \int f(x')G(x', x)\mathrm{d}x'.$$

Example 3.5.1. (a) Determine the eigenfunction expansion of the Green's function $G(x', x)$ for

$$y'' + y = x$$

$$y(0) = 0, \quad y(1) = 0.$$

(b) Find the solution $y(x)$ of the inhomogeneous differential equation through

$$y(x) = \int_0^1 x'G(x', x)\mathrm{d}x'.$$

Solution 3.5.1. (a) To solve this inhomogeneous differential equation, let us first look at the related eigenvalue problem,

$$y'' + y + \lambda y = 0.$$
$$y(0) = 0, \quad y(1) = 0,$$

which is a regular Sturm–Liouville problem, with $p(x) = 1$, $q(x) = 1$, $w(x) = 1$. The solution of the equation

$$y'' = -(1 + \lambda)y$$

is

$$y(x) = A\cos\sqrt{1 + \lambda}\,x + B\sin\sqrt{1 + \lambda}\,x.$$

The boundary condition $y(0) = 0$ requires

$$y(0) = A = 0,$$

so

$$y(1) = B \sin \sqrt{1 + \lambda}.$$

Thus the other boundary condition $y(1) = 0$ makes it necessary that

$$\sqrt{1 + \lambda} = n\pi, \quad n = 1, 2, 3, \ldots.$$

It follows that the eigenvalues are

$$\lambda_n = n^2\pi^2 - 1,$$

and the corresponding eigenfunctions are

$$y_n(x) = \sin n\pi x.$$

Therefore the normalized eigenfunctions are

$$\phi_n(x) = \frac{\sin n\pi x}{\left[\int_0^1 \sin^2 n\pi x \, dx\right]^{1/2}} = \sqrt{2} \sin n\pi x.$$

Hence the Green's function can be written as

$$G(x', x) = \sum_{n=1}^{\infty} \frac{\phi_n(x')\phi_n(x)}{0 - \lambda_n} = 2 \sum_{n=1}^{\infty} \frac{\sin n\pi x' \sin n\pi x}{1 - n^2\pi^2}.$$

(b) The solution $y(x)$ is therefore given by

$$y(x) = \int_0^1 x' G(x', x) dx' = 2 \sum_{n=1}^{\infty} \frac{\sin n\pi x}{1 - n^2\pi^2} \int_0^1 x' \sin n\pi x' \, dx'.$$

Since

$$\int_0^1 x' \sin n\pi x' \, dx' = \left[x' \left(-\frac{1}{n\pi} \cos n\pi x' \right) \right]_0^1 + \frac{1}{n\pi} \int_0^1 \cos n\pi x' \, dx'$$

$$= -\frac{1}{n\pi} \cos n\pi = \frac{(-1)^{n+1}}{n\pi},$$

the solution can be expressed as:

$$y(x) = \frac{2}{\pi} \sum_{n=1}^{\infty} \frac{(-1)^{n+1} \sin n\pi x}{n(1 - n^2\pi^2)}.$$

In this example, with the eigenfunction expansion of the Green's function, we have found the solution of the problem expressed in a Fourier series of sine functions. To show that the Green's function is the response of the system to a unit delta function, it is instructive to solve the same problem with a Green's function obtained directly from the equation

$$\frac{d^2}{dx^2} G(x', x) + G(x', x) = \delta(x' - x). \tag{3.25}$$

This we will do in the following example.

Example 3.5.2. (a) Solve the problem of the previous example with a Green's function obtained from the fact that it is the response of the system to a unit delta function. (b) Solve the inhomogeneous differential equation of the previous example, with the Green's function obtained in (a).

Solution 3.5.2. (a) Since the Green's function is the response of the system to a delta function, we require it to be continuous and bounded in the interval of interest. For $x \neq x'$, the Green's function satisfies the equation

$$\frac{d^2}{dx^2} G(x',x) + G(x',x) = 0.$$

The solution of this equation is given by

$$G(x',x) = A(x') \cos x + B(x') \sin x.$$

As far as x is concerned, $A(x')$ and $B(x')$ are two arbitrary constants. But there is no reason that these constants are the same for $x < x'$ as for $x > x'$, in fact they are not. So let us write $G(x',x)$ as

$$G(x',x) = \begin{cases} a \cos x + b \sin x & x < x', \\ c \cos x + d \sin x & x > x'. \end{cases}$$

Since the Green's function must satisfy the same boundary conditions as the original differential equation. At $x = 0$, $G(x',0) = 0$. Since $x = 0$ is certainly less than x', therefore we require

$$G(x',0) = a \cos 0 + b \sin 0 = a = 0.$$

Furthermore, because at $x = 1$, $G(x',1) = 0$, we have

$$G(x',1) = c \cos 1 + d \sin 1 = 0.$$

It follows that

$$d = -c \frac{\cos 1}{\sin 1}.$$

Thus, for $x > x'$,

$$G(x',x) = c \cos x - c \frac{\cos 1}{\sin 1} \sin x = c \frac{1}{\sin 1} (\sin 1 \cos x - \cos 1 \sin x)$$

$$= c \frac{1}{\sin 1} \sin(1 - x).$$

Hence, with boundary conditions, we are left with two constants in the Green's function to be determined,

$$G(x',x) = \begin{cases} b \sin x & x < x', \\ c \frac{1}{\sin 1} \sin(1 - x) & x > x'. \end{cases}$$

To determine b and c, we invoke the condition that $G(x', x)$ must be continuous at $x = x'$, so

$$b \sin x' = c \frac{1}{\sin 1} \sin(1 - x')$$

Thus,

$$G(x', x) = \begin{cases} G^-(x', x) = b \sin x & x < x', \\ G^+(x', x) = b \frac{\sin x'}{\sin(1-x')} \sin(1 - x) & x > x'. \end{cases}$$

Next we integrate both sides of (3.25) over a small interval across x',

$$\int_{x'-\epsilon}^{x'+\epsilon} \frac{d^2}{dx^2} G(x', x) dx + \int_{x'-\epsilon}^{x'+\epsilon} G(x', x) dx = \int_{x'-\epsilon}^{x'+\epsilon} \delta(x' - x) dx.$$

The integral on the right-hand side is equal to 1, by the definition of the delta function. As $\epsilon \to 0$,

$$\lim_{\epsilon \to 0} \int_{x'-\epsilon}^{x'+\epsilon} G(x', x) dx = 0.$$

This integral is equal to 2ϵ times the average value of $G(x', x)$ over 2ϵ at $x = x'$. Since $G(x', x)$ is bounded, this integral is equal to zero as ϵ goes to zero. Now

$$\int_{x'-\epsilon}^{x'+\epsilon} \frac{d^2}{dx^2} G(x', x) dx = \frac{dG(x', x)}{dx} \bigg|_{x'-\epsilon}^{x'+\epsilon}.$$

It follows that as $\epsilon \to 0$,

$$\lim_{\epsilon \to 0} \int_{x'-\epsilon}^{x'+\epsilon} \frac{d^2}{dx^2} G(x', x) dx = \frac{dG^+(x', x)}{dx} \bigg|_{x=x'} - \frac{dG^-(x', x)}{dx} \bigg|_{x=x'}.$$

Hence

$$-b \frac{\sin x'}{\sin(1 - x')} \cos(1 - x') - b \cos x' = 1,$$

or

$$-\frac{b}{\sin(1 - x')} [\sin(1 - x') \cos x' + \cos(1 - x') \sin x'] = 1.$$

Since

$$[\sin(1 - x') \cos x' + \cos(1 - x') \sin x'] = \sin(1 - x' + x') = \sin 1,$$

so

$$b = -\frac{\sin(1 - x')}{\sin 1}.$$

Thus, the Green's function is given by

$$G(x', x) = \begin{cases} -\frac{\sin(1-x')}{\sin 1} \sin x & x < x', \\ -\frac{\sin x'}{\sin 1} \sin(1 - x) & x > x'. \end{cases}$$

(b)

$$y(x) = \int_0^1 x' G(x', x) dx'$$

$$= -\int_0^x x' \frac{\sin x'}{\sin 1} \sin(1 - x) dx' - \int_x^1 x' \frac{\sin(1 - x')}{\sin 1} \sin x \, dx'$$

$$= -\frac{\sin(1 - x)}{\sin 1} \int_0^x x' \sin x' \, dx' - \frac{\sin x}{\sin 1} \int_x^1 x' \sin(1 - x') dx'.$$

Since

$$\int_0^x x' \sin x' \, dx' = [-x' \cos x' + \sin x']_0^x = -x \cos x + \sin x,$$

$$\int_x^1 x' \sin(1 - x') dx' = [x' \cos(1 - x') + \sin(1 - x')]_x^1$$

$$= 1 - x \cos(1 - x) - \sin(1 - x),$$

so

$$y(x) = -\frac{1}{\sin 1} [-x \sin(1 - x) \cos x + \sin x - x \sin x \cos(1 - x)]$$

$$= -\frac{1}{\sin 1} [-x \sin(1 - x + x) + \sin x] = x - \frac{1}{\sin 1} \sin x.$$

To see if this result is the same as the solution obtained in the previously example, we can expand it in terms of the Fourier sine series in the range of $0 \le x \le 1$,

$$x - \frac{1}{\sin 1} \sin x = \sum_{n=1}^{\infty} a_n \sin n\pi x,$$

$$a_n = 2 \int_0^1 \left(x - \frac{1}{\sin 1} \sin x \right) \sin n\pi x \, dx.$$

It can be readily shown that

$$\int_0^1 x \sin n\pi x \, dx = \frac{(-1)^{n+1}}{n\pi},$$

$$\int_0^1 \sin x \sin n\pi x \, dx = \frac{1}{2} \left[\frac{1}{n\pi - 1} \sin(n\pi - 1) - \frac{1}{n\pi + 1} \sin(n\pi + 1) \right]$$

$$= \frac{1}{2} \left[\frac{(-1)^{n+1}}{n\pi - 1} \sin 1 + \frac{(-1)^{n+1}}{n\pi + 1} \sin 1 \right] = \frac{(-1)^{n+1} n\pi}{n^2 \pi^2 - 1} \sin 1.$$

Thus

$$a_n = 2\left[\frac{(-1)^{n+1}}{n\pi} - \frac{(-1)^{n+1}n\pi}{n^2\pi^2 - 1}\right] = \frac{2(-1)^{n+1}}{n\pi(1 - n^2\pi^2)},$$

and

$$x - \frac{1}{\sin 1}\sin x = \frac{2}{\pi}\sum_{n=1}^{\infty}\frac{(-1)^{n+1}}{n(1 - n^2\pi^2)}\sin n\pi x,$$

which is identical to the result of the previous example.

This problem can be simply solved by the "ordinary method." Clearly x is a particular solution, and the complementary function is $y_c = A\cos x + B\sin x$. Applying the boundary conditions $y(0) = 0$ and $y(1) = 1$ to the solution

$$y(x) = y_p + y_c = x + A\cos x + B\sin x,$$

we get

$$y(x) = x - \frac{1}{\sin 1}\sin x.$$

We used this problem to illustrate how the Green's function works. For such simple problems, the Green's function may not offer any advantage, but the idea of Green's function is a powerful one in dealing with boundary conditions and introducing approximations in solving partial differential equations. We shall see these aspects of the Green's function in a later chapter.

Exercises

1. (a) Use the explicit expressions of the first six Legendre polynomials

$$P_0(x) = 1,\ P_1(x) = x,\ P_2(x) = \frac{1}{2}(3x^2 - 1),\ P_3(x) = \frac{1}{2}(5x^3 - 3x),$$

$$P_4(x) = \frac{1}{8}(35x^4 - 30x^2 + 3),\quad P_5(x) = \frac{1}{8}(63x^5 - 70x^3 + 15x),$$

to show that the conditions

$$P_n(1) = 1,$$

$$\int_{-1}^{1}P_n(x)P_m(x)dx = \frac{2}{2n+1}\delta_{nm},$$

are satisfied by $P_n(x)$ at least for $n = 0$ to $n = 5$.

(b) Show that if $y_n = P_n(x)$ for $n = 0, 1, \ldots, 5$, then

$$(1 - x^2)y_n'' - 2xy_n' + n(n+1)y_n = 0.$$

2. Express the "ramp" function $f(x)$

$$f(x) = \begin{cases} 0 & -1 \le x \le 0 \\ x & 0 \le x \le 1 \end{cases},$$

in terms of the Legendre polynomials in the interval $-1 \le x \le 1$. Find the first four nonvanishing terms explicitly.

Ans. $f(x) = \sum_{n=0}^{\infty} a_n P_n(x)$, $a_n = \frac{2n+1}{2} \int_0^1 x P_n(x) dx$,

$f(x) = \frac{1}{4} P_0(x) + \frac{1}{2} P_1(x) + \frac{5}{16} P_2(x) - \frac{3}{32} P_4(x) + \cdots$.

3. *Laguerre polynomial.* (a) Use the Gram–Schmidt procedure to generate from the set $\{x^n\}$ $(n = 0, 1, \ldots)$ the first three polynomials $L_n(x)$ that are orthogonal over the interval $0 \le x < \infty$ with the weight function e^{-x}. Use the convention that $L_n(0) = 1$.
   (b) Show, by direct integration, that

   $$\int_0^{\infty} L_n(x) L_m(x) e^{-x} dx = \delta_{nm}.$$

   (c) Show that if $y_n = L_n(x)$, then y_n satisfies the Laguerre differential equation
   $$x y_n'' + (1 - x) y_n' + n y_n = 0.$$
   (you may need $\int_0^{\infty} x^n e^{-x} dx = n!$)

   Ans. $L_0(x) = 1$, $L_1(x) = 1 - x$, $L_2(x) = 1 - 2x + \frac{1}{2} x^2$.

4. *Hermite polynomial.* (a) Use the Gram–Schmidt procedure to generate from the set $\{x^n\}$ $(n = 0, 1, \ldots)$ the first three polynomials $H_n(x)$ that are orthogonal over the interval $-\infty \le x < \infty$ with the weight function e^{-x^2}. Fix the multiplicative constant by the requirement

   $$\int_{-\infty}^{\infty} H_n(x) H_m(x) e^{-x^2} dx = \delta_{nm} n! 2^n \sqrt{\pi}.$$

   (b) Show that if $y_n = H_n(x)$, then y_n satisfies the Hermite differential equation
   $$y_n'' + -2x y_n' + 2n y_n = 0.$$
   (you may need $\int_{-\infty}^{\infty} e^{-x^2} dx = \sqrt{\pi}$)

   Ans. $H_0(x) = 1$, $H_1(x) = 2x$, $H_2(x) = 4x^2 - 2$.

5. *Associated laguerre equation.* (a) Express associated Laguerre's differential equation
   $$x y''(x) + (k + 1 - x) y'(x) + n y(x) = 0$$
   in the form of a Sturm–Liouville equation.

(b) Show that in the interval $0 \leq x < \infty$, it is a singular Sturm–Liouville problem.

(c) Find the orthogonality condition of its eigenfunctions.

Ans. (a) $\left[x^{k+1}e^{-x}y'(x)\right]' + nx^{k}e^{-x}y(x) = 0$

(c) $\int_0^\infty x^{k}e^{-x}y_n(x)y_m(x)dx = 0, \quad n \neq m.$

6. *Associated laguerre polynomial.* (a) Use the Gram–Schmidt procedure to generate from the set $\{x^n\}$ $(n = 0, 1, \ldots)$ the first three polynomials $L_n^1(x)$ that are orthogonal over the interval $0 \leq x < \infty$ with the weight function $x\,e^{-x}$. Fix the multiplicative constant by the requirement

$$\int_0^\infty L_n^1(x)L_m^1(x)x\,e^{-x}dx = \delta_{nm}.$$

(b) Show that if $y_n = L_n^1(x)$, then y_n satisfies the Associated Laguerre differential equation with $k = 1$

$$xy_n'' + (k+1+x)y_n' + ny_n = 0.$$

Ans. $L_0^1(x) = 1, \ L_1^1(x) = \frac{1}{\sqrt{2}}(x-2), \ L_2^1(x) = \frac{1}{2\sqrt{3}}(x^2 - 6x + 6).$

7. *Chebyshev polynomial.* (a) Show that the Chebyshev equation

$$(1-x^2)y''(x) - xy'(x) + \lambda y(x) = 0, \quad -1 \leq x \leq 1$$

can be converted into

$$\frac{d^2}{d\theta^2}\Theta(\theta) + \lambda\Theta(\theta) = 0, \quad (0 \leq \theta \leq \pi)$$

with a change of variable $x = \cos\theta$. $(\Theta(0) = y(x(0)) = y(\cos\theta)).$

(b) Show that in terms of $\theta, dy/dx$ can be expressed as

$$\frac{dy}{dx} = \left[A\sqrt{\lambda}\sin\sqrt{\lambda}\theta - B\sqrt{\lambda}\cos\sqrt{\lambda}\theta\right]\frac{1}{\sin\theta}.$$

Hint: $\Theta(\theta) = A\cos\sqrt{\lambda}\theta + B\sin\sqrt{\lambda}\theta;$

$$\frac{dy}{dx} = \frac{d\Theta}{dx} = \frac{d\Theta}{d\theta}\frac{d\theta}{dx}; \quad \frac{d\theta}{dx} = -\frac{1}{\sin\theta}.$$

(c) Show that the conditions for y and dy/dx to be bounded are

$$B = 0, \quad \lambda = n^2, \ n = 0, 1, 2, \ldots.$$

Therefore the eigenvalues and eigenfunctions are

$$\lambda_n = n^2, \quad \Theta_n(\theta) = \cos n\theta$$

(d) The eigenfunctions of the Chebyshev equation are known as Chebyshev polynomial, usually labeled as $T_n(x)$. Find $T_n(x)$ with the condition $T_n(1) = 1$ for $n = 0, 1, 2, 3, 4$.

Hint: $T_n(x) = y_n(x) = \Theta_n(\theta) = \cos n\theta$; $\cos 2\theta = 2\cos^2\theta - 1$, $\cos 3\theta = 4\cos^3\theta - 3\cos\theta$, $\cos 4\theta = 8\cos^4\theta - 8\cos^2\theta + 1$.

(e) Show that for any integer n and m,

$$\int_{-1}^{1} T_n(x) T_m(x) \frac{1}{\sqrt{1-x^2}} dx = \begin{cases} 0 & n \neq m \\ \pi & n = m = 0 \\ \pi/2 & n = m \neq 0 \end{cases}.$$

Ans. (d) $T_0 = 1$, $T_1 = x$, $T_2 = 2x^2 - 1$, $T_3 = 4x^3 - 3x$, $T_4 = 8x^4 - 8x^2 + 1$.

8. *Hypergeometric equation.* Express the hypergeometric equation

$$(x - x^2)y'' + [c - (1 + a + b)x]y' - aby = 0$$

in a Sturm–Liouville form. For it to be a singular Sturm–Liouville problem in the range of $0 \leq x \leq 1$, what conditions must be imposed on a, b and c, if the weight function is required to satisfy the conditions $w(0) = 0$ and $w(1) = 0$?

Hint: Use partial fraction of $\frac{c-(1+a+b)x}{x(1-x)}$ to evaluate $\exp \int^x \frac{c-(1+a+b)x}{x(1-x)} dx$.

Ans. $\left[x^c(1-x)^{1+a+b-c}y'\right]' - abx^{c-1}(1-x)^{a+b-c}y = 0$, $c > 1$, $a + b > c$.

9. Show that if L is a linear operator and

$$\langle h \,|Lh\rangle = \langle Lh \,|h\rangle$$

for all functions h in the complex function space, then

$$\langle f \,|Lg\rangle = \langle Lf \,|g\rangle$$

for all f and g.

Hint: First let $h = f + g$, then let $h = f + ig$.

10. Consider the set of functions $f(x)$ defined in the interval $-\infty < x < \infty$, that goes to zero at least as quickly as x^{-1}, as $x \to \pm\infty$. For a unit weight function, determine whether each of the following linear operators is hermitian when acting upon $\{f(x)\}$.

$$\text{(a)} \frac{d}{dx} + x, \quad \text{(b)} \frac{d^2}{dx^2}, \quad \text{(c)} -i\frac{d}{dx} + x^2, \quad \text{(d)} ix\frac{d}{dx}.$$

Ans. (a) no, (b) yes, (c) yes, (d) no.

11. (a) Express the bounded solution of the following inhomogeneous differential equation

$$(1 - x^2)y'' - 2xy' + ky = f(x), \quad -1 \leq x \leq 1,$$

in terms of Legendre polynomials with the help of a Green's function.
(b) If $k = 14$ and $f(x) = 5x^3$, find the explicit solution.

Ans. $G(x', x) = \sum_{n=0}^{\infty} \frac{2n+1}{2} \frac{P_n(x')P_n(x)}{k-\lambda_n}$

(a) $y(x) = \sum_{n=0}^{\infty} a_n P_n(x)$, $a_n = \frac{2n+1}{2} \frac{1}{k-n(n+1)} \int_{-1}^{1} f(x') P_n(x') \, dx'$.

(b) $y(x) = (10x^3 - 5x)/4$.

12. Determine the eigenvalues and corresponding eigenfunctions for the following problems.

$$\text{(a)} \quad y'' + \lambda y = 0, \quad y(0) = 0, \quad y'(1) = 0.$$
$$\text{(b)} \quad y'' + \lambda y = 0, \quad y'(0) = 0, \quad y'(\pi) = 0.$$
$$\text{(c)} \quad y'' + \lambda y = 0, \quad y(0) = y(2\pi), \quad y'(0) = y'(2\pi).$$

Ans. (a) $\lambda_n = [(2n+1)\pi/2]^2$, $n = 0, 1, 2, \ldots$; $y_n(x) = \sin \frac{\pi}{2}(2n+1)x$.
(b) $\lambda_n = n^2$, $n = 0, 1, 2, \ldots$; $y_n(x) = \cos nx$.
(c) $\lambda_n = n^2$, $n = 0, 1, 2, \ldots$; $y_n(x) = \cos nx$, $\sin nx$.

13. (a) Show that the following differential equation together with the boundary conditions is a Sturm–Liouville problem. What is the weight function?

$$y'' - 2y' + \lambda y = 0, \quad 0 \leq x \leq 1,$$
$$y(0) = 0, \quad y(1) = 0.$$

(b) Determine the eigenvalues and corresponding eigenfunctions of the problem. Fix the multiplication constant by the requirement

$$\int_0^1 y_n(x)y_m(x)w(x)dx = \frac{1}{2}\delta_{nm}.$$

Ans. (a) $[e^{-2x}y']' + \lambda e^{-2x}y = 0$, $w(x) = e^{-2x}$.
(b) $\lambda_n = n^2\pi^2 + 1$, $y_n(x) = e^x \sin n\pi x$.

14. (a) Show that if $\alpha_1, \alpha_2, \alpha_3, \ldots$ are positive roots of

$$\tan \alpha = \frac{h}{\alpha},$$

then $\lambda_n = \alpha_n^2$ and $y_n(x) = \cos \alpha_n x$, $n = 0, 1, 2, \ldots$ are the eigenvalues and eigenfunctions of the following Sturm–Liouville problem:

$$y'' + \lambda y = 0, \qquad 0 \leq x \leq 1$$
$$y'(0) = 0, \quad y'(1) + hy(1) = 0.$$

(b) Show that

$$\int_0^1 \cos \alpha_n x \cos \alpha_m x \, dx = \beta_n^2 \, \delta_{nm}.$$

$$\beta^2 = \frac{\alpha_n^2 + h^2 + h}{2(\alpha_n^2 + h^2)}$$

Hint: $\beta^2 = \frac{1}{2} + \frac{2 \sin 2\alpha_n}{4\alpha_n}$, $\quad \sin 2\alpha_n = 2 \sin \alpha_n \cos \alpha_n = \dfrac{2\alpha_n h}{\alpha_n^2 + h^2}$.

15. Find the eigenfunction expansion for the solution with boundary conditions $y(0) = y(\pi) = 0$ of the inhomogeneous differential equation

$$y'' + ky = f(x),$$

where k is a constant and

$$f(x) = \begin{cases} x & 0 \leq x \leq \pi/2 \\ \pi - x & \pi/2 \leq x \leq \pi. \end{cases}$$

Ans. $y(x) = \frac{4}{\pi} \sum_{n=\text{odd}} \frac{(-1)^{(n-1)/2}}{n^2(k-n^2)} \sin nx.$

16. (a) Find the normalized eigenfunctions $y_n(x)$ of the Hermitian operator d^2/dx^2 that satisfy the boundary conditions $y_n(0) = y_n(\pi) = 0$. Construct the Green's function of this operator $G(x', x)$.
    (b) Show that the Green's function obtained from

$$\frac{d^2}{dx^2} G(x'x) = \delta(x' - x)$$

is

$$G(x', x) = \begin{cases} x(x' - \pi)/\pi & 0 \leq x \leq x' \\ x'(x - \pi)/\pi & x' \leq x \leq \pi. \end{cases}$$

(c) By expanding the function given in (b) in terms of the eigenfunctions $y_n(x)$, verify that it is the same function as that derived in (a).

Ans. (a) $y_n(x) = \left(\frac{2}{\pi}\right)^{1/2} \sin nx$, $n = 1, 2, \ldots$.
$G(x', x) = -\frac{2}{\pi} \sum_{n=1}^{\infty} \frac{1}{n^2} \sin nx' \sin nx.$

4

Bessel and Legendre Functions

In the last chapter we have seen a number of named differential equations. These equations are of considerable importance in engineering and sciences because they occur in numerous applications. In Chap. 6, we will discuss a variety of physical problems which lead to these equations. Unfortunately these equations cannot be solved in terms of elementary functions. To solve them, we have to resort to power series expansions. Functions represented by these series solutions are called special functions.

An enormous amount of details are known about these special functions. Evaluations of these functions and formulas involving them can be found in many books and computer programs. We will mention some of them in the last section.

In order to be able to work with these functions and to have a feeling of understanding when results are expressed in terms of them, we need to know not only their definitions, but also some of their properties. Certain amount of familiarity with these special functions is necessary for us to deal with problems in mathematical physics.

In this chapter, we will first introduce the power series solutions of second-order differential equations, known as the Frobenius method. Next we will apply this method to finding the series solutions of Bessel's and Legendre's equations.

Undoubtedly, the most frequently encountered functions in solving second-order differential equations are trigonometric, hyperbolic, Bessel, and Legendre functions. Since the reader is certainly familiar with trigonometric and hyperbolic functions, we will not include them in our discussions. Our discussions are mostly about the characteristics and properties of Bessel and Legendre functions.

In exercises, we will present some other special functions mentioned in the last chapter. Their properties can be derived by similar methods discussed in this chapter.

4.1 Frobenius Method of Differential Equations

4.1.1 Power Series Solution of Differential Equation

A second-order linear homogeneous differential equation in the form of

$$\frac{d^2y}{dx^2} + p(x)\frac{dy}{dx} + q(x)y = 0 \tag{4.1}$$

can be solved by expressing $y(x)$ in a power series

$$y(x) = \sum_{n=0}^{\infty} a_n x^n, \tag{4.2}$$

if $p(x)$ and $q(x)$ are analytic at $x = 0$.

The idea of this method is simple. If $p(x)$ and $q(x)$ are analytic at $x = 0$, then they can be expressed in terms of Taylor series

$$p(x) = p_0 + p_1 x + p_2 x^2 + \cdots$$
$$q(x) = q_0 + q_1 x + q_2 x^2 + \cdots.$$

Around the point $x = 0$, the differential equation becomes

$$y'' + p_0 y' + q_0 y = 0.$$

This is a differential equation of constant coefficients. The solution is given by either an exponential function or a power of x times an exponential function. Both of these functions can be expressed in terms of a power series around $x = 0$. Therefore it is natural for us to use (4.2) as a trial solution. After (4.2) is substituted into (4.1), we determine the coefficients a_n in such a way that the differential equation (4.1) is identically satisfied. If the series with coefficients so determined is convergent, then it is indeed a solution. The following example illustrates how this procedure works.

Example 4.1.1. Solve the differential equation

$$y'' + y = 0$$

by expanding $y(x)$ in a power series.

Solution 4.1.1. With

$$y = \sum_{n=0}^{\infty} a_n x^n,$$

$$y' = \sum_{n=0}^{\infty} a_n n x^{n-1},$$

$$y'' = \sum_{n=0}^{\infty} a_n n (n-1) x^{n-2},$$

the differential equation can be written as

$$\sum_{n=0}^{\infty} a_n n (n-1) x^{n-2} + \sum_{n=0}^{\infty} a_n x^n = 0.$$

The first two terms of the first summation $[a_0 (0) (-1) x^{-2}$ and $a_1 (1) (0) x^{-1}]$ are zero, so the summation is starting from $n = 2$,

$$\sum_{n=2}^{\infty} a_n n (n-1) x^{n-2} + \sum_{n=0}^{\infty} a_n x^n = 0. \tag{4.3}$$

In order to collect terms, let us write the index in the first summation as

$$n = k + 2,$$

so the first summation can be written as

$$\sum_{n=2}^{\infty} a_n n (n-1) x^{n-2} = \sum_{k=0}^{\infty} a_{k+2} (k+2) (k+1) x^k.$$

Now k is a running index, it does not matter what name it is called. So we can rename it back to n, that is

$$\sum_{k=0}^{\infty} a_{k+2} (k+2) (k+1) x^k = \sum_{n=0}^{\infty} a_{n+2} (n+2) (n+1) x^n.$$

Thus (4.3) can be written as

$$\sum_{n=0}^{\infty} a_{n+2}(n+2) (n+1) x^n + \sum_{n=0}^{\infty} a_n x^n = \sum_{n=0}^{\infty} [a_{n+2}(n+2) (n+1) + a_n] x^n = 0.$$

For this series to vanish, the coefficients of x^n have to be zero for all n. Therefore

$$a_{n+2}(n+2) (n+1) + a_n = 0,$$

or

$$a_{n+2} = -\frac{1}{(n+2) (n+1)} a_n.$$

This is known as the recurrence relation. This equation relates all even coefficients to a_0 and all odd coefficients to a_1. For

$$n = 0, \quad a_2 = -\frac{1}{2 \cdot 1} a_0,$$

$$n = 2, \quad a_4 = -\frac{1}{4 \cdot 3} a_2 = -\frac{1}{4 \cdot 3} \left(-\frac{1}{2 \cdot 1} a_0 \right) = \frac{1}{4!} a_0,$$

$$n = 4, \quad a_6 = -\frac{1}{6!} a_0,$$

. . . .

$$n = 1, \qquad a_3 = -\frac{1}{3 \cdot 2} a_1,$$

$$n = 3, \qquad a_5 = -\frac{1}{5 \cdot 4} a_3 = -\frac{1}{5 \cdot 4}\left(-\frac{1}{3 \cdot 2} a_1\right) = \frac{1}{5!} a_1,$$

$$n = 5, \qquad a_7 = -\frac{1}{7!} a_1,$$

. . . .

It follows that

$$y(x) = \sum_{n=0}^{\infty} a_n x^n = a_0 \left(1 - \frac{1}{2!}x^2 + \frac{1}{4!}x^4 - \frac{1}{6!}x^6 + \cdots\right)$$
$$+ a_1 \left(x - \frac{1}{3!}x^3 + \frac{1}{5!} - \frac{1}{7!} + \cdots\right).$$

These two series are readily recognized as cosine and sine functions,

$$y(x) = a_0 \cos x + a_1 \sin x.$$

4.1.2 Classifying Singular Points

Now the question is if $p(x)$ and $q(x)$ are not analytic at $x = 0$, can we still use the power series method? In other words, if $x = 0$ is a singular point for $p(x)$ and/or $q(x)$, is there a general method to solve the equation? To answer the question, we must distinguish two kinds of singular points.

Definition. Let x_0 be a singular point of $p(x)$ and/or $q(x)$. We call it a regular singular (or nonessential singular) point of the differential equation (4.1) if $(x - x_0)\,p(x)$ and $(x - x_0)^2\,q(x)$ are analytic at x_0. We call it an irregular singular (or essential singular) point of the equation if it is not a regular singular point.

By this definition, $x = 0$ is a regular singular point of the equation

$$y'' + \frac{f(x)}{x}y' + \frac{g(x)}{x^2}y = 0,$$

if $f(x)$ and $g(x)$ are analytic at $x = 0$. When we say that they are analytic, we mean that they can be expanded in terms of Taylor series

$$f(x) = \sum_{n=0}^{\infty} f_n x^n, \qquad g(x) = \sum_{n=0}^{\infty} g_n x^n,$$

including cases that $f(x)$ and $g(x)$ are polynomials of finite orders. For example, the equation

$$xy'' + 2y' + xy = 0$$

has a regular singular point at $x = 0$, since written in the form of

$$y'' + \frac{2}{x} y' + \frac{x^2}{x^2} y = 0,$$

we can see that 2 and x^2 are both analytic at $x = 0$.

If the singularity is only regular singular, we can use the following Frobenius series to solve the equation. Fortunately, almost all singular points we encounter in mathematical physics are regular singular points.

For the convenience of our discussion, we will assume that the singular point x_0 is at 0. In the case that it is not zero, all we need to do is to make a change of variable $\xi = x - x_0$, and solve the equation in ξ. At the end, ξ is changed back to x, so that the series is expanded in terms of $(x - x_0)$.

4.1.3 Frobenius Series

A differential equation with a regular singular point at $x = 0$ in the form of

$$y'' + \frac{f(x)}{x} y' + \frac{g(x)}{x^2} y = 0$$

can be solved by expression $y(x)$ in the following series

$$y(x) = x^p \sum_{n=0}^{\infty} a_n x^n. \tag{4.4}$$

if $f(x)$ and $g(x)$ are analytic at $x = 0$.

The idea of this method is simple. If $f(x)$ and $g(x)$ are analytic at $x = 0$, then they can be expressed in terms of Taylor series

$$f(x) = f_0 + f_1 x + f_2 x^2 + \cdots$$
$$g(x) = g_0 + g_1 x + g_2 x^2 + \cdots.$$

Around the point $x = 0$, the differential equation can be written as

$$y'' + \frac{1}{x} f_0 y' + \frac{1}{x^2} g_0 y = 0.$$

or

$$x^2 y'' + f_0 x y' + g_0 y = 0 \tag{4.5}$$

This is an Euler–Cauchy differential equation which has a solution in the form of

$$y(x) = x^p.$$

This is the case because, after this function is put in (4.5)

$$p(p-1) x^p + f_0 p x^p + g_0 x^p = 0,$$

we can always find a p from the quadratic equation

$$p(p-1) + f_0 p + g_0 = 0,$$

so that x^p is a solution of (4.5).

Thus it is natural for us to use (4.4) as a trial solution. In fact there is a mathematical theorem known as Fuchs' theorem which says that if $x = 0$ is a regular singular point, then at least one solution can be found this way. We will be satisfied in learning how to find this solution rather than to prove this theorem.

After (4.4) is substituted into the differential equation, we determine the coefficients a_n in such a way that the equation is identically satisfied. If the series with coefficients so determined is convergent, then it is indeed a solution. In using (4.4), we can assume $a_0 \neq 0$, because if a_0 is equal to zero, we can increase p by one and rename a_1 as a_0. The following example illustrates how this procedure works.

Example 4.1.2. Solve the differential equation

$$xy'' + 2y' + xy = 0$$

by expanding $y(x)$ in a Frobenius series.

Solution 4.1.2. With

$$y = \sum_{n=0}^{\infty} a_n x^{n+p},$$

$$y' = \sum_{n=0}^{\infty} a_n (n+p) x^{n+p-1},$$

$$y'' = \sum_{n=0}^{\infty} a_n (n+p)(n+p-1) x^{n+p-2},$$

the differential equation becomes

$$\sum_{n=0}^{\infty} a_n (n+p)(n+p-1) x^{n+p-1} + 2 \sum_{n=0}^{\infty} a_n (n+p) x^{n+p-1}$$

$$+ \sum_{n=0}^{\infty} a_n x^{n+p+1} = 0,$$

or

$$x^p \left\{ \sum_{n=0}^{\infty} a_n [(n+p)(n+p-1) + 2(n+p)] x^{n-1} + \sum_{n=0}^{\infty} a_n x^{n+1} \right\} = 0.$$

Since
$$(n+p)(n+p-1) + 2(n+p) = (n+p)(n+p+1),$$
and x^p cannot be identically equal to zero for all x, so
$$\sum_{n=0}^{\infty} a_n (n+p)(n+p+1)x^{n-1} + \sum_{n=0}^{\infty} a_n x^{n+1} = 0.$$

In order to collect terms, we separate out the $n=0$ and $n=1$ terms in the first summation,

$$a_0 p (p+1) x^{-1} + a_1 (p+1)(p+2) + \sum_{n=2}^{\infty} a_n (n+p)(n+p+1)x^{n-1}$$

$$+ \sum_{n=0}^{\infty} a_n x^{n+1} = 0.$$

Furthermore,

$$\sum_{n=2}^{\infty} a_n (n+p)(n+p+1)x^{n-1} = \sum_{n=0}^{\infty} a_{n+2}(n+p+2)(n+p+3)x^{n+1},$$

therefore

$$a_0 p (p+1) x^{-1} + a_1 (p+1)(p+2)$$

$$+ \sum_{n=0}^{\infty} [a_{n+2}(n+p+2)(n+p+3) + a_n] x^{n+1} = 0.$$

For this to vanish, all coefficients have to be zero,

$$a_0 p (p+1) = 0, \qquad (4.6)$$
$$a_1 (p+1)(p+2) = 0, \qquad (4.7)$$
$$a_{n+2}(n+p+2)(n+p+3) + a_n = 0. \qquad (4.8)$$

Since $a_0 \neq 0$, it follows from (4.6)

$$p(p+1) = 0.$$

This equation is called the indicial equation. Clearly

$$p = -1, \qquad p = 0.$$

There are three possibilities that (4.7) is satisfied,

$$\begin{array}{lll} \text{case 1:} & p = -1, & a_1 \neq 0, \\ \text{case 2:} & p = -1, & a_1 = 0, \\ \text{case 3:} & p = 0, & a_1 = 0. \end{array}$$

From here on we solve the problem in these three separate cases. In case 1, $p = -1$, it follows from (4.8) that

$$a_{n+2} = \frac{-1}{(n+2)(n+1)}a_n.$$

This kind of relation is known as recurrence relation. From this relation, we have

$$n = 0: \quad a_2 = \frac{-1}{2\cdot 1}a_0$$

$$n = 2: \quad a_4 = \frac{-1}{4\cdot 3}a_2 = \frac{-1}{4\cdot 3}\left(\frac{-1}{2\cdot 1}a_0\right) = \frac{(-1)^2}{4!}a_0$$

$$n = 4: \quad a_6 = \frac{-1}{6\cdot 5}a_4 = \frac{-1}{6\cdot 5}\left[\frac{(-1)^2}{4!}a_0\right] = \frac{(-1)^3}{6!}a_0$$

. . . .

$$n = 1: \quad a_3 = \frac{-1}{3\cdot 2}a_1$$

$$n = 3: \quad a_5 = \frac{-1}{5\cdot 4}a_3 = \frac{-1}{5\cdot 4}\left(\frac{-1}{3\cdot 2}a_1\right) = \frac{(-1)^2}{5!}a_1$$

$$n = 5: \quad a_7 = \frac{-1}{7\cdot 6}a_5 = \frac{-1}{7\cdot 6}\left[\frac{(-1)^2}{5!}a_1\right] = \frac{(-1)^3}{7!}a_1$$

. . . . .

It is thus clear that the solution of the differential equation can be written as

$$y(x) = x^{-1}a_0\left(1 - \frac{1}{2!}x^2 + \frac{1}{4!}x^4 - \frac{1}{6!}x^5 + \cdots\right)$$
$$+x^{-1}a_1\left(x - \frac{1}{3!}x^3 + \frac{1}{5!}x^5 - \frac{1}{7!}x^7 + \cdots\right),$$

which we recognize as

$$y(x) = a_0\frac{1}{x}\cos x + a_1\frac{1}{x}\sin x.$$

In this we have found both linearly independent solutions of this second-order differential equation.

In case 2, $p = -1$, and $a_1 = 0$. Because of the recurrence relation, all odd coefficients are zero,

$$a_1 = a_3 = a_5 = \cdots = 0.$$

Therefore we are left with

$$y(x) = a_0\frac{1}{x}\cos x,$$

which is one of the solutions.

In case 3, $p = 0$, and $a_1 = 0$. In this case, all odd coefficients are again equal to zero, and for the even coefficients, the recurrence relation becomes

$$a_{n+2} = \frac{-1}{(n+3)(n+2)} a_n.$$

So the solution can be written as

$$
\begin{aligned}
y(x) &= x^0 a_0 \left(1 - \frac{1}{3!} x^2 + \frac{1}{5!} x^4 - \frac{1}{7!} x^6 + \cdots \right) \\
&= a_0 \frac{1}{x} \left(x - \frac{1}{3!} x^3 + \frac{1}{5!} x^5 - \frac{1}{7!} x^7 + \cdots \right) \\
&= a_0 \frac{1}{x} \sin x,
\end{aligned}
$$

which is the other solution. Note that the a_0 in case 2 is not necessarily equal to the a_0 in case 3. They are arbitrary constants. We recover the general solution by a linear combination of the solutions in case 2 and in case 3,

$$y(x) = c_1 \frac{1}{x} \cos x + c_2 \frac{1}{x} \sin x.$$

The Frobenius series is a generalized power series

$$y = x^p \sum_{n=0}^{\infty} a_n x^n.$$

If the exponent p is a positive integer, it becomes a Taylor series. If p is a negative integer, it becomes a Laurent series. Any equation that can be solved by Taylor or Laurent series, it can also be solved by Frobenius series. Frobenius series is even more general than that because p may be a fraction number, in fact it may even be a complex number. Therefore if one is trying to solve a differential equation by series expansion, instead of first trying to determine if the expansion center is an ordinary point or a regular singular point, one can just try to solve it with the Frobenius series. However, before accepting the series as the solution of the equation, one must determine whether the series is convergent or divergent.

4.2 Bessel Functions

Bessel function is one of the most important special functions in mathematical physics. It occurs, mostly but not exclusively, in problems with cylindrical symmetry. It is the solution of the equation

$$x^2 y''(x) + x y'(x) + (x^2 - n^2) y(x) = 0, \tag{4.9}$$

where n is a given number. This linear homogeneous differential equation is known as the Bessel's equation, named after Wilhelm Bessel (1752–1833), a great German astronomer and mathematician.

4.2.1 Bessel Functions $J_n(x)$ of Integer Order

Although n can be any real number, but we will first limit our attention primarily to the case where n is an integer $(n = 0, 1, 2, \cdots)$. We seek a solution of the Bessel's equation in the form of Frobenius series

$$y = x^p \sum_{j=0} a_j x^j = \sum_{j=0} a_j x^{j+p}, \qquad (4.10)$$

where p is some constant, and

$$a_0 \neq 0.$$

Assume for the present that the function is differentiable, so

$$y' = \sum_{j=0} a_j (j + p) x^{j+p-1},$$

$$y'' = \sum_{j=0} a_j (j + p)(j + p - 1) x^{j+p-2}.$$

Substituting them into (4.9), we obtain

$$\sum_{j=0} \left[(j + p)(j + p - 1) + (j + p) + (x^2 - n^2) \right] a_j x^{j+p} = 0,$$

or

$$x^p \left[\sum_{j=0} [(j + p)^2 - n^2] a_j x^j + \sum_{i=0} a_i x^{i+2} \right] = 0. \qquad (4.11)$$

After $j = 0$ and $j = 1$ terms are written out explicitly, the first summation becomes

$$\sum_{j=0} [(j + p)^2 - n^2] a_j x^j = [p^2 - n^2] a_0 + [(p + 1)^2 - n^2] a_1 x$$

$$+ \sum_{j=2} [(j + p)^2 - n^2] a_j x^j,$$

and the second summation can be written as

$$\sum_{i=0} a_i x^{i+2} = \sum_{j=2} a_{j-2} x^j,$$

The quantity in the bracket of (4.11) must be equal to zero, therefore

$$[p^2 - n^2]a_0 + [(p+1)^2 - n^2]a_1 x + \sum_{j=2}\{[(j+p)^2 - n^2]a_j + a_{j-2}\}x^j = 0.$$

For this equation to hold, the coefficient of each power of x must vanish. Thus,

$$[p^2 - n^2]a_0 = 0, \tag{4.12}$$

$$[(p+1)^2 - n^2]a_1 = 0, \tag{4.13}$$

$$[(j+p)^2 - n^2]a_j + a_{j-2} = 0. \tag{4.14}$$

Since $a_0 \neq 0$, (4.12) requires

$$p = \pm n,$$

we will first proceed with a choice of $+n$. Clearly (4.13) requires

$$a_1 = 0.$$

From (4.14), we have the recurrence relation

$$a_j = \frac{-a_{j-2}}{(j+n)^2 - n^2} = \frac{-1}{j(j+2n)}a_{j-2}. \tag{4.15}$$

Since $a_1 = 0$, this recurrence relation requires $a_3 = 0$, then $a_5 = 0$, etc.; thus

$$a_{2j-1} = 0 \quad j = 1, 2, 3, \dots$$

Since all nonvanishing coefficients have even indices, we set

$$j = 2k, \quad k = 0, 1, 2, \dots,$$

thus the recurrence relation (4.15) becomes

$$a_{2k} = \frac{-1}{2^2 k(k+n)}a_{2(k-1)}. \tag{4.16}$$

This relation holds for any k, specifically we have

$$a_2 = -\frac{1}{2^2 \cdot 1 \cdot (n+1)}a_0,$$

$$a_4 = -\frac{1}{2^2 \cdot 2 \cdot (n+2)}a_2 = \frac{(-1)^2}{2^4 \cdot 2! \cdot (n+2)(n+1)}a_0,$$

$$a_{2k} = \frac{(-1)^k}{2^{2k} k!(n+k)(n+k-1)\cdots\cdots(n+1)}a_0. \tag{4.17}$$

Thus a_0 is a common factor in all terms of the series. Therefore it is a multiplicative constant and can be set to any value. However, by convention, if a_0 is chosen to be

$$a_0 = \frac{1}{2^n n!}, \qquad (4.18)$$

the resulting series for $y(x)$ is designated as $J_n(x)$, called Bessel function of the first kind of order n. With this choice, (4.17) becomes

$$a_{2k} = \frac{(-1)^k}{k!(k+n)!} \frac{1}{2^{n+2k}}, \qquad k = 0, 1, 2, \ldots \qquad (4.19)$$

and

$$
\begin{aligned}
J_n(x) &= \sum_{k=0}^{\infty} a_{2k} x^{n+2k} = \sum_{k=0}^{\infty} \frac{(-1)^k}{k!(k+n)!} \left(\frac{x}{2}\right)^{n+2k} \\
&= \frac{x^n}{2^n n!} \left(1 - \frac{x^2}{2^2(n+1)} + \frac{x^4}{2^4 2!(n+1)(n+2)} - \cdots \right). \qquad (4.20)
\end{aligned}
$$

By ratio test, this series is absolutely convergent for all x. Hence $J_n(x)$ is bounded everywhere from $x = 0$ to $x \to \infty$.

The results for J_0, J_1, J_2 are shown in Fig. 4.1. They are alternating series. The error in cutting off after n terms is less than the first term dropped. The Bessel functions oscillate but are not periodic. The amplitude of $J_n(x)$ is not constant but decreases asymptotically.

4.2.2 Zeros of the Bessel Functions

As it is seen in Fig. 4.1 for each n, there are a series of x values for which $J_n(x) = 0$. These x values are the zeros of Bessel functions. They are very important in practical applications. They can be found in tables, such as "Table of First 700 Zeros of Bessel Functions" by C.L. Beattie, Bell Tech. J. **37**, 689 (1958) and Bell Monograph 3055. The first few are listed in Table 4.1.

As an example of how to use this table, let us answer the following question. If λ_{nj} is the jth root of $J_n(\lambda c) = 0$ where $c = 2$, find λ_{01}, λ_{23}, λ_{53}. The answer should be

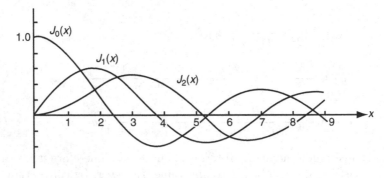

Fig. 4.1. Bessel functions, $J_0(x)$, $J_1(x)$, $J_2(x)$

Table 4.1. Zeros of the Bessel function

Number of zeros	$J_0(x)$	$J_1(x)$	$J_2(x)$	$J_3(x)$	$J_4(x)$	$J_5(x)$
1	2.4048	3.8317	5.1356	6.3802	7.5883	8.7715
2	5.5201	7.0156	8.4172	9.7610	11.0647	12.3386
3	8.6537	10.1735	11.6198	13.0152	14.3725	15.7002
4	11.7915	13.3237	14.7960	16.2235	17.6160	18.9801
5	14.9309	16.4706	17.9598	19.4094	20.8269	22.2178

$$\lambda_{01} = \frac{2.4048}{2} = 1.2024,$$

$$\lambda_{23} = \frac{11.6198}{2} = 5.8099,$$

$$\lambda_{53} = \frac{15.7002}{2} = 7.8501.$$

4.2.3 Gamma Function

For Bessel function of noninteger order we need an extension of the factorials. This can be done via gamma function.

The gamma function is defined by the integral

$$\Gamma(\alpha) = \int_0^\infty e^{-t} t^{\alpha-1}\, dt. \tag{4.21}$$

With integration by parts, we obtain

$$\Gamma(\alpha+1) = \int_0^\infty e^{-t} t^\alpha\, dt = -e^{-t} t^\alpha \Big|_0^\infty + \alpha \int_0^\infty e^{-t} t^{\alpha-1}\, dt.$$

The first expression on the right is zero, and the integral on the right is $\alpha\Gamma(\alpha)$. This gives the basic relation

$$\Gamma(\alpha+1) = \alpha\Gamma(\alpha).$$

Since

$$\Gamma(1) = \int_0^\infty e^{-t}\, dt = 1,$$

we conclude for integer n,

$$\Gamma(n+1) = n\Gamma(n) = n(n-1)\Gamma(n-1)$$
$$= n(n-1)\cdots 1\Gamma(1) = n! \tag{4.22}$$

For a noninteger α, the integral of (4.21) can be evaluated. The gamma functions $\Gamma(\alpha)$ for both positive and negative α are shown in Fig. 4.2.

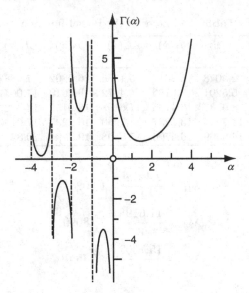

Fig. 4.2. Gamma function $\Gamma(\alpha)$

It follows from (4.22) that

$$0! = \Gamma(1) = 1. \tag{4.23}$$

Since $n\Gamma(n) = \Gamma(n+1)$, thus $\Gamma(n) = \Gamma(n+1)/n$. It follows that

$$\Gamma(0) = \frac{\Gamma(1)}{0} \to \infty,$$

$$\Gamma(-1) = \frac{\Gamma(0)}{-1} \to \infty,$$

and for any negative integer

$$\Gamma(-n) = \frac{\Gamma(-n+1)}{-n} \to \infty.$$

The special case of $\Gamma(1/2)$ is of particular interest,

$$\Gamma\left(\frac{1}{2}\right) = \int_0^\infty e^{-t} t^{-1/2} \, \mathrm{d}t. \tag{4.24}$$

Let $t = x^2$, so $\mathrm{d}t = 2x \, \mathrm{d}x$ *and* $t^{-1/2} = x^{-1}$

$$\Gamma\left(\frac{1}{2}\right) = \int_0^\infty e^{-x^2} \frac{1}{x} 2x \, \mathrm{d}x = 2 \int_0^\infty e^{-x^2} \, \mathrm{d}x.$$

For a definite integral, the name of the integration variable is immaterial

$$\int_0^\infty e^{-x^2}\,dx = \int_0^\infty e^{-y^2}\,dy,$$

$$\left[\Gamma\left(\frac{1}{2}\right)\right]^2 = 4\int_0^\infty e^{-x^2}\,dx \int_0^\infty e^{-y^2}\,dy$$

$$= 4\int_0^\infty \int_0^\infty e^{-x^2-y^2}\,dx\,dy.$$

The double integral can be considered as a surface integral over the first quadrant of the entire plane. Change to the polar coordinates,

$$x^2 + y^2 = \rho^2$$
$$da = \rho\,d\theta\,d\rho,$$

we have

$$\left[\Gamma\left(\frac{1}{2}\right)\right]^2 = 4\int_0^\infty \int_0^{\pi/2} e^{-\rho^2}\rho\,d\theta\,d\rho$$

$$= 4\frac{\pi}{2}\int_0^\infty e^{-\rho^2}\rho\,d\rho = \pi\left[-e^{-\rho^2}\right]_0^\infty = \pi.$$

Thus

$$\Gamma\left(\frac{1}{2}\right) = \sqrt{\pi}. \tag{4.25}$$

4.2.4 Bessel Function of Noninteger Order

In our development of Bessel function of integer order, we had in (4.18) $a_0 = 1/(2^n n!)$, which can be written as

$$a_0 = \frac{1}{2^n \Gamma(n+1)}.$$

This suggests that, for noninteger α, we choose

$$a_0 = \frac{1}{2^\alpha \Gamma(\alpha+1)}.$$

Following exactly the same procedure as for the integer order, we find the noninteger order Bessel function is given by

$$J_\alpha(x) = \sum_{k=0}^\infty \frac{(-1)^k}{k!\,\Gamma(k+\alpha+1)}\left(\frac{x}{2}\right)^{\alpha+2k}. \tag{4.26}$$

In fact this formula can be used for both integer and noninteger α.

Example 4.2.1. Show that

$$J_{1/2}(x) = \sqrt{\frac{2}{\pi x}} \sin x, \qquad J_{-1/2}(x) = \sqrt{\frac{2}{\pi x}} \cos x.$$

Solution 4.2.1. By definition,

$$J_{1/2}(x) = \sum_{k=0}^{\infty} \frac{(-1)^k}{k!\Gamma(k + \frac{1}{2} + 1)} \left(\frac{x}{2}\right)^{(1/2)+2k}$$

$$= \left(\frac{x}{2}\right)^{-1/2} \sum_{k=0}^{\infty} \frac{(-1)^k}{k!\Gamma(k + \frac{1}{2} + 1)} \frac{x^{2k+1}}{2^{2k+1}}.$$

$$\Gamma\left(k + \frac{1}{2} + 1\right) = \left(k + \frac{1}{2}\right) \Gamma\left(k + \frac{1}{2}\right)$$

$$= \left(k + \frac{1}{2}\right)\left(k + \frac{1}{2} - 1\right)\left(k + \frac{1}{2} - 2\right) \cdots \frac{1}{2}\Gamma\left(\frac{1}{2}\right)$$

$$= \frac{(2k + 1)(2k - 1)(2k - 3)\cdots 1}{2^{k+1}}\Gamma\left(\frac{1}{2}\right).$$

It follows that

$$k!\Gamma(k + \frac{1}{2} + 1)2^{2k+1} = k!\left[(2k + 1)(2k - 1)\cdots 1\right]\Gamma\left(\frac{1}{2}\right)2^k$$

$$= \left[2k\left(2k - 2\right)\cdots 2\right]\left[(2k + 1)(2k - 1)\cdots 1\right]\Gamma\left(\frac{1}{2}\right)$$

$$= (2k + 1)!\Gamma\left(\frac{1}{2}\right) = (2k + 1)!\sqrt{\pi}.$$

Thus

$$J_{1/2}(x) = \sqrt{\frac{2}{\pi x}} \sum_{k=0}^{\infty} \frac{(-1)^k \, x^{2k+1}}{(2k + 1)!}.$$

But

$$\sin x = x - \frac{1}{3!}x^3 + \frac{1}{5!}x^5 + \cdots = \sum_{k=0}^{\infty} \frac{(-1)^k \, x^{2k+1}}{(2k + 1)!},$$

therefore

$$J_{1/2}(x) = \sqrt{\frac{2}{\pi x}} \sin x. \tag{4.27}$$

Similarly,

$$J_{-1/2}(x) = \sum_{k=0}^{\infty} \frac{(-1)^k}{k!\Gamma(k - \frac{1}{2} + 1)} \left(\frac{x}{2}\right)^{-(1/2)+2k}$$

$$= \left(\frac{x}{2}\right)^{-1/2} \sum_{k=0}^{\infty} \frac{(-1)^k}{(2k)!\Gamma(\frac{1}{2})}x^{2k} = \sqrt{\frac{2}{\pi x}} \cos x. \tag{4.28}$$

4.2.5 Bessel Function of Negative Order

If α is not an integer, the Bessel function of the negative order of $J_{-\alpha}(x)$ is very simple. All we have to do is to replace α by $-\alpha$ in the expression of $J_\alpha(x)$, that is

$$J_{-\alpha}(x) = \sum_{k=0}^{\infty} \frac{(-1)^k}{k!\Gamma(k-\alpha+1)} \left(\frac{x}{2}\right)^{-\alpha+2k}. \tag{4.29}$$

Since the first term of J_α and $J_{-\alpha}$ is a finite nonzero multiple of x^α and $x^{-\alpha}$, respectively, clearly J_α and $J_{-\alpha}$ are linearly independent. Therefore the general solution of Bessel's equation of order α is

$$y(x) = c_1 J_\alpha(x) + c_2 J_{-\alpha}(x).$$

However, if α is an integer, the negative order $J_{-n}(x)$ and the positive order Bessel function $J_n(x)$ are not linearly independent. This can be seen as follows. Starting with the definition

$$J_{-n}(x) = \sum_{k=0}^{\infty} \frac{(-1)^k}{k!\Gamma(k-n+1)} \left(\frac{x}{2}\right)^{-n+2k}.$$

If $k < n$, then $\Gamma(k-n+1) \to \infty$ and all the corresponding terms will be zero. Therefore the series actually starts with $k = n$,

$$J_{-n}(x) = \sum_{k=n}^{\infty} \frac{(-1)^k}{k!\Gamma(k-n+1)} \left(\frac{x}{2}\right)^{-n+2k}.$$

Let us define $j = k - n$, then $k = n + j$, thus

$$J_{-n}(x) = \sum_{j=0}^{\infty} \frac{(-1)^{n+j}}{(n+j)!\Gamma(j+1)} \left(\frac{x}{2}\right)^{-n+2(j+n)}$$

$$= (-1)^n \sum_{j=0}^{\infty} \frac{(-1)^j}{\Gamma(n+j+1)j!} \left(\frac{x}{2}\right)^{2j+n} = (-1)^n J_n(x).$$

Therefore $J_{-n}(x)$ and $J_n(x)$ are linearly dependent. So in this case, there must be another linearly independent solution of Bessel's equation of order n.

4.2.6 Neumann Functions and Hankel Functions

To determine the second linearly independent solution of the Bessel function when $\alpha = n$ and n is an integer, it is customary to form a particular linear combination of $J_\alpha(x)$ and $J_{-\alpha}(x)$ and then letting $\alpha \to n$. The combination

$$N_\alpha(x) = \frac{\cos(\alpha\pi) J_\alpha(x) - J_{-\alpha}(x)}{\sin(\alpha\pi)} \tag{4.30}$$

is called the Bessel function of the second kind of order α. It is also known as the Neumann function. In some literature, it is denoted as $Y_\alpha(x)$.

For noninteger α, $N_\alpha(x)$ is clearly a solution of the Bessel equation, since it is a linearly combination of two linearly independent solutions $J_\alpha(x)$ and $J_{-\alpha}(x)$.

For integer α, $\alpha = n$ and $n = 0, 1, 2, \ldots$, (4.30) becomes

$$N_n(x) = \frac{\cos(n\pi) J_n(x) - J_{-n}(x)}{\sin(n\pi)},$$

which gives an indeterminate form of $0/0$, since $\cos(n\pi) = (-1)^n$, $\sin(n\pi) = 0$ and $J_n(x) = (-1)^n J_{-n}(x)$. We can use l'Hôspital's rule to evaluate this ratio. If we define the Neumann function $N_n(x)$ as

$$N_n(x) = \lim_{\alpha \to n} \frac{\cos(\alpha\pi) J_\alpha(x) - J_{-\alpha}(x)}{\sin(\alpha\pi)}.$$

Then

$$\begin{aligned}
N_n(x) &= \left[\frac{\frac{\partial}{\partial\alpha}(\cos(\alpha\pi) J_\alpha(x) - J_{-\alpha}(x))}{\frac{\partial}{\partial\alpha}\sin(\alpha\pi)} \right]_{\alpha=n} \\
&= \left[\frac{-\pi\sin(\alpha\pi) J_\alpha(x) + \cos(\alpha\pi)\frac{\partial}{\partial\alpha}J_\alpha - \frac{\partial}{\partial\alpha}J_{-\alpha}(x)}{\pi\cos(\alpha\pi)} \right]_{\alpha=n} \\
&= \frac{1}{\pi}\left[\frac{\partial}{\partial\alpha}J_\alpha(x) - (-1)^n \frac{\partial}{\partial\alpha}J_{-\alpha}(x) \right]_{\alpha=n}.
\end{aligned} \tag{4.31}$$

Now we will show that $N_n(x)$ so defined is indeed a solution of the Bessel's equation. By definition, J_α and $J_{-\alpha}$, respectively, satisfy the following differential equations

$$x^2 J_\alpha''(x) + x J_\alpha'(x) + (x^2 - \alpha^2) J_\alpha(x) = 0,$$
$$x^2 J_{-\alpha}''(x) + x J_{-\alpha}'(x) + (x^2 - \alpha^2) J_{-\alpha}(x) = 0.$$

Differentiate with respect to α,

$$x^2 \frac{d^2}{dx^2}\left(\frac{\partial J_\alpha}{\partial\alpha}\right) + x\frac{d}{dx}\left(\frac{\partial J_\alpha}{\partial\alpha}\right) + (x^2 - \alpha^2)\frac{\partial J_\alpha}{\partial\alpha} - 2\alpha J_\alpha = 0,$$

$$x^2 \frac{d^2}{dx^2}\left(\frac{\partial J_{-\alpha}}{\partial\alpha}\right) + x\frac{d}{dx}\left(\frac{\partial J_{-\alpha}}{\partial\alpha}\right) + (x^2 - \alpha^2)\frac{\partial J_{-\alpha}}{\partial\alpha} - 2\alpha J_{-\alpha} = 0.$$

Multiplying the second equation by $(-1)^n$ and subtracting it from the first equation, we have

$$x^2 \frac{d^2}{dx^2}\left(\frac{\partial J_\alpha}{\partial\alpha} - (-1)^n \frac{\partial J_{-\alpha}}{\partial\alpha}\right) + x\frac{d}{dx}\left(\frac{\partial J_\alpha}{\partial\alpha} - (-1)^n \frac{\partial J_{-\alpha}}{\partial\alpha}\right)$$

$$+ (x^2 - \alpha^2)\left(\frac{\partial J_\alpha}{\partial\alpha} - (-1)^n \frac{\partial J_{-\alpha}}{\partial\alpha}\right) - 2\alpha\left(J_\alpha - (-1)^n J_{-\alpha}\right) = 0.$$

Taking the limit $\alpha \to n$, the last term drops out because

$$J_n - (-1)^n J_{-n} = 0.$$

Clearly the Neumann function expressed in (4.31) satisfies the Bessel's equation.

Neumann function has a logarithmic term, since

$$\frac{\partial J_\alpha}{\partial \alpha} = \frac{\partial}{\partial \alpha} \left[x^\alpha \sum_{k=0}^{\infty} \frac{(-1)^n}{k! \Gamma(k+\alpha+1)} \left(\frac{x}{2}\right)^{2k} \right]$$

$$= \frac{\partial x^\alpha}{\partial \alpha} \sum_{k=0}^{\infty} \frac{(-1)^n}{k! \Gamma(k+\alpha+1)} \left(\frac{x}{2}\right)^{2k} + x^\alpha \frac{\partial}{\partial \alpha} \sum_{k=0}^{\infty} \frac{(-1)^n}{k! \Gamma(k+\alpha+1)} \left(\frac{x}{2}\right)^{2k},$$

and

$$\frac{\partial x^\alpha}{\partial \alpha} = \frac{\partial}{\partial \alpha} e^{\alpha \ln x} = e^{\alpha \ln x} \ln x = x^\alpha \ln x.$$

Thus $N_n(x)$ contains a term $J_n(x) \ln x$. Clearly it is linearly independent of $J_n(x)$.

Neumann functions diverge as $x \to 0$. For $\alpha \neq 0$, it diverges because the series for $J_{-\alpha}$ starts with a term $x^{-\alpha}$. For $\alpha = 0$, the term $J_0(x) \ln x$ goes to $-\infty$ as x goes to zero.

Like Bessel functions, the values and zeros of Neumann functions have been extensively tabulated. The first three orders of Neumann functions are shown in Fig. 4.3.

Since $J_\alpha(x)$ and $N_\alpha(x)$ always constitute a pair of linearly independent solutions, the general solution of the Bessel's equation can be written as

$$y(x) = c_1 J_\alpha(x) + c_2 N_\alpha(x).$$

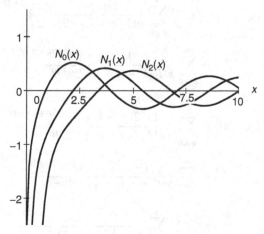

Fig. 4.3. Neumann functions $N_0(x)$, $N_1(x)$, $N_2(x)$

This expression is valid for all α.

In physical applications, we often require the solution to be bounded at the origin. Since $N_\alpha(x)$ diverges at $x = 0$, the coefficient c_2 must be set to zero, and the solution is just a constant times $J_\alpha(x)$. However, there are problems where the origin is excluded. In such cases, both $J_\alpha(x)$ and $N_\alpha(x)$ have to be used.

Hankel Functions. The following linear combinations are useful in the study of wave propagation, especially in the asymptotic region where they have pure complex exponential behavior. They are called Hankel functions of the first kind

$$H_n^{(1)}(x) = J_n(x) + \mathrm{i}N_n(x),$$

and Hankel functions of the second kind

$$H_n^{(2)}(x) = J_n(x) - \mathrm{i}N_n(x).$$

These functions are also known as Bessel functions of the third kind.

4.3 Properties of Bessel Function

4.3.1 Recurrence Relations

Starting with the series representations, we can deduce the following properties of Bessel functions

1.
$$\frac{\mathrm{d}}{\mathrm{d}x}[x^{n+1} J_{n+1}(x)] = x^{n+1} J_n(x). \tag{4.32}$$

Proof:

$$x^{n+1} J_{n+1}(x) = x^{n+1} \sum_{k=0}^{\infty} \frac{(-1)^k}{k!\Gamma(k+n+2)} \left(\frac{x}{2}\right)^{n+2k+1}$$

$$= \sum_{k=0}^{\infty} \frac{(-1)^k}{k!\Gamma(k+n+2)} \frac{x^{2n+2k+2}}{2^{n+2k+1}}.$$

$$\frac{\mathrm{d}}{\mathrm{d}x}[x^{n+1} J_{n+1}(x)] = \sum_{k=0}^{\infty} \frac{(-1)^k 2(n+k+1)}{k!\Gamma(k+n+2)} \frac{x^{2n+2k+1}}{2^{n+2k+1}}$$

$$= \sum_{k=0}^{\infty} \frac{(-1)^k}{k!\Gamma(k+n+1)} \frac{x^{2n+2k+1}}{2^{n+2k}}$$

$$= x^{n+1} \sum_{k=0}^{\infty} \frac{(-1)^k}{k!\Gamma(k+n+1)} \frac{x^{n+2k}}{2^{n+2k}} = x^{n+1} J_n(x).$$

2.

$$\frac{\mathrm{d}}{\mathrm{d}x}[x^{-n}J_n(x)] = -x^{-n}J_{n+1}(x).$$ (4.33)

Proof:

$$x^{-n}J_n(x) = x^{-n}\sum_{k=0}^{\infty}\frac{(-1)^k}{k!\Gamma(k+n+1)}\frac{x^{n+2k}}{2^{n+2k}} = \sum_{k=0}^{\infty}\frac{(-1)^k}{k!\Gamma(k+n+1)}\frac{x^{2k}}{2^{n+2k}},$$

$$\frac{\mathrm{d}}{\mathrm{d}x}[x^{-n}J_n(x)] = \sum_{k=0}^{\infty}\frac{(-1)^k 2k}{k!\Gamma(k+n+1)}\frac{x^{2k-1}}{2^{n+2k}}.$$

Since the first term with $k = 0$ is zero, it follows that the summation starts with $k = 1$.

$$\frac{\mathrm{d}}{\mathrm{d}x}[x^{-n}J_n(x)] = \sum_{k=1}^{\infty}\frac{(-1)^k 2k}{k!\Gamma(k+n+1)}\frac{x^{2k-1}}{2^{n+2k}}$$

$$= \sum_{k=1}^{\infty}\frac{(-1)^k}{(k-1)!\Gamma(k+n+1)}\frac{x^{2k-1}}{2^{n+2k-1}}.$$

Now let $j = k - 1$, so $k = j + 1$ and

$$\frac{\mathrm{d}}{\mathrm{d}x}[x^{-n}J_n(x)] = \sum_{j=0}^{\infty}\frac{(-1)^{j+1}}{(j)!\Gamma(j+n+2)}\frac{x^{2j+1}}{2^{n+2j+1}}$$

$$= -x^{-n}\sum_{j=0}^{\infty}\frac{(-1)^j}{(j)!\Gamma(j+n+2)}\frac{x^{n+2j+1}}{2^{n+2j+1}} = -x^{-n}J_{n+1}(x).$$

With these properties, we can derive the following recurrence relations

3.

$$J_{n+1}(x) = \frac{2n}{x}J_n(x) - J_{n-1}(x),$$ (4.34)

4.

$$J_n'(x) = \frac{1}{2}[J_{n-1}(x) - J_{n+1}(x)],$$ (4.35)

5.

$$J_0'(x) = -J_1(x).$$ (4.36)

Starting with

$$\frac{\mathrm{d}}{\mathrm{d}x}[x^{n+1}J_{n+1}(x)] = (n+1)x^n J_{n+1}(x) + x^{n+1}J_{n+1}'(x),$$

it follows from (4.32) that

$$(n+1)x^n J_{n+1}(x) + x^{n+1}J_{n+1}'(x) = x^{n+1}J_n(x),$$

which can be written as

$$J'_{n+1}(x) = J_n(x) - \frac{n+1}{x} J_{n+1}(x),$$

or

$$J'_n(x) = J_{n-1}(x) - \frac{n}{x} J_n(x). \tag{4.37}$$

From (4.33), we have

$$-nx^{-n-1} J_n(x) + x^{-n} J'_n(x) = -x^{-n} J_{n+1}(x),$$

or

$$J'_n(x) = \frac{n}{x} J_n(x) - J_{n+1}(x). \tag{4.38}$$

It follows from (4.37), (4.38) that

$$J_{n-1}(x) - \frac{n}{x} J_n(x) = \frac{n}{x} J_n(x) - J_{n+1}(x),$$

or

$$J_{n+1}(x) = \frac{2n}{x} J_n(x) - J_{n-1}(x)$$

which is the recurrence relation 3.

Adding (4.37)–(4.38), we have

$$J'_n(x) = \frac{1}{2} [J_{n-1}(x) - J_{n+1}(x)].$$

which is the recurrence relation 4.

The particular case $J'_0(x)$ follows directly from property 2. Put $n = 0$ in (4.33), we have

$$J'_0(x) = -J_1(x).$$

which is the recurrence relation 5.

These recurrence relations are very useful. It means that as long as we know $J_0(x)$ and $J_1(x)$, all higher orders Bessel functions and their derivatives can be generated from these relations.

Another interesting relation follows from property 1 is

$$\int_0^r d[x^{n+1} J_{n+1}(x)] = \int_0^r x^{n+1} J_n(x) \, dx.$$

Therefore

6.

$$\int_0^r x^{n+1} J_n(x) \, dx = r^{n+1} J_{n+1}(r). \tag{4.39}$$

An important special case is for $n = 0$,

7.

$$\int_0^r x J_0(x) \, dx = r J_1(r). \tag{4.40}$$

4.3.2 Generating Function of Bessel Functions

Although Bessel functions are of interest primarily as solutions of differential equations, it is intructive and convenient to develop them from a completely different approach, that of the generating function.

Recall

$$\exp(x) = \sum_{n=0}^{\infty} \frac{1}{n!} x^n,$$

so

$$\exp\left(\frac{xt}{2}\right) = \sum_{n=0}^{\infty} \frac{1}{n!} \left(\frac{xt}{2}\right)^n, \qquad \exp\left(-\frac{x}{2t}\right) = \sum_{n=0}^{\infty} \frac{1}{n!} \left(-\frac{x}{2t}\right)^n.$$

It follows

$$\exp\left(\frac{xt}{2} - \frac{x}{2t}\right) = \sum_{l=0}^{\infty} \frac{1}{l!} \left(\frac{xt}{2}\right)^l \sum_{m=0}^{\infty} \frac{1}{m!} \left(-\frac{x}{2t}\right)^m$$

$$= \sum_{l=0}^{\infty} \sum_{m=0}^{\infty} \frac{(-1)^m}{l!m!} \left(\frac{x}{2}\right)^{l+m} t^{l-m}. \qquad (4.41)$$

Let

$$l - m = n, \quad \text{then} \quad l = m + n \quad \text{and} \quad l + m = 2m + n,$$

so (4.41) can be written as

$$\exp\left(\frac{xt}{2} - \frac{x}{2t}\right) = \sum_{n=-\infty}^{\infty} \left\{ \sum_{m=0}^{\infty} \frac{1}{(n+m)!} \frac{(-1)^m}{m!} \left(\frac{x}{2}\right)^{2m+n} \right\} t^n.$$

Since the Bessel function $J_n(x)$ is given by

$$J_n(x) = \sum_{m=0}^{\infty} (-1)^m \frac{1}{(n+m)!m!} \left(\frac{x}{2}\right)^{2m+n},$$

it is clear

$$\exp\left(\frac{xt}{2} - \frac{x}{2t}\right) = \sum_{n=-\infty}^{\infty} J_n(x) t^n. \qquad (4.42)$$

The left-hand side of this equation is known as the generating function of the Bessel functions, sometimes designated as $G(x, t)$,

$$G(x, t) = \exp\left(\frac{xt}{2} - \frac{x}{2t}\right).$$

4.3.3 Integral Representation

A particularly useful and powerful way of treating Bessel functions employs integral representations. If we substitute

$$t = e^{i\theta},$$

then

$$t - \frac{1}{t} = e^{i\theta} - e^{-i\theta} = 2i\sin\theta.$$

Therefore (4.42) can be written as

$$e^{(ix\sin\theta)} = \sum_{n=-\infty}^{\infty} J_n(x)e^{in\theta} = J_0(x) + \sum_{n=1}^{\infty}\left[J_n(x)e^{in\theta} + J_{-n}(x)e^{-in\theta}\right]$$

$$= J_0(x) + \sum_{n=1}^{\infty} J_n(x)(\cos n\theta + i\sin n\theta)$$

$$+ \sum_{n=1}^{\infty}(-1)^n J_n(x)(\cos n\theta - i\sin n\theta)$$

$$= J_0(x) + 2[J_2(x)\cos 2\theta + J_4(x)\cos 4\theta + \cdots]$$

$$+ 2i[J_1(x)\sin\theta + J_3(x)\sin 3\theta + J_5(x)\sin 5\theta + \cdots].$$

But

$$e^{(ix\sin\theta)} = \cos(x\sin\theta) + i\sin(x\sin\theta),$$

thus

$$\cos(x\sin\theta) = J_0(x) + 2[J_2(x)\cos 2\theta + J_4(x)\cos 4\theta + \cdots], \qquad (4.43)$$
$$\sin(x\sin\theta) = 2[J_1(x)\sin\theta + J_3(x)\sin 3\theta + J_5(x)\sin 5\theta + \cdots]. \qquad (4.44)$$

These are Fourier type representations. The coefficients $J_n(x)$ can be readily obtained. For example, multiply (4.43) by $\cos(n\theta)$ and integrate from 0 to π, we obtain

$$\frac{1}{\pi}\int_0^\pi \cos(x\sin\theta)\cos n\theta \, d\theta = \begin{cases} J_n(x), & n \text{ even} \\ 0, & n \text{ odd} \end{cases}.$$

Similarly from (4.44)

$$\frac{1}{\pi}\int_0^\pi \sin(x\sin\theta)\sin n\theta \, d\theta = \begin{cases} 0, & n \text{ even} \\ J_n(x), & n \text{ odd} \end{cases}.$$

Adding these two equations, we get

$$J_n(x) = \frac{1}{\pi}\int_0^\pi [\cos(x\sin\theta)\cos n\theta + \sin(x\sin\theta)\sin n\theta]d\theta$$

$$= \frac{1}{\pi}\int_0^\pi \cos(x\sin\theta - n\theta)d\theta. \qquad (4.45)$$

As a special case,

$$J_0(x) = \frac{1}{\pi} \int_0^{\pi} \cos(x \sin \theta) \mathrm{d}\theta. \tag{4.46}$$

On the other hand, cosine is an even function and sine is an odd function

$$\int_0^{\pi} \cos(x \sin \theta) \mathrm{d}\theta = \frac{1}{2} \int_0^{2\pi} \cos(x \sin \theta) \mathrm{d}\theta,$$

$$\int_0^{2\pi} \sin(x \sin \theta) \mathrm{d}\theta = 0.$$

Therefore (4.46) can be written as

$$J_0(x) = \frac{1}{2\pi} \int_0^{2\pi} \cos(x \sin \theta) \mathrm{d}\theta = \frac{1}{2\pi} \int_0^{2\pi} [\cos(x \sin \theta) + \mathrm{i} \sin(x \sin \theta)] \mathrm{d}\theta$$

$$= \frac{1}{2\pi} \int_0^{2\pi} \exp(\mathrm{i}\, x \sin \theta) \mathrm{d}\theta = \frac{1}{2\pi} \int_0^{2\pi} \exp(\mathrm{i}\, x \cos \theta) \mathrm{d}\theta.$$

This form is very useful in the Fraunhofer diffraction with a circular aperture.

4.4 Bessel Functions as Eigenfunctions of Sturm–Liouville Problems

4.4.1 Boundary Conditions of Bessel's Equation

As discussed in the last chapter, by itself Bessel's equation is not a Sturm–Liouville equation. There is no way for it to satisfy any given boundary condition. However, the closely related equation

$$x^2 \frac{\mathrm{d}^2 y}{\mathrm{d}x^2} + x \frac{\mathrm{d}y}{\mathrm{d}x} + \left(\lambda^2 x^2 - n^2\right) y = 0 \tag{4.47}$$

is a Sturm–Liouville equation. It can easily be shown that

$$y(x) = J_n(\lambda x)$$

is a solution of this equation. Let $z = \lambda x$, then

$$\frac{\mathrm{d}y}{\mathrm{d}x} = \frac{\mathrm{d}J_n(z)}{\mathrm{d}z} \frac{\mathrm{d}z}{\mathrm{d}x} = \lambda \frac{\mathrm{d}J_n(z)}{\mathrm{d}z},$$

$$\frac{\mathrm{d}^2 y}{\mathrm{d}x^2} = \frac{\mathrm{d}}{\mathrm{d}x} \left[\lambda \frac{\mathrm{d}J_n(z)}{\mathrm{d}z} \right] = \frac{\mathrm{d}}{\mathrm{d}z} \left[\lambda \frac{\mathrm{d}J_n(z)}{\mathrm{d}z} \right] \frac{\mathrm{d}z}{\mathrm{d}x} = \lambda^2 \frac{\mathrm{d}^2 J_n(z)}{\mathrm{d}z^2}.$$

Substituting them into (4.47), we have

$$x^2 \lambda^2 \frac{\mathrm{d}^2 J_n(z)}{\mathrm{d}z^2} + x\lambda \frac{\mathrm{d}J_n(z)}{\mathrm{d}z} + \left(\lambda^2 x^2 - n^2\right) J_n(z) =$$

$$z^2 \frac{\mathrm{d}^2 J_n(z)}{\mathrm{d}z^2} + z \frac{\mathrm{d}J_n(z)}{\mathrm{d}z} + \left(z^2 - n^2\right) J_n(z) = 0.$$

The second line is just the regular Bessel's equation. Therefore we have established that $J_n(\lambda x)$ is the solution of (4.47).

We have shown in the last chapter that (4.47) written in the form

$$\frac{\mathrm{d}}{\mathrm{d}x}\left(x\frac{\mathrm{d}y}{\mathrm{d}x}\right) + \left(\lambda^2 x - \frac{n^2}{x}\right) y = 0 \tag{4.48}$$

together with a boundary condition at $x = c$ constitute a Sturm–Liouville problem in the interval of $0 \le x \le c$. The general boundary condition is of the form

$$Ay(c) + By'(c) = 0,$$

where A and B are two constants. If $B = 0$, it is known as the Dirichlet condition. If $A = 0$, it is known as the Neumann condition.

The problem also requires that the solution be regular (bounded) at $x = 0$. This precludes Neumann function as a solution.

This means that only those values of λ that satisfy the equation

$$AJ_n(\lambda c) - B \left.\frac{\mathrm{d}J_n(\lambda x)}{\mathrm{d}x}\right|_{x=c} = 0 \tag{4.49}$$

are acceptable. Since Bessel functions have oscillatory character, there are infinite number of λ that satisfy this equation. These values of λ are the eigenvalues of the problem. For example, if $B = 0$, $n = 0$, $c = 2$, then

$$J_0(2\lambda) = 0.$$

The jth root of this equation, labeled λ_{0j}, can be found from the table of zeros of $J_n(x)$ (Table 4.1) as

$$\lambda_{01} = \frac{2.4048}{2} = 1.2924, \quad \lambda_{02} = \frac{5.5201}{2} = 2.7601, \quad \text{etc.}$$

The zeros of $J'(x)$ are also tabulated. So if $A = 0$, λ_{nj} can also be read from the table. In general, if both A and B are nonzero, then λ_{nj} has to be numerically calculated.

4.4.2 Orthogonality of Bessel Functions

Corresponding to the set of eigenvalues $\{\lambda_{nj}\}$, the eigenfunctions are $\{J_n(\lambda_{nj}x)\}$. These eigenfunctions form a complete set and they are orthogonal to each other with respect to the weight function x, that is

$$\int_0^c J_n\left(\lambda_{ni}x\right) J_n\left(\lambda_{nk}x\right) x\,\mathrm{d}x = 0 \quad\text{if}\quad \lambda_{ni} \neq \lambda_{nk}.$$

Therefore any well-behaved function $f(x)$ in the interval $0 \leq x \leq c$ can be expanded into a Fourier–Bessel series

$$f(x) = \sum_{j=1}^{\infty} a_j J_n\left(\lambda_{nj}x\right),$$

where

$$a_j = \frac{1}{\int_0^c \left[J_n\left(\lambda_{nj}x\right)\right]^2 x\,\mathrm{d}x} \int_0^c f(x) J_n(\lambda_{nj}x) x\,\mathrm{d}x.$$

In Sect. 4.4.3, we will evaluate the normalization integral

$$\beta_{nj}^2 = \int_0^c \left[J_n\left(\lambda_{nj}x\right)\right]^2 x\,\mathrm{d}x$$

under various boundary conditions.

4.4.3 Normalization of Bessel Functions

One way to find the value of the normalization integral β_{nj}^2 is to substitute $y = J_n\left(\lambda x\right)$ into (4.48) and multiply it by $2x(\mathrm{d}/\mathrm{d}x)J_n(\lambda x)$:

$$2x\frac{\mathrm{d}}{\mathrm{d}x}J_n(\lambda x)\left\{\frac{\mathrm{d}}{\mathrm{d}x}\left(x\frac{\mathrm{d}}{\mathrm{d}x}J_n(\lambda x)\right) + \left(\lambda^2 x - \frac{n^2}{x}\right)J_n(\lambda x)\right\} = 0.$$

It is not difficult to see that this equation can be written as

$$\frac{\mathrm{d}}{\mathrm{d}x}\left(x\frac{\mathrm{d}}{\mathrm{d}x}J_n(\lambda x)\right)^2 + (\lambda^2 x^2 - n^2)J_n(\lambda x)2\frac{\mathrm{d}}{\mathrm{d}x}J_n(\lambda x) = 0,$$

or

$$\frac{\mathrm{d}}{\mathrm{d}x}\left(x\frac{\mathrm{d}}{\mathrm{d}x}J_n(\lambda x)\right)^2 + (\lambda^2 x^2 - n^2)\frac{\mathrm{d}}{\mathrm{d}x}\left(J_n(\lambda x)\right)^2 = 0.$$

Furthermore,

$$\lambda^2 x^2 \frac{\mathrm{d}}{\mathrm{d}x}\left(J_n(\lambda x)\right)^2 = \frac{\mathrm{d}}{\mathrm{d}x}\left[\lambda^2 x^2 \left(J_n(\lambda x)\right)^2\right] - 2\lambda^2 x \left(J_n(\lambda x)\right)^2.$$

Thus

$$\frac{\mathrm{d}}{\mathrm{d}x}\left[\left(x\frac{\mathrm{d}}{\mathrm{d}x}J_n(\lambda x)\right)^2 + \lambda^2 x^2 \left(J_n(\lambda x)\right)^2 - n^2 \left(J_n(\lambda x)\right)^2\right] = 2\lambda^2 x \left(J_n(\lambda x)\right)^2.$$

$$(4.50)$$

From (4.38), we have

$$x \frac{\mathrm{d}}{\mathrm{d}x} J_n(x) = n J_n(x) - x J_{n+1}(x).$$

Replace x by λx leading to

$$x \frac{\mathrm{d}}{\mathrm{d}x} J_n(\lambda x) = n J_n(\lambda x) - \lambda x J_{n+1}(\lambda x). \tag{4.51}$$

Square it,

$$\left(x \frac{\mathrm{d}}{\mathrm{d}x} J_n(\lambda x) \right)^2 = n^2 J_n^2(\lambda x) - 2n\lambda x J_n(\lambda x) J_{n+1}(\lambda x) + \lambda^2 x^2 J_{n+1}^2(\lambda x),$$

and substitute it into (4.50)

$$\frac{\mathrm{d}}{\mathrm{d}x} \left[\lambda^2 x^2 J_{n+1}^2(\lambda x) + \lambda^2 x^2 J_n^2(\lambda x) - 2n\lambda x J_n(\lambda x) J_{n+1}(\lambda x) \right] = 2\lambda^2 x \left(J_n(\lambda x) \right)^2.$$

By integrating with respect to x, we obtain

$$[\lambda^2 x^2 J_{n+1}^2(\lambda x) + \lambda^2 x^2 J_n^2(\lambda x) - 2n\lambda x J_n(\lambda x) J_{n+1}(\lambda x)]_0^c = 2\lambda^2 \int_0^c x (J_n(\lambda x))^2 \mathrm{d}x,$$

or

$$\int_0^c x \left(J_n(\lambda x) \right)^2 \mathrm{d}x = \frac{c^2}{2} \left[J_{n+1}^2(\lambda c) + J_n^2(\lambda c) \right] - \frac{nc}{\lambda} J_n(\lambda c) J_{n+1}(\lambda c). \tag{4.52}$$

Now if the boundary condition is such that $B = 0$ in (4.49), then

$$J_n(\lambda_{nj} c) = 0.$$

In this case, the normalization constant is given by

$$\beta_{nj}^2 = \int_0^c x \left(J_n(\lambda_{nj} x) \right)^2 \mathrm{d}x = \frac{1}{2} c^2 J_{n+1}^2(\lambda_{nj} c), \qquad B = 0. \tag{4.53}$$

If $B \neq 0$, then (4.49) can be written as

$$\frac{A}{B} J_n(\lambda c) = \left. \frac{\mathrm{d} J_n(\lambda x)}{\mathrm{d}x} \right|_{x=c}$$

which, according to (4.51), is given by

$$\left. \frac{\mathrm{d} J_n(\lambda x)}{\mathrm{d}x} \right|_{x=c} = \frac{n}{c} J_n(\lambda c) - \lambda J_{n+1}(\lambda c).$$

It follows that

$$J_{n+1}(\lambda c) = \frac{1}{\lambda}\left(\frac{n}{c} - \frac{A}{B}\right)J_n(\lambda c).$$ (4.54)

Putting it into (4.52), we have

$$\int_0^c x\left(J_n(\lambda x)\right)^2 dx = \frac{c^2}{2}J_n^2(\lambda c)\left[1 + \frac{1}{\lambda^2}\left(\frac{n}{c} - \frac{A}{B}\right)^2\right] - \frac{nc}{\lambda^2}\left(\frac{n}{c} - \frac{A}{B}\right)J_n^2(\lambda c)$$

$$= \frac{1}{2\lambda^2}J_n^2(\lambda c)\left[(\lambda c)^2 - n^2 + \left(\frac{Ac}{B}\right)^2\right].$$

Therefore, in the case $\lambda = \lambda j$, where λ_j is the root of (4.54),

$$\beta_{nj}^2 = \int_0^c x\left(J_n(\lambda_j x)\right)^2 dx = \frac{1}{2\lambda_j^2}J_n^2(\lambda_j c)\left[(\lambda_j c)^2 - n^2 + \left(\frac{Ac}{B}\right)^2\right], \quad B \neq 0.$$

(4.55)

4.5 Other Kinds of Bessel Functions

4.5.1 Modified Bessel Functions

Besides the Bessel equation of order n, one also encounter the modified Bessel equation

$$x^2 y'' + xy' - \left(x^2 + n^2\right)y = 0.$$

The only difference between this equation and the Bessel equation is the minus sign in front of the second x^2 term. If we change x to ix, this equation reduces to the Bessel equation. Therefore $J_n(ix)$ and $N_n(ix)$ are solutions of this equation.

It is customary to define

$$I_n(x) = i^{-n}J_n(ix)$$

as the modified Bessel function of the first kind. Since

$$J_n(ix) = \sum_{k=0}^{\infty}\frac{(-1)^k}{k!\,\Gamma(k+n+1)}\left(\frac{ix}{2}\right)^{2k+n},$$

the modified Bessel function $I_n(x)$

$$I_n(x) = \frac{1}{i^n}J_n(ix) = \sum_{k=0}^{\infty}\frac{1}{k!\,\Gamma(k+n+1)}\left(\frac{x}{2}\right)^{2k+n}$$

is a real and monotonically increasing function.

The second modified Bessel functions of the second kind are usually defined as $K_n(x)$

$$K_n(x) = \frac{\pi}{2} i^{n+1} H_n^{(1)}(ix) = \frac{\pi}{2} i^{n+1} [J_n(ix) + i N_n(ix)].$$

These functions do not have multiple zeros and are not orthogonal functions. They should be compared with $\sinh(x) = -i\sin(ix)$ and $\cosh(x) = \cos(ix)$. Because of this analogy, $I_n(x)$ and $K_n(x)$ are also called hyperbolic Bessel functions. The i factors are adjusted to make them real for real x. The first three $I_n(x)$ and $K_n(x)$ are shown in Fig. 4.4. Note that the first modified Bessel functions I_n are well behaved at the origin but diverge at infinity, the second modified Bessel functions K_n diverge at the origin but well behaved at infinity.

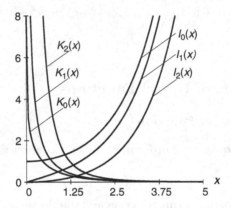

Fig. 4.4. Modified Bessel functions. The functions $I_n(x)$ diverge as $x \to \infty$, and the functions $K_n(x)$ diverge at the origin

4.5.2 Spherical Bessel Functions

The equation

$$x^2 y'' + 2xy' + [x^2 - l(l+1)]y = 0 \tag{4.56}$$

with integer l, arises as the radial part of the wave equation in spherical coordinates. It is called spherical Bessel's equation because it can be transformed into a Bessel equation by a change of variables. Let

$$y = \frac{1}{(x)^{1/2}} z(x).$$

So

$$y' = \frac{z'}{x^{1/2}} - \frac{1}{2}\frac{z}{x^{3/2}},$$

$$y'' = \frac{z''}{x^{1/2}} - \frac{z'}{x^{3/2}} + \frac{3}{4}\frac{z}{x^{5/2}}.$$

Substituting them into (4.56) and multiplying it by $x^{1/2}$, we have

$$x^2[z'' - \frac{1}{x}z' + \frac{3}{4}\frac{1}{x^2}z] + 2x[z' - \frac{1}{2}\frac{1}{x}z] + [x^2 - l(l+1)]z = 0$$

or

$$x^2 z'' + x z' + [x^2 - l(l+1) - \frac{1}{4}]z = 0. \tag{4.57}$$

Since

$$l(l+1) + \frac{1}{4} = (l+\frac{1}{2})^2,$$

clearly (4.57) is a Bessel equation of the order $l + 1/2$, therefore

$$z(x) = C_1 J_{l+1/2}(x) + C_2 J_{-(l+1/2)}(x),$$

and

$$y(x) = C_1 \sqrt{\frac{1}{x}} J_{l+1/2}(x) + C_2 \sqrt{\frac{1}{x}} J_{-(l+1/2)}(x).$$

The two linearly independent solutions are known as spherical Bessel function $j_l(x)$ and spherical Neumann function $n_l(x)$. They are, respectively, defined as

$$j_l(x) = \sqrt{\frac{\pi}{2x}} J_{l+1/2}(x),$$

$$n_l(x) = \sqrt{\frac{\pi}{2x}} N_{l+1/2}(x).$$

Since

$$N_{l+1/2}(x) = \frac{\cos\left[(l+1/2)\pi\right]J_{l+1/2}(x) - J_{-(l+1/2)}(x)}{\sin\left[(l+1/2)\pi\right]} = (-1)^{l+1}J_{-(l+1/2)}(x),$$

so $n_l(x)$ can also be written as

$$n_l(x) = (-1)^{l+1}\sqrt{\frac{\pi}{2x}} J_{-(l+1/2)}(x).$$

The spherical Hankel functions are defined as

$$h_l^{(1)}(x) = j_l(x) + i n_l(x),$$

$$h_l^{(2)}(x) = j_l(x) - i n_l(x).$$

All these functions can be expressed in closed forms of elementary functions. In (4.27) and (4.28), we have showed that

$$J_{1/2}(x) = \sqrt{\frac{2}{\pi x}} \sin x, \qquad J_{-(1/2)}(x) = \sqrt{\frac{2}{\pi x}} \cos x.$$

With the recurrence relation (4.34)

$$J_{n+1}(x) = \frac{2n}{x} J_n(x) - J_{n-1}(x),$$

we can generate all Bessel functions of half-integer orders. For example, with $n = 1/2$, we have

$$J_{3/2}(x) = \frac{1}{x} J_{1/2}(x) - J_{-1/2}(x) = \frac{1}{x}\sqrt{\frac{2}{\pi x}} \sin x - \sqrt{\frac{2}{\pi x}} \cos x.$$

With $n = -1/2$, we have

$$J_{1/2}(x) = \frac{-1}{x} J_{-1/2}(x) - J_{-3/2}(x),$$

or

$$J_{-3/2}(x) = \frac{-1}{x} J_{-1/2}(x) - J_{1/2}(x) = -\frac{1}{x}\sqrt{\frac{2}{\pi x}} \cos x - \sqrt{\frac{2}{\pi x}} \sin x.$$

Therefore,

$$j_0(x) = \sqrt{\frac{\pi}{2x}} J_{1/2}(x) = \frac{1}{x} \sin x,$$

$$j_1(x) = \sqrt{\frac{\pi}{2x}} J_{3/2}(x) = \frac{1}{x^2} \sin x - \frac{1}{x} \cos x,$$

and

$$n_0(x) = -\sqrt{\frac{\pi}{2x}} J_{-1/2}(x) = -\frac{1}{x} \cos x,$$

$$n_1(x) = \sqrt{\frac{\pi}{2x}} J_{-3/2}(x) = -\frac{1}{x^2} \cos x - \frac{1}{x} \sin x.$$

The higher order of spherical Bessel functions can be generated by the recurrence relation

$$f_{l+1} = \frac{2l+1}{x} f_l - f_{l-1}, \tag{4.58}$$

where f_l can be j_l, n_l, $h_l^{(1)}$, or $h_l^{(2)}$. This recurrence relation is obtained by multiplying (4.34) by $\sqrt{\pi/(2x)}$ and set $n = l + 1/2$.

The asymptotic expressions of j_l, n_l, $h_l^{(1)}$, and $h_l^{(2)}$ are of considerable interests. In the asymptotic region, the $1/x$ term dominates all other terms. Since $j_0(x) = (1/x)\sin x$, and

$$\lim_{x \to \infty} j_1(x) \to -\frac{1}{x}\cos x = \frac{1}{x}\sin\left(x - \frac{\pi}{2}\right),$$

so for $l = 0$ and $l = 1$, asymptotically we can write

$$j_l(x) \to \frac{1}{x}\sin\left(x - \frac{l\pi}{2}\right), \qquad l = 0, 1.$$

For higher order of l, we see from the recurrence relation (4.58) that

$$\lim_{x \to \infty} j_{l+1}(x) = \lim_{x \to \infty} [-j_{l-1}(x)] \to -\frac{1}{x}\sin\left(x - \frac{(l-1)\pi}{2}\right)$$

$$= \frac{1}{x}\sin\left(x - \frac{(l-1)\pi}{2} - \pi\right) = \frac{1}{x}\sin\left(x - \frac{(l+1)\pi}{2}\right).$$

Therefore for any integer l, asymptotically

$$j_l(x) \to \frac{1}{x}\sin\left(x - \frac{l\pi}{2}\right).$$

Similarly, we can show that asymptotically

$$n_l(x) \to -\frac{1}{x}\cos\left(x - \frac{l\pi}{2}\right),$$

for all l. Furthermore,

$$h_l^{(1)}(x) \to \frac{1}{x}\sin\left(x - \frac{l\pi}{2}\right) + i\frac{-1}{x}\cos\left(x - \frac{l\pi}{2}\right) = \frac{1}{x}e^{i(x - ((l+1)/2)\pi)},$$

$$h_l^{(2)}(x) \to \frac{1}{x}\sin\left(x - \frac{l\pi}{2}\right) - i\frac{-1}{x}\cos\left(x - \frac{l\pi}{2}\right) = \frac{1}{x}e^{-i(x - ((l+1)/2)\pi)}.$$

These asymptotic expressions are very useful for scattering problems.

We should mention that from $j_0(x)$, $j_1(x)$, and (4.58), one can show with mathematical induction that

$$j_l(x) = x^l \left(-\frac{1}{x}\frac{d}{dx}\right)^l \frac{\sin x}{x}.$$

Similarly,

$$n_l(x) = x^l \left(-\frac{1}{x}\frac{d}{dx}\right)^l \frac{\cos x}{x}.$$

These are known as Rayleigh's formulas.

4.6 Legendre Functions

The differential equation

$$(1 - x^2)\frac{d^2}{dx^2}y - 2x\frac{d}{dx}y + \lambda y = 0 \qquad (4.59)$$

is known as Legendre's equation, after the French mathematician Adrien-Marie Legendre (1785–1833). This equation occurs in numerous physical applications. The solution of this equation is called Legendre function which is one of the most important special functions.

Legendre function ariss in the solution of partial differential equations when the Laplacian is expressed in the spherical coordinates. In that usage, the variable x is the cosine of the polar angle $(x = \cos\theta)$. Therefore x is limited in the range of $-1 \le x \le 1$.

4.6.1 Series Solution of Legendre Equation

We start with the series solution of (4.59)

$$y(x) = x^k \sum_{n=0}^{\infty} a_n x^n, \quad (a_0 \ne 0). \qquad (4.60)$$

$$\frac{dy}{dx} = \sum_{n=0}^{\infty} a_n (n+k)x^{n+k-1},$$

$$\frac{d^2 y}{dx^2} = \sum_{n=0}^{\infty} a_n (n+k)(n+k-1)x^{n+k-2}.$$

Put them into (4.59), we have

$$(1 - x^2)\sum_{n=0}^{\infty} a_n (n+k)(n+k-1)x^{n+k-2} - 2x\sum_{n=0}^{\infty} a_n (n+k)x^{n+k-1}$$

$$+ \lambda \sum_{n=0}^{\infty} a_n x^{n+k} = 0,$$

or

$$\sum_{n=0}^{\infty} a_n (n+k)(n+k-1)x^{n+k-2} - \sum_{n=0}^{\infty} a_n (n+k)(n+k-1)x^{n+k}$$

$$- 2\sum_{n=0}^{\infty} a_n (n+k)x^{n+k} + \lambda \sum_{n=0}^{\infty} a_n x^{n+k} = 0.$$

Combining the last three summations

$$\sum_{n=0}^{\infty} a_n (n+k)(n+k-1)x^{n+k-2} - \sum_{n=0}^{\infty} a_n [(n+k)(n+k-1)+2(n+k)-\lambda]x^{n+k} = 0.$$

Explicitly writing out the first two terms in the first summation, amd using

$$(n+k)(n+k-1) + 2(n+k) - \lambda = (n+k)(n+k+1) - \lambda,$$

we have

$$a_0 k(k-1)x^{k-2} + a_1(k+1)kx^{k-1} + \sum_{n=2} a_n(n+k)(n+k-1)x^{n+k-2}$$
$$- \sum_{n=0} a_n[(n+k)(n+k+1) - \lambda]x^{n+k} = 0. \tag{4.61}$$

Shifting the index by 2 units, we can write the summation starting with $n=2$ as

$$\sum_{n=2} a_n(n+k)(n+k-1)x^{n+k-2} = \sum_{n=0} a_{n+2}(n+k+2)(n+k+1)x^{n+k}.$$

Thus (4.61) becomes

$$a_0 k(k-1)x^{k-2} + a_1(k+1)kx^{k-1}+$$
$$\sum_{n=0} \{a_{n+2}(n+k+2)(n+k+1) - a_n[(n+k)(n+k+1) - \lambda]\} x^{n+k} = 0,$$

which implies

$$a_0 k(k-1) = 0, \tag{4.62}$$
$$a_1(k+1)k = 0, \tag{4.63}$$
$$a_{n+2}(n+k+2)(n+k+1) - a_n[(n+k)(n+k+1) - \lambda] = 0. \tag{4.64}$$

Since $a_0 \neq 0$, from (4.62) we have $k=0$ or $k=1$. If $k=1$, then from (4.63) a_1 must be 0. If $k=0$, a_1 can either be 0, or not equal to 0. So we have three cases:

case 1 : $k=0$ and $a_1 = 0$,
case 2 : $k=1$ and $a_1 = 0$,
case 3 : $k=0$ and $a_1 \neq 0$.

We will first take up case 1. Since $k=0$, (4.64) becomes

$$a_{n+2}(n+2)(n+1) - a_n[n^2 + n - \lambda] = 0,$$

or

$$a_{n+2} = \frac{n(n+1) - \lambda}{(n+2)(n+1)} a_n. \tag{4.65}$$

Since $a_1 = 0$, from this equation, it is seen that $a_3 = a_5 = \cdots = 0$. Therefore we have only even terms left. Let us write this equation as

$$a_{n+2} = f(n)a_n,$$

where

$$f(n) = \frac{n(n+1) - \lambda}{(n+2)(n+1)}. \tag{4.66}$$

$$y(x) = a_0(1 + f(0)x^2 + f(2)f(0)x^4 + f(4)f(2)f(0)x^6 + \cdots), \tag{4.67}$$

which is an infinite series. We must now determine if this is a converging series. For this purpose, it is helpful to write the series in the form of

$$y(x) = \sum_{j=0}^{\infty} b_j(x^2)^j,$$

where

$$b_0 = a_0,$$
$$b_1 = a_0 f(0),$$
$$b_j = a_0 f(2j - 2)f(2j - 4) \cdots f(0)$$
$$b_{j+1} = a_0 f(2j)f(2j - 2) \cdots f(0).$$

For the ratio test, we examine

$$R = \lim_{j \to \infty} \frac{b_{j+1}(x^2)^{j+1}}{b_j(x^2)^j} = \lim_{j \to \infty} f(2j)x^2.$$

By (4.66),

$$f(2j) = \frac{2j(2j + 1) - \lambda}{(2j + 2)(2j + 1)}, \tag{4.68}$$

so

$$\lim_{j \to \infty} f(2j)x^2 = x^2.$$

Therefore for $-1 < x < 1$, the series converges. However, for $x = 1$, the ordinary ratio test gives no information. We must go to the second-order ratio test (also known as Gauss's test or Raabe's test) which says with

$$R \to 1 - \frac{s}{j}, \tag{4.69}$$

if $s > 1$, the series converges, if $s \leq 1$, the series diverges. This test is based on comparison with the Riemann Zeta function $\xi(s)$, defined as

$$\xi(s) = \sum_{j=1}^{\infty} \frac{1}{j^s}.$$

Since

$$\int_1^\infty x^{-s}\, dx = \begin{cases} \left.\dfrac{x^{-s+1}}{-s+1}\right|_1^\infty & s \neq 1 \\ \left.\ln x\right|_1^\infty & s = 1 \end{cases},$$

The integral is divergent for $s \leq 1$ and convergent for $s > 1$. Thus, according to integral test, $\xi(s)$ is divergent for $s \leq 1$ and convergent for $s > 1$. For $\xi(s)$,

$$R = \lim_{j \to \infty} \frac{1/(j+1)^s}{1/j^s} = \lim_{j \to \infty} \frac{j^s}{(j+1)^s} = \lim_{j \to \infty} \left(\frac{j+1}{j}\right)^{-s} \to 1 - \frac{s}{j},$$

which is in the form of (4.69). It can be shown that the same convergence criteria can be applied to any series that asymptotically behaves in the same way as the Riemann Zeta function $\xi(s)$ (see, for example, John M.H. Olmsted, Advance Calculus, Prentice Hall, 1961).

Now in (4.68)

$$R = \lim_{j \to \infty} f(2j) = \frac{j}{j+1} = \frac{j}{j(1+1/j)} = 1 - \frac{1}{j} \cdots .$$

Since $s = 1$, the series will diverge. Therefore for an arbitrary λ, the solution will not be bounded at $x = \pm 1$. However, if

$$\lambda = l(l+1), \quad l = \text{even integer},$$

then the series will terminate and become a polynomial and we will not have the convergence problem. It is clear from (4.65)

$$a_{n+2} = \frac{n(n+1) - l(l+1)}{(n+2)(n+1)} a_n, \tag{4.70}$$

that $a_{l+2} = 0$. It follows that $a_{l+4} = a_{l+6} = \cdots = 0$. For example, if $l = 0$, then $a_2 = a_4 = \cdots = 0$. The solution, according to (4.60), is simply $y = a_0$. For any l, the solution can be systematically generated from (4.66) and (4.67). Since

$$f(n) = \frac{n(n+1) - l(l+1)}{(n+2)(n+1)}, \tag{4.71}$$

$$y(x) = a_0(1 + f(0)x^2 + f(2)f(0)x^4 + f(4)f(2)f(0)x^6 + \cdots),$$

thus

$l = 0: \; f(0) = 0, \quad y = a_0.$ $\qquad\qquad$ (4.72)

$l = 2: \; f(0) = -3; \; f(2) = 0, \quad y = a_0(1 - 3x^2).$ $\qquad$ (4.73)

$l = 4: \; f(0) = -10; \; f(2) = -\dfrac{7}{6}; \; f(4) = 0, \quad y = a_0(1 - 10x^2 + \dfrac{70}{6}x^4).$

For case 2, $k = 1$, $a_1 = 0$, (4.64) becomes

$$a_{n+2}(n+3)(n+2) - a_n[(n+1)(n+2) - \lambda] = 0,$$

or

$$a_{n+2} = \frac{(n+1)(n+2) - \lambda}{(n+3)(n+2)} a_n. \tag{4.74}$$

Again because of $a_1 = 0$ and this recurrence relation, all a_n with odd n will be zero. The solution is then given by

$$y = x(a_0 + a_2 x^2 + a_4 x^4 + \cdots)$$
$$= a_0 x + a_2 x^3 + a_4 x^5 + \cdots, \tag{4.75}$$

which is a series with odd powers of x. With a similar argument as in case 1, one can show this series diverges for $x = 1$, unless

$$\lambda = l(l+1) \quad l = \text{odd integer}.$$

In that case,

$$a_{n+2} = \frac{(n+1)(n+2) - l(l+1)}{(n+3)(n+2)} a_n = f(n+1)a_n,$$

and

$$y = a_0[x + f(1)x^3 + f(3)f(1)x^5 + f(5)f(3)f(1)x^7 + \cdots].$$

It follows

$$l = 1, \quad f(1) = 0, \quad y = a_0 x, \tag{4.76}$$

$$l = 3, \quad f(1) = -\frac{5}{3}; \quad f(3) = 0, \quad y = a_0\left(x - \frac{5}{3}x^3\right). \tag{4.77}$$

For case 3, $k = 0$, $a_1 \neq 0$. Since a_0 is not equal to zero, it can be easily shown that the solution is the sum of two infinite series, one with even powers of x, and the other with odd powers of x. In this case, the series solution diverges at $x = \pm 1$.

4.6.2 Legendre Polynomials

In the last section we found that the solution of the Legendre Equation

$$\frac{d}{dx}(1 - x^2)\frac{d}{dx}y(x) + l(l+1)y(x) = 0$$

is given by a polynomial of order l. Furthermore the polynomial contains only even orders of x if l is even, and only odd orders of x if l is odd. These solutions are determined up to a multiplicative constant a_0. Now, by convention, if a_0 is chosen in such a way that $y(1) = 1$, then these polynomials are known as Legendre polynomials $P_l(x)$. For example, from (4.73)

$$l = 2, \quad y_2(x) = a_0(1 - 3x^2),$$

requiring

$$y_2(1) = a_0(-2) = 1$$

we have to choose

$$a_0 = -\frac{1}{2}.$$

With this choice $y_2(x)$ is known as $P_2(x)$, that is

$$y_2(x) = -\frac{1}{2}(1 - 3x^2) = \frac{1}{2}(3x^2 - 1) = P_2(x).$$

Similarly, from (4.77)

$$l = 3, \quad y_3(x) = a_0\left(x - \frac{5}{3}x^3\right),$$

with

$$a_0 = -\frac{3}{2}, \quad y_3(1) = 1.$$

Therefore

$$P_3(x) = \frac{1}{2}(5x^3 - 3x).$$

The first few Legendre polynomials are shown in Fig. 4.5 and they are listed as follows

$$P_0(x) = 1, \quad P_1(x) = x, \quad P_2(x) = \frac{1}{2}(3x^2 - 1)$$

$$P_3(x) = \frac{1}{2}(5x^3 - 3x), \quad P_4(x) = \frac{1}{8}(35x^4 - 30x^2 + 3)$$

$$P_5(x) = \frac{1}{8}(63x^5 - 70x^3 + 15x).$$

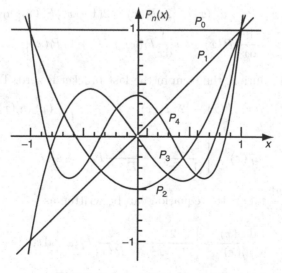

Fig. 4.5. Legendre polynomials

Clearly
$$P_l(1) = 1, \quad \text{for all } l.$$

If l is even, then $P_l(x)$ is an even function, symmetric with respect to zero, and if l is odd, then $P_l(x)$ is an odd function, antisymmetric with respect to zero. That is
$$P_l(-x) = (-1)^l P_l(x),$$
and in particular
$$P_l(-1) = (-1)^l.$$

4.6.3 Legendre Functions of the Second Kind

By choosing $\lambda = l(l+1)$, we force one of the two infinite series to become a polynomial. The second solution is still an infinite series. This infinite series can also be expressed in a closed form, although it diverges at $x = \pm 1$.

Another way to obtain the second solution is the so called "method of reduction of order." With $P_l(x)$ being a solution of the Legendre equation of order l, we write the second solution as
$$y_2(x) = u_l(x) P_l(x).$$

Requiring $y_2(x)$ to satisfy the Legendre equation of the same order, we can determine $u_l(x)$. Substituting $y_2(x)$ into the Legendre equation, we have
$$\left(1 - x^2\right) \frac{\mathrm{d}^2}{\mathrm{d}x^2} [u_l(x) P_l(x)] - 2x \frac{\mathrm{d}}{\mathrm{d}x} [u_l(x) P_l(x)] + l(l+1) [u_l(x) P_l(x)] = 0.$$

This equation can be written as
$$\left(1 - x^2\right) P_l(x) u_l''(x) - \left[2x P_l(x) - 2(1 - x^2) P_l'(x)\right] u_l'(x)$$
$$+ \left[\left(1 - x^2\right) \frac{\mathrm{d}^2}{\mathrm{d}x^2} P_l(x) - 2x \frac{\mathrm{d}}{\mathrm{d}x} P_l(x) + l(l+1) P_l(x)\right] u_l(x) = 0.$$

Since $P_l(x)$ is a solution, the term in the last bracket is zero. Therefore
$$\left(1 - x^2\right) P_l(x) u_l''(x) - \left[2x P_l(x) - 2(1 - x^2) P_l'(x)\right] u_l'(x) = 0,$$
or
$$u_l''(x) = \left[\frac{2x}{(1 - x^2)} - \frac{2}{P_l(x)} P_l'(x)\right] u_l'(x).$$

Since $u_l''(x) = \dfrac{\mathrm{d}}{\mathrm{d}x} u_l'(x)$, this equation can be written as
$$\frac{\mathrm{d}u_l'(x)}{u_l'(x)} = \left[\frac{2x}{(1 - x^2)} - \frac{2}{P_l(x)} P_l'(x)\right] \mathrm{d}x.$$

Integrating both sides, we have

$$\ln u_l'(x) = -\ln\left(1 - x^2\right) - 2\ln P_l(x) + C,$$

or

$$u_l'(x) = \frac{c}{(1 - x^2) P_l^2(x)}.$$

Thus

$$u_l(x) = \int^x \frac{c\, dx'}{(1 - x'^2) P_l^2(x')}.$$

The arbitrary constant c contributes nothing to the property of the function. With c chosen to be 1, the Legendre function of the second kind is designated as $Q_l(x)$.

With $P_0(x) = 1$,

$$u_0(x) = \int^x \frac{dx'}{(1 - x'^2)} = \frac{1}{2}\int^x \left(\frac{dx'}{1 + x'} + \frac{dx'}{1 - x'}\right) = \frac{1}{2}\ln\frac{1 + x}{1 - x}.$$

The additive constant of this integral is chosen to be zero for convenience. With this choice

$$Q_0(x) = u_0(x) P_0(x) = \frac{1}{2}\ln\frac{1 + x}{1 - x}. \tag{4.78}$$

Similarly, with $P_1(x) = x$,

$$u_1(x) = \int^x \frac{dx'}{(1 - x'^2)\, x'^2} = \int^x \left(\frac{dx'}{1 - x'^2} + \frac{dx'}{x'^2}\right) = \frac{1}{2}\ln\frac{1 + x}{1 - x} - \frac{1}{x},$$

and

$$Q_1(x) = u_1(x) P_1(x) = \frac{x}{2}\ln\frac{1 + x}{1 - x} - 1 = P_1(x) Q_0(x) - 1.$$

Higher order $Q_l(x)$ can be obtained with the recurrence relation

$$l Q_l(x) = (2l - 1)\, x Q_{l-1}(x) - (l - 1)\, Q_{l-2}(x),$$

which is also satisfied by P_l, as we shall show in Sect. 4.7. The first few $Q_l(x)$ are as follows:

$$Q_0 = \frac{1}{2}\ln\frac{1 + x}{1 - x}, \quad Q_1 = P_1 Q_0 - 1, \quad Q_2 = P_2 Q_0 - \frac{3}{2}x,$$

$$Q_3 = P_3 Q_0 - \frac{5}{2}x^2, \quad Q_4 = P_4 Q_0 - \frac{35}{8}x^3 + \frac{55}{24}x,$$

$$Q_5 = P_5 Q_0 - \frac{63}{8}x^4 + \frac{49}{8}x^2 - \frac{8}{15}.$$

It can be shown that these expressions are the infinite series solutions obtained from the Frobenius method. For example, the coefficients in the second solution of (4.75)

$$y = a_0 x + a_2 x^3 + a_4 x^5 + \cdots,$$

are given in (4.74)

$$a_{n+2} = \frac{(n+1)(n+2) - \lambda}{(n+3)(n+2)} a_n.$$

With $\lambda = 0$,

$$a_{n+2} = \frac{(n+1)}{(n+3)} a_n.$$

Therefore

$$y = a_0 \left(x + \frac{1}{3}x^3 + \frac{1}{5}x^5 + \cdots \right) = \frac{a_0}{2} \ln \frac{1+x}{1-x}.$$

This expression is identical to $a_0 Q_0(x)$.

Therefore the general solution of the Legendre equation for integer order l is

$$y(x) = c_1 P_l(x) + c_2 Q_l(x),$$

where $P_l(x)$ is a polynomial which converges for all x, and $Q_l(x)$ diverges at $x = \pm 1$.

4.7 Properties of Legendre Polynomials

4.7.1 Rodrigues' Formula

The Legendre polynomials can be summarized by the Rodrigues formula

$$P_l(x) = \frac{1}{2^l l!} \frac{d^l}{dx^l} (x^2 - 1)^l. \qquad (4.79)$$

Clearly this formula will give a polynomial of order l. We will prove this polynomial is the Legendre polynomial. There are two parts of the proof. First we will show that $d^l/dx^l (x^2 - 1)^l$ satisfies the Legendre equation. Then we will show that $P_l(1)$ with $P_l(x)$ given by (4.79) is equal to 1.

Before we start the proof, let us first recall the Leibniz's rule for differentiating products:

$$\frac{d^m}{dx^m} A(x)B(x) = \sum_{k=0}^{m} \frac{m!}{k!(m-k)!} \frac{d^{m-k}}{dx^{m-k}} A(x) \frac{d^k}{dx^k} B(x)$$

$$= \left[\frac{d^m}{dx^m} A(x) \right] B(x) + m \left[\frac{d^{m-1}}{dx^{m-1}} A(x) \right] \left[\frac{d}{dx} B(x) \right] + \cdots + A(x) \left[\frac{d^m}{dx^m} B(x) \right].$$

$$(4.80)$$

To prove the first part of (4.79), let

$$v = (x^2 - 1)^l,$$

$$\frac{dv}{dx} = l(x^2 - 1)^{l-1}2x$$

$$(x^2 - 1)\frac{dv}{dx} = l(x^2 - 1)^l 2x = 2lxv. \tag{4.81}$$

Differentiating the left-hand side of this equation $l + 1$ times by Leibniz' rule [with $A(x) = \dfrac{dv}{dx}$ and $B(x) = (x^2 - 1)$]:

$$\frac{d^{l+1}}{dx^{l+1}}(x^2 - 1)\frac{dv}{dx} = (x^2 - 1)\frac{d^{l+2}v}{dx^{l+2}} + (l+1)2x\frac{d^{l+1}v}{dx^{l+1}} + \frac{(l+1)l}{2!}2\frac{d^l v}{dx^l},$$

and differentiating the right-hand side of (4.81) $l + 1$ times by Leibniz' rule [with $A(x) = v$ and $B(x) = 2lx$]:

$$\frac{d^{l+1}}{dx^{l+1}}2lvx = 2lx\frac{d^{l+1}v}{dx^{l+1}} + (l+1)2l\frac{d^l v}{dx^l},$$

we have

$$(x^2 - 1)\frac{d^{l+2}v}{dx^{l+2}} + (l+1)2x\frac{d^{l+1}v}{dx^{l+1}} + \frac{(l+1)l}{2!}2\frac{d^l v}{dx^l} = 2lx\frac{d^{l+1}v}{dx^{l+1}} + (l+1)2l\frac{d^l v}{dx^l}.$$

Simplifying this equation, we get

$$(x^2 - 1)\frac{d^{l+2}v}{dx^{l+2}} + 2x\frac{d^{l+1}v}{dx^{l+1}} - l(l+1)\frac{d^l v}{dx^l} = 0,$$

or

$$\frac{d}{dx}\left\{(1 - x^2)\frac{d}{dx}\left[\frac{d^l v}{dx^l}\right]\right\} + l(l+1)\left[\frac{d^l v}{dx^l}\right] = 0.$$

Clearly

$$\frac{d^l v}{dx^l} = \frac{d^l}{dx^l}(x^2 - 1)^l$$

satisfies the Legendre equation. Now if we can show

$$\left[\frac{1}{2^l l!}\frac{d^l}{dx^l}(x^2 - 1)^l\right]_{x=1} = 1,$$

then

$$P_l(x) = \frac{1}{2^l l!}\frac{d^l}{dx^l}(x^2 - 1)^l$$

must be the Legendre polynomial.

This can be done by writing

$$\frac{d^l}{dx^l}(x^2-1)^l = \frac{d^l}{dx^l}(x-1)^l(x+1)^l$$

and using the Leibniz' rule with $A(x) = (x-1)^l$ and $B(x) = (x+1)^l$

$$\frac{d^l}{dx^l}(x-1)^l(x+1)^l = \left[\frac{d^l}{dx^l}(x-1)^l\right](x+1)^l + l\left[\frac{d^{l-1}}{dx^{l-1}}(x-1)^l\right]\left[\frac{d}{dx}(x+1)^l\right] + \cdots.$$

(4.82)

Notice

$$\frac{d^{l-1}}{dx^{l-1}}(x-1)^l = l!(x-1),$$

at $x = 1$, this term is equal to zero. In fact as long as the number of times $(x-1)^l$ is differentiated is less than l, the result will contain the factor $(x-1)$. At $x = 1$, the derivative is equal to zero. Therefore all terms on the right-hand side of (4.82) except the first term are equal to zero at $x = 1$, thus

$$\left[\frac{d^l}{dx^l}(x-1)^l(x+1)^l\right]_{x=1} = \left\{\left[\frac{d^l}{dx^l}(x-1)^l\right](x+1)^l\right\}_{x=1}.$$

Now

$$\frac{d^l}{dx^l}(x-1)^l = l!, \quad \left[(x+1)^l\right]_{x=1} = 2^l,$$

therefore

$$\frac{1}{2^l l!}\frac{d^l}{dx^l}(x^2-1)^l\bigg|_{x=1} = 1.$$

This completes our proof.

4.7.2 Generating Function of Legendre Polynomials

We will prove a very important identity

$$\frac{1}{\sqrt{1-2xz+z^2}} = \sum_{n=0}^{\infty} P_n(x)z^n, \quad |z| < 1. \tag{4.83}$$

This relation has an advantage of summarizing P_n into a single function. This enables us to derive relationships between Legendre polynomials of different orders without using explicit forms. Besides its many applications in physics, it is also very useful in statistics.

To prove this relationship, we will make use of the fact that the power series

$$f = \frac{1}{\sqrt{1-2xz+z^2}} = \sum_{n=0}^{\infty} F_n(x)z^n \tag{4.84}$$

exists. Then we will show that the coefficient $F_n(x)$ satisfies the Legendre equation, and that $F_n(1) = 1$. That will enable us to identify $F_n(x)$ as $P_n(x)$. Now

$$\frac{\partial f}{\partial x} = -\frac{1}{2}(1 - 2xz + z^2)^{-3/2}(-2z) = zf^3, \tag{4.85}$$

$$\frac{\partial f}{\partial z} = -\frac{1}{2}(1 - 2xz + z^2)^{-3/2}(-2x + 2z) = (x - z)f^3, \tag{4.86}$$

$$f = \frac{f^3}{f^2} = f^3(1 - 2xz + z^2) = (1 - x^2)f^3 + (x - z)^2 f^3.$$

It follows from the last equation

$$(1 - x^2)f^3 = f - (x - z)^2 f^3.$$

Thus

$$(1 - x^2)\frac{\partial f}{\partial x} = (1 - x^2)zf^3 = z\left[f - (x - z)^2 f^3\right]$$

$$= z\left[f - (x - z)\frac{\partial f}{\partial z}\right] = -z\frac{\partial}{\partial z}\left[(x - z)f\right].$$

Differentiate both sides of the last equation with respect to x

$$\frac{\partial}{\partial x}\left[(1 - x^2)\frac{\partial f}{\partial x}\right] = -z\frac{\partial}{\partial z}\frac{\partial}{\partial x}\left[(x - z)f\right] = -z\frac{\partial}{\partial z}\left[f + (x - z)\frac{\partial f}{\partial x}\right]$$

$$= -z\frac{\partial}{\partial z}\left[f + (x - z)zf^3\right] = -z\frac{\partial}{\partial z}\left[f + z\frac{\partial f}{\partial z}\right]$$

$$= -z\frac{\partial}{\partial z}\frac{\partial}{\partial z}(zf) = -z\frac{\partial^2}{\partial z^2}(zf).$$

With the series expansion of f, we have

$$\frac{\partial}{\partial x}\left[(1 - x^2)\frac{\partial}{\partial x}\sum_{n=0}^{\infty}F_n(x)z^n\right] = -z\frac{\partial^2}{\partial z^2}\left[z\sum_{n=0}^{\infty}F_n(x)z^n\right],$$

or

$$\sum_{n=0}^{\infty}\left[\frac{\partial}{\partial x}(1 - x^2)\frac{\partial}{\partial x}F_n(x)\right]z^n = -z\sum_{n=0}^{\infty}F_n(x)\frac{\partial^2}{\partial z^2}z^{n+1}$$

$$= -z\sum_{n=0}^{\infty}F_n(x)(n + 1)nz^{n-1} = -\sum_{n=0}^{\infty}F_n(x)(n + 1)nz^n.$$

Thus

$$\sum_{n=0}^{\infty}\left[\frac{\partial}{\partial x}(1 - x^2)\frac{\partial}{\partial x}F_n(x) + n(n + 1)F_n(x)\right]z^n = 0.$$

Therefore $F_n(x)$ satisfies the Legendre equation

$$\frac{\partial}{\partial x}(1-x^2)\frac{\partial}{\partial x}F_n(x) + n(n+1)F_n(x) = 0.$$

Furthermore, for $x = 1$

$$\frac{1}{\sqrt{1-2z+z^2}} = \sum_{n=0}^{\infty} F_n(1)z^n.$$

Since

$$\frac{1}{\sqrt{1-2z+z^2}} = \frac{1}{\sqrt{(1-z)^2}} = \frac{1}{1-z},$$

and

$$\frac{1}{1-z} = \sum_{n=0}^{\infty} z^n,$$

hence

$$F_n(1) = 1.$$

Thus, (4.83) is established. The left-hand side of that equation is known as the generating function $G(x, z)$ of Legendre polynomials,

$$G(x, z) = \frac{1}{\sqrt{1-2xz+z^2}} = \sum_{n=0}^{\infty} P_n(x)z^n.$$

4.7.3 Recurrence Relations

The following recurrence formulas are very useful in handling Legendre polynomials and their derivatives.

1.
$$(n+1)P_{n+1}(x) - (2n+1)xP_n(x) + nP_{n-1}(x) = 0. \qquad (4.87)$$

Proof:

$$\frac{\partial}{\partial z}\left[(1-2xz+z^2)^{-1/2}\right] = \frac{\partial}{\partial z}\sum_{n=0}^{\infty} P_n(x)z^n,$$

$$(1-2xz+z^2)^{-3/2}(x-z) = \sum_{n=0}^{\infty} P_n(x)nz^{n-1},$$

$$(x-z)\frac{(1-2xz+z^2)^{-1/2}}{1-2xz+z^2} = \sum_{n=0}^{\infty} P_n(x)nz^{n-1},$$

$$(x-z)\sum_{n=0}^{\infty} P_n(x)z^n = (1-2xz+z^2)\sum_{n=0}^{\infty} P_n(x)nz^{n-1},$$

$$\sum_{n=0}^{\infty} x P_n(x) z^n - \sum_{n=0}^{\infty} P_n(x) z^{n+1}$$

$$= \sum_{n=0}^{\infty} n P_n(x) z^{n-1} - \sum_{n=0}^{\infty} 2xn P_n(x) z^n + \sum_{n=0}^{\infty} n P_n(x) z^{n+1}.$$

Collecting the terms, we have

$$\sum_{n=0}^{\infty} n P_n(x) z^{n-1} - \sum_{n=0}^{\infty} (2n+1)x P_n(x) z^n + \sum_{n=0}^{\infty} (n+1) P_n(x) z^{n+1} = 0.$$

Since

$$\sum_{n=0}^{\infty} n P_n(x) z^{n-1} = \sum_{n=1}^{\infty} n P_n(x) z^{n-1} = \sum_{n=0}^{\infty} (n+1) P_{n+1}(x) z^n,$$

$$\sum_{n=0}^{\infty} (n+1) P_n(x) z^{n+1} = \sum_{n=1}^{\infty} n P_{n-1}(x) z^n = \sum_{n=0}^{\infty} n P_{n-1}(x) z^n,$$

thus

$$\sum_{n=0}^{\infty} \left[(n+1) P_{n+1}(x) - (2n+1)x P_n(x) + n P_{n-1}(x) \right] z^n = 0.$$

Hence

$$P_{n+1}(x) = \frac{1}{n+1} \left[(2n+1)x P_n(x) - n P_{n-1}(x) \right]. \tag{4.88}$$

This means as long as we know $P_{n-1}(x)$, $P_n(x)$, we can generate $P_{n+1}(x)$. For example with $P_0(x) = 1$, and $P_1(x) = x$, this recurrence relation with $n = 1$ will give $P_2(x) = 1/2(3x^2 - 1)$. With $P_1(x)$ and $P_2(x)$, we can generate $P_3(x)$ and so on. In other words, just with the first two orders, a Do-Loop in the computer code can automatically generate all orders of the polynomials.

2.

$$n P_n(x) - x P_n'(x) + P_{n-1}'(x) = 0, \quad \text{where} \quad P_n'(x) = \mathrm{d}/\mathrm{d}x P_n(x). \tag{4.89}$$

Proof: Taking the derivative of the generating function with respect to z and x, we get, respectively,

$$\frac{x - z}{(1 - 2xz + z^2)^{3/2}} = \sum_{n=0}^{\infty} n P_n(x) z^{n-1},$$

$$\frac{z}{(1 - 2xz + z^2)^{3/2}} = \sum_{n=0}^{\infty} P_n'(x) z^n.$$

Taking the ratio, we get

$$\frac{x - z}{z} = \frac{\sum_{n=0} n P_n(x) z^{n-1}}{\sum_{n=0} P'_n(x) z^n}$$

It follows that

$$(x - z) \sum_{n=0}^{\infty} P'_n(x) z^n = z \sum_{n=0}^{\infty} P_n(x) n z^{n-1}$$

$$\sum_{n=0}^{\infty} x P'_n(x) z^n - \sum_{n=0}^{\infty} P'_n(x) z^{n+1} = \sum_{n=0}^{\infty} n P_n(x) z^n. \qquad (4.90)$$

Now

$$\sum_{n=0}^{\infty} x P'_n(x) z^n = \sum_{n=1}^{\infty} x P'_n(x) z^n,$$

since $P'_0(x) = 0$, and

$$\sum_{n=0}^{\infty} n P_n(x) z^n = \sum_{n=1}^{\infty} n P_n(x) z^n,$$

since $n = 0$ term is equal to zero. It can also easily be shown that

$$\sum_{n=0}^{\infty} P'_n(x) z^{n+1} = \sum_{n=1}^{\infty} P'_{n-1}(x) z^n.$$

Thus (4.90) can be written as

$$\sum_{n=1}^{\infty} \left[x P'_n(x) - P'_{n-1}(x) - n P_n(x) \right] z^n = 0.$$

It follows

$$x P'_n(x) - P'_{n-1}(x) - n P_n(x) = 0. \qquad (4.91)$$

This relation together with the recurrence relation of (4.87) can generate the derivative of all orders of the Legendre polynomial based on the knowledge of $P_0 = 1$ and $P_1 = x$.

Many other relations can be derived from these two recurrence relations. For example, from (4.88) we have

$$P'_{n+1} = \frac{2n+1}{n+1}(P_n + x P'_n) - \frac{n}{n+1} P'_{n-1},$$

or

$$x P'_n = \frac{n+1}{2n+1} P'_{n+1} - P_n + \frac{n}{2n+1} P'_{n-1}.$$

Putting it into (4.91)

$$\frac{n+1}{2n+1}P'_{n+1} - P_n + \frac{n}{2n+1}P'_{n-1} - P'_{n-1} - nP_n = 0,$$

and collecting terms

$$\frac{n+1}{2n+1}P'_{n+1} - \frac{n+1}{2n+1}P'_{n-1} - (n+1)P_n = 0,$$

we obtain another very important relation

$$P_n = \frac{1}{2n+1}\left[P'_{n+1} - P'_{n-1}\right]. \tag{4.92}$$

This equation enables us to do the following integral

$$\int_x^1 P_n(\zeta)d\zeta = \frac{1}{2n+1}\left[P_{n-1}(x) - P_{n+1}(x)\right].$$

4.7.4 Orthogonality and Normalization of Legendre Polynomials

As we have discussed in the last chapter, the Legendre's equation by itself in the form of

$$\frac{\mathrm{d}}{\mathrm{d}x}\left[(1-x^2)\frac{\mathrm{d}}{\mathrm{d}x}y\right] + \lambda y = 0$$

constitutes a Sturm–Liouville problem in the interval of $-1 \leq x \leq 1$ with a unit weight function. The requirements that solutions must be regular (bounded) at $x = \pm 1$ restrict the acceptable eigenvalues λ to $l(l+1)$, where l is an integer. The corresponding eigenfunctions are the Legendre polynomials $P_l(x)$. Since eigenfunctions of a Sturm–Liouville problem must be orthogonal to each other with respect to the weight function, therefore

$$\int_{-1}^1 P_l(x)P_{l'}(x)\mathrm{d}x = 0, \quad \text{if} \quad l \neq l'.$$

An extension to this fact is that

$$\int_{-1}^1 x^n P_l(x)\mathrm{d}x = 0, \quad \text{if} \quad n < l,$$

since x^n can be expressed as a linear combination of Legendre polynomials, the highest order of which is n.

Any well-behaved function $f(x)$ in the interval $-1 \leq x \leq 1$ can be expanded as a Fourier–Legendre series

$$f(x) = \sum_{n=0}^{\infty} a_n P_n(x),$$

where

$$a_n = \frac{1}{\int_{-1}^{1} P_n^2(x)\mathrm{d}x} \int_{-1}^{1} f(x)P_n(x)\mathrm{d}x.$$

To carry out this expansion, we need to know the value of the normalization integral

$$\beta_n^2 = \int_{-1}^{1} P_n^2(x)\mathrm{d}x.$$

There are many ways to evaluate this integral. The simplest is to make use of (4.92)

$$P_n(x) = \frac{1}{2n+1} \left[P_{n+1}'(x) - P_{n-1}'(x) \right].$$

Multiply both sides by $P_n(x)$ and integrate

$$\int_{-1}^{1} P_n^2(x)\mathrm{d}x = \frac{1}{2n+1} \left[\int_{-1}^{1} P_n(x)P_{n+1}'(x)\mathrm{d}x - \int_{-1}^{1} P_n(x)P_{n-1}'(x)\mathrm{d}x \right].$$

The second integral on the right-hand side is equal to zero because $P_{n-1}'(x)$ is a polynomial of order $n-2$. The first integral on the right-hand side can be written as

$$\int_{-1}^{1} P_n(x)P_{n+1}'(x)\mathrm{d}x = \int_{-1}^{1} P_n(x)\frac{\mathrm{d}P_{n+1}(x)}{\mathrm{d}x}\mathrm{d}x = \int_{-1}^{1} P_n(x)\mathrm{d}P_{n+1}(x)$$

$$= [P_n(x)P_{n+1}(x)]_{-1}^{1} - \int_{-1}^{1} P_{n+1}(x)P_n'(x)\mathrm{d}x.$$

Again, the last integral is zero because $P_n'(x)$ is a polynomial of order $n-1$. The integrated part is

$$[P_n(x)P_{n+1}(x)]_{-1}^{1} = P_n(1)P_{n+1}(1) - P_n(-1)P_{n+1}(-1)$$

$$= 1 - (-1)^n(-1)^{n+1} = 2.$$

Thus

$$\beta_n^2 = \int_{-1}^{1} P_n^2(x)\mathrm{d}x = \frac{2}{2n+1}.$$

4.8 Associated Legendre Functions and Spherical Harmonics

4.8.1 Associated Legendre Polynomials

The equation

$$\frac{\mathrm{d}}{\mathrm{d}x}(1-x^2)\frac{\mathrm{d}}{\mathrm{d}x}y(x) + \left[\lambda - \frac{m^2}{1-x^2}\right]y(x) = 0. \qquad (4.93)$$

is known as the associated Legendre equation. This equation is of Sturm–Liouville form. It becomes an eigenvalue equation for λ if we require that the solution be bounded (finite) at the singular points $x = \pm 1$. If $m = 0$, it reduces to the Legendre equation and $\lambda = l(l+1)$. For m equal to nonzero integer, we could solve this equation with a series expansion as we did for the Legendre equation. However, it is more efficient and interesting to relate the $m \neq 0$ case to the $m = 0$ case.

One way of doing this to start with the regular Legendre equation

$$(1 - x^2)\frac{d^2}{dx^2}P_l(x) - 2x\frac{d}{dx}P_l(x) + l(l+1)P_l(x) = 0 \qquad (4.94)$$

and convert it into the associated Legendre equation by multiple differentiations. With the Leibnitz's formula (4.80), we can write

$$\frac{d^m}{dx^m}\left[(1 - x^2)\frac{d^2}{dx^2}P_l\right] = (1 - x^2)\frac{d^{m+2}}{dx^{m+2}}P_l - 2mx\frac{d^{m+1}}{dx^{m+1}}P_l - m(m-1)\frac{d^m}{dx^m}P_l,$$

$$\frac{d^m}{dx^m}\left[2x\frac{d}{dx}P_l\right] = 2x\frac{d^{m+1}}{dx^{m+1}}P_l + 2m\frac{d^m}{dx^m}P_l.$$

Therefore differentiating (4.94) m times and collecting terms, we have

$$(1-x^2)\frac{d^{m+2}}{dx^{m+2}}P_l - 2x(m+1)\frac{d^{m+1}}{dx^{m+1}}P_l + [l(l+1) - m(m+1)]\frac{d^m}{dx^m}P_l = 0. \quad (4.95)$$

Denoting

$$u = \frac{d^m}{dx^m}P_l(x),$$

the above equation becomes

$$(1 - x^2)u'' - 2x(m+1)u' + [l(l+1) - m(m+1)]u = 0. \qquad (4.96)$$

Now we will show that with

$$u(x) = \left(1 - x^2\right)^{-m/2}y(x),$$

$y(x)$ satisfies the associated Legendre equation. Inserting $u(x)$ and its derivatives

$$u' = (1 - x^2)^{-m/2}y' + \frac{m}{2}(1 - x^2)^{-m/2-1}2xy = \left(y' + \frac{mx}{1 - x^2}y\right)(1 - x^2)^{-m/2},$$

$$u'' = \left[y'' + \frac{2mx}{1 - x^2}y' + \frac{m}{1 - x^2}y + \frac{m(m+2)x^2}{(1 - x^2)^2}y\right](1 - x^2)^{-m/2},$$

into (4.96) and simplify, we see that

$$(1 - x^2)y'' - 2xy' + \left[l(l+1) - \frac{m^2}{1 - x^2}\right]y == 0,$$

which is the associated Legendre equation. Thus

$$y(x) = (1 - x^2)^{m/2} u(x) = (1 - x^2)^{m/2} \frac{\mathrm{d}^m}{\mathrm{d}x^m} P_l(x)$$

must be the solution of the associated Legendre equation. A negative value for m does not change m^2 in the equation, so this solution is also a solution for the corresponding negative m.

This solution is called the associated Legendre function, customarily designated as $P_l^m(x)$. For both positive and negative m, it is defined as

$$P_l^m(x) = (1 - x^2)^{|m|/2} \frac{\mathrm{d}^{|m|}}{\mathrm{d}x^{|m|}} P_l(x). \tag{4.97}$$

Using the Rodrigues formula, it can be written as

$$P_l^m(x) = \frac{1}{2^l l!} (1 - x^2)^{m/2} \frac{\mathrm{d}^{l+m}}{\mathrm{d}x^{l+m}} (x^2 - 1)^l. \tag{4.98}$$

Since the highest power term in $(x^2 - 1)^l$ is x^{2l}, clearly $-l \le m \le l$.

A constant times (4.97) is also a solution, this enables some authors to define P_l^m differently by multiplying a factor $(-1)^m$. Still others use (4.98) to define $P_l^{-m}(x)$ which differs from (4.97) by a constant factor (see exercise 25). Unless otherwise stated, we shall assume that the associated Legendre polynomial is defined in (4.97). The first few polynomials are listed below. In applications most often $x = \cos\theta$, therefore $P_l^m(\cos\theta)$ are also listed side by side,

$$P_1^1(x) = (1 - x^2)^{1/2} \qquad\qquad P_1^1(\cos\theta) = \sin\theta$$
$$P_2^1(x) = 3x(1 - x^2)^{1/2} \qquad\qquad P_2^1(\cos\theta) = 3\cos\theta\sin\theta$$
$$P_2^2(x) = 3(1 - x^2) \qquad\qquad P_2^2(\cos\theta) = 3\sin^2\theta$$
$$P_3^1(x) = (3/2)(5x^2 - 1)(1 - x^2)^{1/2} \quad P_3^1(\cos\theta) = (3/2)(5\cos^2\theta - 1)\sin\theta$$
$$P_3^2(x) = 15x(1 - x^2) \qquad\qquad P_3^2(\cos\theta) = 15\cos\theta\sin^2\theta$$
$$P_3^3(x) = 15(1 - x^2)^{3/2} \qquad\qquad P_3^3(\cos\theta) = 15\sin^3\theta.$$

4.8.2 Orthogonality and Normalization of Associated Legendre Functions

The associated Legendre equation is of the Sturm–Liouville form, therefore its eigenfunctions, the associated Legendre functions, are orthogonal to each other.

Let us recall the standard Sturm–Liouville equation

$$\frac{\mathrm{d}}{\mathrm{d}x}\left[r(x)\frac{\mathrm{d}}{\mathrm{d}x}y\right] + q(x)y + \lambda w(x)y = 0.$$

If $r(a) = r(b) = 0$, then this equation, by itself, is a singular Sturm–Liouville problem in the range of $a \leq x \leq b$, provided the solution is bounded at $x = a$ and $x = b$. This means that its eigenfunctions are orthogonal with respect to the weight function $w(x)$.

With the associated Legendre equation written in the form

$$\frac{d}{dx}\left[(1 - x^2)\frac{d}{dx}P_l^m(x)\right] + l(l+1)P_l^m(x) - \frac{m^2}{1-x^2}P_l^m(x) = 0,$$

we can identify

$$r(x) = x^2 - 1, \quad q(x) = -l(l+1), \quad w(x) = \frac{1}{1-x^2}, \quad \lambda = m^2,$$

and conclude that

$$\int_{-1}^{1} P_l^m(x)P_l^{m'}(x)\frac{1}{1-x^2}dx = 0, \qquad m' \neq m. \tag{4.99}$$

On the other hand, we can identify

$$r(x) = 1 - x^2, \quad q(x) = -\frac{m^2}{1-x^2}, \quad w(x) = 1, \quad \lambda = l(l+1),$$

so we see that $P_l^m(x)$ also satisfies the orthogonality condition

$$\int_{-1}^{1} P_l^m(x)P_{l'}^m(x)dx = 0, \quad l \neq l'. \tag{4.100}$$

In practical applications, (4.100) is of great importance, while (4.99) is only a mathematical curiosity.

To use $\{P_l^m(x)\}$ as a basis set in the generalized Fourier series, we must evaluate the normalization integral,

$$\int_{-1}^{1} [P_l^m(x)]^2\, dx = \beta_{lm}^2.$$

By definition

$$\int_{-1}^{1} [P_l^m(x)]^2\, dx = \int_{-1}^{1} (1-x^2)^m \frac{d^m P_l(x)}{dx^m}\frac{d^m P_l(x)}{dx^m}dx$$

$$= \int_{-1}^{1} (1-x^2)^m \frac{d^m P_l(x)}{dx^m}d\left[\frac{d^{m-1}P_l(x)}{dx^{m-1}}\right].$$

Integrating by parts we have

$$\int_{-1}^{1} (1-x^2)^m \frac{d^m P_l(x)}{dx^m}d\left[\frac{d^{m-1}P_l(x)}{dx^{m-1}}\right] = (1-x^2)^m \frac{d^m P_l(x)}{dx^m}\frac{d^{m-1}P_l(x)}{dx^{m-1}}\Bigg|_{-1}^{1}$$

$$- \int_{-1}^{1} \frac{d^{m-1}P_l(x)}{dx^{m-1}}\frac{d}{dx}\left[(1-x^2)^m\frac{d^m P_l(x)}{dx^m}\right]dx.$$

The integrated part vanishes at both upper and lower limits, and

$$\frac{\mathrm{d}}{\mathrm{d}x}\left[(1-x^2)^m \frac{\mathrm{d}^m P_l(x)}{\mathrm{d}x^m}\right] = (1-x^2)^m \frac{\mathrm{d}^{m+1} P_l(x)}{\mathrm{d}x^{m+1}} - 2mx(1-x^2)^{m-1}\frac{\mathrm{d}^m P_l(x)}{\mathrm{d}x^m}.$$

Replacing m with $m-1$ in (4.95), the equation becomes

$$(1-x^2)\frac{\mathrm{d}^{m+1}}{\mathrm{d}x^{m+1}}P_l - 2xm\frac{\mathrm{d}^m}{\mathrm{d}x^m}P_l + [l(l+1)-(m-1)m]\frac{\mathrm{d}^{m-1}}{\mathrm{d}x^{m-1}}P_l = 0.$$

Multiplying this equation by $(1-x^2)^{m-1}$, we see that

$$(1-x^2)^m \frac{\mathrm{d}^{m+1} P_l(x)}{\mathrm{d}x^{m+1}} - 2mx(1-x^2)^{m-1}\frac{\mathrm{d}^m P_l(x)}{\mathrm{d}x^m}$$

$$= -(1-x^2)^{m-1}(l+m)(l-m+1)\frac{\mathrm{d}^{m-1} P_l(x)}{\mathrm{d}x^{m-1}}.$$

Therefore

$$\int_{-1}^1 [P_l^m(x)]^2\,\mathrm{d}x = (l+m)(l-m+1)\int_{-1}^1 \frac{\mathrm{d}^{m-1} P_l(x)}{\mathrm{d}x^{m-1}}(1-x^2)^{m-1}\frac{\mathrm{d}^{m-1} P_l(x)}{\mathrm{d}x^{m-1}}$$

$$= (l+m)(l-m+1)\int_{-1}^1 \left[P_l^{m-1}(x)\right]^2\,\mathrm{d}x.$$

Clearly this process can be continued, after m times

$$\int_{-1}^1 [P_l^m(x)]^2\,\mathrm{d}x = k_{lm}\int_{-1}^1 [P_l(x)]^2\,\mathrm{d}x,$$

where

$$k_{lm} = (l+m)(l-m+1)(l+m-1)(l-m+2)\cdots(l+1)l$$

$$= (l+m)(l+m-1)\cdots(l+1)l\cdots(l-m+2)(l-m+1)$$

$$= (l+m)(l+m-1)\cdots(l-m+1)\frac{(l-m)!}{(l-m)!} = \frac{(l+m)!}{(l-m)!}.$$

Since

$$\int_{-1}^1 [P_l(x)]^2\,\mathrm{d}x = \frac{2}{2l+1},$$

it follows that the normalization constant β_{lm}^2 is given by

$$\beta_{lm}^2 = \int_{-1}^1 [P_l^m(x)]^2\,\mathrm{d}x = \frac{2}{2l+1}\frac{(l+m)!}{(l-m)!}.$$

4.8.3 Spherical Harmonics

The major use of the associated Legendre polynomials is in conjunction with the spherical harmonics $Y_l^m(\theta, \varphi)$ which is the angular part of the solution of the Laplace equation expressed in the spherical coordinates,

$$Y_l^m(\theta, \varphi) = (-1)^m \sqrt{\frac{1}{2\pi} \frac{2l+1}{2} \frac{(l-m)!}{(l+m)!}} P_l^m(\cos\theta) e^{im\varphi}. \qquad (4.101)$$

where θ is the polar angle and φ is the azimuthal angle and $m \geq 0$. For $m \leq 0$,

$$Y_l^{-|m|}(\theta, \varphi) = (-1)^m \left[Y_l^{|m|}(\theta, \varphi) \right]^*. \qquad (4.102)$$

Over the surface of a sphere, $\{Y_l^m\}$ forms a complete orthonormal set,

$$\int_{\varphi=0}^{2\pi} \int_{\theta=0}^{\pi} Y_{l_1}^{m_1*}(\theta, \varphi) Y_{l_2}^{m_2}(\theta, \varphi) \sin\theta \, d\theta \, d\varphi = \delta_{l_1,l_2} \delta_{m_1,m_2}.$$

The orthogonality with respect to (m_1, m_2) comes from the φ-dependent part $\exp(im\varphi)$

$$\int_{\varphi=0}^{2\pi} e^{-im_1\varphi} e^{im_2\varphi} \, d\varphi = \int_{\varphi=0}^{2\pi} e^{i(m_2-m_1)\varphi} \, d\varphi = 2\pi \delta_{m_1,m_2},$$

and the orthogonality with respect to (l_1, l_2) is due to the associated Legendre function $P_l^m(\cos\theta)$,

$$\int_{\theta=0}^{\pi} P_{l_1}^m(\cos\theta) P_{l_2}^m(\cos\theta) \sin\theta \, d\theta \, d\varphi = \frac{2}{2l+1} \frac{(l+m)!}{(l-m)!} \delta_{l_1,l_2}.$$

The first few spherical harmonics are listed as follows

$$Y_0^0(\theta, \varphi) = \frac{1}{\sqrt{4\pi}} \qquad Y_2^0(\theta, \varphi) = \sqrt{\frac{5}{16\pi}} (3\cos^2\theta - 1)$$

$$Y_1^0(\theta, \varphi) = \sqrt{\frac{3}{4\pi}} \cos\theta \qquad Y_2^{\pm 1}(\theta, \varphi) = \mp\sqrt{\frac{5}{24\pi}} 3\sin\theta\cos\theta e^{i\varphi}$$

$$Y_1^{\pm 1}(\theta, \varphi) = \mp\sqrt{\frac{3}{8\pi}} \sin\theta e^{i\varphi} \qquad Y_2^{\pm 2}(\theta, \varphi) = \sqrt{\frac{5}{96\pi}} 3\sin^2\theta e^{2i\varphi}.$$

The factor $(-1)^m$ in (4.101) is a phase factor. Although it is not necessary, it is most convenient in the quantum theory of angular momentum. It is called the Condon–Shortley phase, after the authors of a classic text on atomic spectroscopy. Some authors define spherical harmonics without it. Still others use $\cos\varphi$ or $\sin\varphi$ instead of $e^{i\varphi}$. So whenever spherical harmonics are used, the phase convention should be specified.

Any well–behaved function of θ and φ can be expanded as a sum of spherical harmonics,

$$f(\theta, \varphi) = \sum_{l=0}^{\infty} \sum_{m=-l}^{l} c_{lm} Y_l^m(\theta, \varphi),$$

where

$$c_{lm} = \int_{-1}^{1} \int_{0}^{2\pi} [Y_l^m(\theta, \varphi)]^* f(\theta, \varphi) \, d\varphi \, d(\cos\theta).$$

This is another example of generalized Fourier series where the basis set is the solutions of a Sturm–Liouville problem.

4.9 Resources on Special Functions

We have touched only a small part of the huge amount of information about special functions. For the wealth of material available, see the three volume set of "Higher Transcendental Functions" by A. Erdélyi, W. Magnus, F. Oberhettinger, and F.G. Triconi (McGraw-Hill, 1981). An excellent summary is given in "Formulas and Theorems for the Special Functions of Mathematical Physics" by W. Magnus, F. Oberhettinger, and R.P. Soni, (Springer, New York, 1966).

Extensive numerical tables of many special functions are presented in "Handbook of Mathematical Function with Formulas, Graphs, and Mathematical Tables" edited by M. Abramowitz and I.A. Stegun (Dover Publications, 1972).

Digital computers have made numerical evaluations of special functions readily available. There are several computer programs for calculating special functions listed in "Numerical Recipes" by William H. Press, Brian P. Flannery, Saul A. Teukolsky and William T. Vetterling (Cambridge University Press, 1986). A more recent comprehensive compilation of special functions, including computer programs, algorithms and tables is given in "Computation of Special Functions" by S. Zhang and J. Jin (John Wiley & Sons, 1996).

It should also be mentioned that a number of commercial computer packages are available to perform algebraic manipulations, including evaluating special functions. They are called computer algebraic systems, some prominent ones are Matlab, Maple, Mathematica, MathCad, and MuPAD.

This book is written with the software "Scientific WorkPlace", which also provides an interface to MuPAD (Before version 5, it also came with Maple). Instead of requiring the user to adhere to a rigid syntax, the user can use natural mathematical notations. For example, the figures of Bessel function J_n, Neumann function N_n, and modified Bessel functions I_n and K_n shown in this chapter are all automatically plotted. To plot $J_n(x)$, all you have to do is (1) from the Insert menu, choose Math Name, (2) type BesselJ in the name box, enter a subscript, enter an argument enclosed in parentheses, (3) choose OK, and (4) click on the 2D plot button. The program will return with a graph of J_0, J_1, or J_3 depending on what subscript you put in.

Unfortunately, as powerful and convenient as these computer algebraic systems are, sometimes they fail without any apparent reason. For example, in plotting $N_n(x)$, after the program successfully returned $N_0(x)$ and $N_1(x)$, with the same scale, the program hanged up on $N_2(x)$. After the scale is changed slightly, the program worked again. Even worse, the intention of the user is sometimes misinterpreted, and the computer returns with an answer to a wrong problem without the user knowing it. Therefore these systems must be used with caution.

Exercises

1. Solve the differential equation

$$\frac{dT(t)}{dt} + \alpha T(t) = 0$$

by expanding $T(t)$ into a Frobenius series

$$T(t) = t^p \sum_{n=0} a_n t^n.$$

Ans. $T(t) = a_0 \left(1 - \alpha t + \frac{\alpha^2 t^2}{2!} - \frac{\alpha^3 t^3}{3!} + \cdots \right) = a_0 \, e^{-\alpha t}.$

2. *Laguerre Polynomials.* Use the Frobenius method to solve the Laguerre equation

$$xy'' + (1-x)y' + \lambda y = 0.$$

Show that if λ is a nonnegative integer n, then the solution is a polynomial of order n. If the polynomial is normalized such that it is equal to 1 at $x = 0$, it is known as the Laguerre polynomial $L_n(x)$. Show that

$$L_n(x) = \sum_{k=0}^{n} \frac{(-1)^k \, n!}{(n-k)! \, (k!)^2} x^k.$$

Find the explicit expression of $L_0(x)$, $L_1(x)$, $L_2(x)$ and show that they are identical to results obtained in Exercise 3 of the last chapter from the Gram–Schmidt procedure.

3. Find the coefficients c_n of the expansion

$$f(x) = \sum_{n=0}^{\infty} c_n L_n(x)$$

in the interval of $0 \le x < \infty$. Let $f(x) = x^2$, find c_n and verify the results with the explicit expressions of $L_0(x)$, $L_1(x)$, $L_2(x)$.

   Hint: Recall Laguerre equation is a Sturm–Liouville problem in the interval of $0 \le x < \infty$. Its eigenfunctions are orthogonal with respect to weight function e^{-x}.

4. *Rodrigues Formula for Laguerre Polynomials.* Show that

$$L_n(x) = \frac{1}{n!}e^x \frac{d^n}{dx^n}\left(x^n e^{-x}\right).$$

Hint: Use Leibnitz's rule to carry out the differentiation of $x^n e^{-x}$.

5. *Associated Laguerre Polynomials.* (a) Show that the solution of the associated Laguerre equation

$$xy'' + (K + 1 - x)y' + ny = 0$$

is given by

$$y = \frac{d^k}{dx^k}L_{n+k}(x).$$

(b) Other than a multiplicative constant, these solutions are known as the associated Laguerre polynomials $L_n^k(x)$. Show that $L_0^1(x)$, $L_1^1(x)$, $L_2^1(x)$ found in Exercise 6 of the last chapter are, respectively, proportional to

$$\frac{d}{dx}L_l(x), \quad \frac{d}{dx}L_2(x), \quad \frac{d}{dx}L_3(x).$$

Hint: (a) Start with the Laguerre equation of order $n + k$. Differentiate it k times.

Warning: There are many different notations used in literature for associated Laguerre polynomials. When dealing with these polynomials, one must be careful with their definition.

6. *Hermite Polynomials.* Use the Frobenius method to show that the following polynomials

$$y = \sum_{k=0}^{} c_k x^k,$$

where

$$\frac{c_{k+2}}{c_k} = \frac{2k - 2n}{(k+1)(k+2)},$$

are solutions to the Hermite equation

$$y'' - 2xy' + 2ny = 0.$$

Here we have a terminating series with $k = 0, 2, \ldots, n$ terms for n even and with $k = 1, 3, \ldots, n$ terms for n odd. If the coefficient with the highest power of x is normalized to 2^n, then these polynomials are known as Hermite polynomials $H_n(x)$. Find the explicit expressions of $H_0(x)$, $H_1(x)$, $H_2(x)$. Show that these are the same polynomials found in Exercise 4 of the last chapter from the Gram–Schmidt procedure.

7. With the product of

$$e^{-t^2} = 1 - t^2 + \frac{t^4}{2!} - \frac{t^6}{3!} + \cdots = \sum_{l=0}^{\infty} \frac{(-1)^l}{l!} x^{2l},$$

$$e^{2xt} = 1 + 2xt + \frac{(2x)^2}{2!} t^2 + \frac{(2x)^3}{3!} t^3 + \cdots = \sum_{k=0}^{\infty} \frac{(2x)^k}{k!} t^k,$$

written as

$$e^{-t^2} e^{2xt} = \sum_{n=0}^{\infty} A_n(x) t^n,$$

show that for $n = 0$, $n = 1$, $n = 2$,

$$A_n(x) = \frac{1}{n!} H_n(x),$$

where $H_n(x)$ are the Hermite polynomials found in the previous problem.

8. *Generating Function of Hermite Polynomials.* Show that (a) $A_n(x)$ of the previous problem can be written as

$$A_n(x) = \sum_{k=0}^{n} c_{k,n} x^k, \quad \text{and} \quad c_{n,n} = \frac{2^n}{n!},$$

where k and n are either both even or both odd.
(b) Show that c_{kn} is given by

$$c_{k,n} = \frac{2^k}{k!} \frac{(-1)^{(n-k)/2}}{[(n-k)/2]!},$$

(c) Show that the ratio of $c_{k+2,n}$ over $c_{k,n}$ is given by

$$\frac{c_{k+2,n}}{c_{k,n}} = \frac{2k - 2n}{(k+2)(k+1)}.$$

(d) Show that in general

$$e^{-t^2} e^{2xt} = \sum_{n=0}^{\infty} \frac{1}{n!} H_n(x) t^n,$$

where $H_n(x)$ is the Hermite polynomial of order n. The left-hand side of this equation is known as the generating function $G(x,t)$ of Hermite polynomials,

$$G(x,t) = e^{-t^2 + 2xt}.$$

Hint: (a) The power of x can come only from the expansion of e^{2xt}.

(b) The coefficient of t^n is the product of the coefficient of t^k in the expansion of e^{2xt} and the coefficient of t^{n-k} in the expansion of e^{-t^2}. Set $(n-k) = 2l$.

(d) The coefficients satisfy the recurrence relation of the Hermite polynomials and the coefficient of the highest power of x is normalized to $(1/n!)2^n$.

9. *Recurrence Relations of Hermite Polynomials.* Show that

$$\text{(a)}\quad 2xH_n(x) = 2nH_{n-1}(x) + H_{n+1}(x), \quad n \geq 1,$$

$$\text{(b)}\quad \frac{\mathrm{d}}{\mathrm{d}x}H_n(x) = 2nH_{n-1}(x), \quad n \geq 1.$$

Hint: (a) Consider $\dfrac{\partial}{\partial t}G(x,t)$. (b) Consider $\dfrac{\partial}{\partial x}G(x,t)$.

10. *Rodrigues Formula for Hermite Polynomials.* Show that

$$\text{(a)}\quad \left.\frac{\partial^n}{\partial t^n}G(x,t)\right|_{t=0} = H_n(x),$$

$$\text{(b)}\quad \frac{\partial^n}{\partial t^n}G(x,t) = e^{x^2}\frac{\partial^n}{\partial t^n}e^{-(t-x)^2},$$

$$\text{(c)}\quad H_n(x) = (-1)^n\, e^{x^2}\frac{\partial^n}{\partial x^n}e^{-x^2}.$$

Hint: (b) $G(x,t) = e^{-t^2+2xt} = e^{x^2-(t-x)^2} = e^{x^2}e^{-(t-x)^2}$

(c) $e^{x^2}\dfrac{\partial^n}{\partial t^n}e^{-(t-x)^2} = (-1)^n\, e^{x^2}\dfrac{\partial^n}{\partial x^n}e^{-(t-x)^2}$.

11. Use the series expression of Bessel functions

$$J_n(x) = \sum_{k=0}^{\infty} \frac{(-1)^k}{k!\,\Gamma(n+k+1)}\left(\frac{x}{2}\right)^{n+2k}$$

to show that

$$\text{(a)}\quad J_0(0) = 1; \quad J_n(0) = 0 \ \text{ for } \ n = 1,\, 2,\, 3,\ldots,$$

$$\text{(b)}\quad J_0'(x) = -J_1(x).$$

12. Let λ_{nj} be the jth root of $J_n(\lambda c) = 0$, where $c = 2$. Find λ_{01}, λ_{12}, λ_{23} with the table of "zeros of the Bessel function."
Ans. 1.2024, 3.5078, 5.8099.

13. Show that

$$(a) \int_0^c J_0(\lambda r)\, r\, dr = \frac{c}{\lambda} J_1(\lambda c).$$

$$(b) \int_0^1 J_1(\lambda r)\, dr = \frac{1}{\lambda}, \quad \text{if} \quad J_0(\lambda) = 0.$$

Hint: (a) Use (4.32). (b) Use (4.36).

14. Show that

$$J_0(x) + 2 \sum_{n=1}^{\infty} J_{2n}(x) = 1.$$

Hint: Use the generating function.

15. Solve

$$y''(x) + y(x) = 0$$

by substituting $y = \sqrt{x}u$ and solving the resulting Bessel's equation. Then show that the solution is equivalent to

$$y(x) = A \sin x + B \cos x.$$

16. (a) Show that an equation of the form

$$x^2 y''(x) + x y'(x) + \left(a x^\beta\right)^2 y - b^2 y = 0$$

is transformed into a Bessel equation

$$z^2 y''(z) + z y'(z) + z^2 y - \left(\frac{b}{\beta}\right)^2 y = 0$$

by a change of variable

$$z = \frac{a x^\beta}{\beta}.$$

(b) Solve

$$x^2 y'' + x y' + 4x^4 y - 16 y = 0.$$

Ans. (b) $y(x) = c_1 J_2\left(x^2\right) + c_2 N_2\left(x^2\right)$.

17. Solve

$$x^2 y''(x) - x y'(x) + x^2 y(x) = 0.$$

Hint: Let $y(x) = x^\alpha u(x)$, and show that the equation becomes

$$x^2 u'' + (2\alpha - 1)\, x u' + \left[x^2 + \left(\alpha^2 - 2\alpha\right)\right] u = 0.$$

then set $\alpha = 1$.
Ans. $y(x) = x\left[c_1 J_1(x) + c_2 N_1(x)\right]$.

18. Use the Rodrigues formula

$$P_l(x) = \frac{1}{2^l l!} \frac{d^l}{dx^l} \left(x^2 - 1\right)^l$$

to develop the following series expressions of the Legendre polynomials

$$P_l(x) = \frac{1}{2^l l!} \sum_{k=0}^{[l/2]} \frac{(-1)^k \, l!}{k! \, (l-k)!} \frac{(2l-2k)!}{(l-2k)!} x^{l-2k},$$

where $[l/2] = l/2$ if l is even, and $[l/2] = (l-1)/2$ if l is odd. Show that

$$P_{2n}(0) = (-1)^n \frac{(2n)!}{2^{2n} (n!)^2}, \qquad P_{2n+1}(0) = 0.$$

19. Use the generating function $G(0,t)$ to find $P_{2n}(0)$.
Hint:

$$\frac{1}{\sqrt{1+t^2}} = 1 - \frac{1}{2}t^2 + \frac{1}{2!} \frac{1}{2} \frac{3}{2} t^4 + \cdots$$

$$= \sum_{n=0}^{2n} \frac{(-1)^n}{n!} \frac{(2n-1)!!}{2^n} t^{2n},$$

where $(2n-1)!! = (2n-1)(2n-3)\cdots 1$, and

$$(2n-1)!! = (2n-1)!! \frac{(2n)!!}{(2n)!!} = \frac{(2n)!}{2^n n!}.$$

20. Use Rodrigues' formula to prove that

$$P'_{n+1}(x) - P'_{n-1}(x) = (2n+1) P_n(x).$$

Hence show that

$$\int_x^1 P_n(x) dx = \frac{1}{2n+1} \left[P_{n-1}(x) - P_{n+1}(x)\right].$$

21. Use the fact that both $P_n(x)$ and $P_m(x)$ satisfy the Legendre equation to show that

$$\int_{-1}^1 P_n(x) P_m(x) dx = 0 \quad \text{if } m \neq n.$$

Hint: Multiply the Legendre equation

$$\frac{d}{dx}\left[(1-x^2)\frac{d}{dx}P_m(x)\right] + m(m+1)P_m(x) = 0$$

by $P_n(x)$ and integrate by parts:

$$\int_{-1}^{1}(1-x^2)\left[\frac{d}{dx}P_m(x)\right]\frac{d}{dx}P_n(x)dx = m(m+1)\int_{-1}^{1}P_n(x)P_m(x)dx.$$

Similarly,

$$\int_{-1}^{1}(1-x^2)\left[\frac{d}{dx}P_n(x)\right]\frac{d}{dx}P_m(x)dx = n(n+1)\int_{-1}^{1}P_m(x)P_n(x)dx.$$

Then get the conclusion from

$$[m(m+1) - n(n+1)]\int_{-1}^{1}P_n(x)P_m(x)dx = 0.$$

22. Use the Rodrigues formula to show

$$\int_{-1}^{1}[P_n(x)]^2\,dx = \frac{2}{2n+1}.$$

Hint: First show that

$$\int_{-1}^{1}[P_n(x)]^2\,dx = \left[\frac{1}{2^n n!}\right]^2\int_{-1}^{1}d\left[\frac{d^{n-1}}{dx^{n-1}}\left(x^2-1\right)^n\right]\frac{d^n}{dx^n}\left(x^2-1\right)^n,$$

then use integration by parts repeatedly to show

$$\int_{-1}^{1}[P_n(x)]^2\,dx = (-1)^n\left[\frac{1}{2^n n!}\right]^2\int_{-1}^{1}\left(x^2-1\right)^n\frac{d^{2n}}{dx^{2n}}\left(x^2-1\right)^n\,dx.$$

Note that $(d^{2n}/dx^{2n})\left(x^2-1\right)^n = (2n)!$, so

$$\int_{-1}^{1}[P_n(x)]^2\,dx = \left[\frac{1}{2^n n!}\right]^2(2n)!\int_{-1}^{1}\left(1-x^2\right)^n\,dx.$$

Evaluate the integral on the right-hand side with a change a variable $x = \cos\theta$. First note

$$\int_{-1}^{1}\left(1-x^2\right)^n\,dx = \int_{-1}^{1}\left(1-\cos^2\theta\right)^n\,d\cos\theta = \int_{0}^{\pi}(\sin\theta)^{2n+1}d\theta,$$

then show that

$$(2n+1)\int_{0}^{\pi}(\sin\theta)^{2n+1}d\theta = 2n\int_{0}^{\pi}(\sin\theta)^{2n-1}\,d\theta,$$

by repeated integration by parts of

$$\int_{-1}^{1}\left(1-\cos^2\theta\right)^n\,d\cos\theta = \int_{-1}^{1}(\sin\theta)^{2n}\,d\cos\theta = -\int_{\pi}^{0}\cos\theta\frac{d}{d\theta}(\sin\theta)^{2n}\,d\theta$$

$$= 2n\int_{0}^{\pi}(\sin\theta)^{2n-1}\,d\theta - 2n\int_{0}^{\pi}(\sin\theta)^{2n+1}\,d\theta.$$

Finally get the result from

$$\int_0^\pi (\sin\theta)^{2n+1}\mathrm{d}\theta = \frac{2n(2n-2)\cdots 2}{(2n+1)(2n-1)\cdots 3}\int_0^\pi \sin\theta\,\mathrm{d}\theta = 2\frac{[2^n n!]^2}{(2n+1)!}.$$

23. If

$$f(x) = \begin{cases} 1 & 0 < x < 1 \\ 0 & -1 < x < 0 \end{cases},$$

show that $f(x)$ can be expressed as

$$f(x) = \frac{1}{2} + \frac{3}{4}x + \sum_{n=1}^\infty (-1)^n \frac{4n+3}{4n+4}\frac{(2n)!}{2^{2n}(n!)^2}P_{2n+1}(x).$$

Hint: Show that

$$f(x) = \frac{1}{2}P_0(x) + \frac{1}{2}\sum_{n=0}^\infty [P_{2n}(0) - P_{2n+2}(0)]P_{2n+1}(x).$$

24. Show that

$$\text{(a)} \quad \int_{-1}^1 x P_n(x) P_m(x)\mathrm{d}x = \begin{cases} \dfrac{2(n+1)}{(2n+1)(2n+3)} & \text{if } m = n+1 \\ \dfrac{2n}{(2n-1)(2n+1)} & \text{if } m = n-1 \\ 0 & \text{otherwise} \end{cases},$$

$$\text{(b)} \quad \int_{-1}^1 x^2 P_n(x) P_m(x)\mathrm{d}x = \begin{cases} \dfrac{2(n+1)(n+2)}{(2n+1)(2n+3)(2n+5)} & m = n+2 \\ \dfrac{2(2n^2+2n-1)}{(2n-1)(2n+1)(2n+3)} & m = n \\ \dfrac{2(n-1)n}{(2n-3)(2n-1)(2n+1)} & m = n-2 \\ 0 & \text{otherwise} \end{cases}.$$

Hint: Use (4.88).

25. Show that if $P_l^{-m}(x)$ and $Y_l^m(\theta,\varphi)$ are, respectively, defined as

$$P_l^{-m}(x) = (-1)^m \frac{(l-m)!}{(l+m)!}P_l^m(x),$$

and

$$Y_l^m(\theta,\varphi) = (-1)^m \sqrt{\frac{1}{2\pi}\frac{2l+1}{2}\frac{(l-m)!}{(l+m)!}}P_l^m(\cos\theta)\mathrm{e}^{\mathrm{i}m\varphi}, \quad \text{for } -l \le m \le l,$$

then $Y_l^{-m}(\theta,\varphi)$ obtained from the last expression is the same as

$$Y_l^{-|m|}(\theta,\varphi) = (-1)^m \left[Y_l^{|m|}(\theta,\varphi)\right]^*,$$

defined in (4.102).

Part III

Partial Differential Equations

5

Partial Differential Equations in Cartesian Coordinates

An equation involving one or more partial derivatives of an unknown function of two or more independent variables is called a partial differential equation. Compared with ordinary differential equations, far more problems in physical sciences lead to partial differential equations. In fact, most of mathematical physics deals with partial differential equations.

In general, the totality of solutions of a partial differential equation is very large. However, a unique solution of a partial differential equation corresponding to a given physical problem can usually be obtained by the use of either boundary and/or initial conditions. In practice, the boundary conditions frequently serve as a guide in choosing a particular form of the solution, which satisfies the partial differential equation as well as the boundary conditions.

The field of partial differential equation is very wide. We are going to focus our attention on the equations that arise most often in physics, namely

$$\nabla^2 u = \frac{1}{a^2} \frac{\partial^2 u}{\partial t^2}, \quad \text{Wave equation}$$

$$\nabla^2 u = \frac{1}{a^2} \frac{\partial u}{\partial t}, \quad \text{Diffusion equation}$$

$$\nabla^2 u = 0, \quad \text{Laplace's equation,}$$

where ∇^2 is the operator

$$\nabla^2 = \frac{\partial^2}{\partial x^2} + \frac{\partial^2}{\partial y^2} + \frac{\partial^2}{\partial z^2}.$$

The Schrödinger equation in quantum mechanics also has a similar form except that an imaginary number is attached to the time derivative.

In this chapter we only consider problems expressible in cartesian coordinates. Problems with curved boundaries will be considered in Chap. 6.

The amazing thing is that there are many problems which are physically unrelated, but they can be described by the same or very similar partial differential equation.

5.1 One-Dimensional Wave Equations

5.1.1 The Governing Equation of a Vibrating String

As an example, we will derive the equation that governs the small vibrations of an elastic string of length L, fixed at both endpoints. The dependent variable $u(x,t)$ represents, at time t, the displacement of the point of the string that is at distance x away from the first end point 0.

We shall assume that the string is homogeneous, that is, the mass of the string per unit length, denoted as ρ, is a constant. We also shall assume that the string undergoes only small vertical displacements from its equilibrium position. (The displacements do not have to be in vertical direction, but for the sake of discussion, we assume it is.)

Let us consider the segment of the string between x and $x + \Delta x$, where Δx is a small increment, as shown in Fig. 5.1. The quantities T_1 and T_2 in the figure are the tensions at the points P and Q of the string. Both T_1 and T_2 are tangential to the curve of the spring. Because there is no horizontal motion of the string, the net horizontal force exerted on the segment must be zero. In other words, the horizontal components of the tensions at P and Q must be equal and opposite. That is

$$T_1 \cos \theta_1 = T_2 \cos \theta_2 = T, \tag{5.1}$$

where T is a constant equal to the horizontal force with which the string is stretched. If the amplitude is small, we can regard T as the tension of the string.

There is a net force in the vertical direction, F_u, that causes the vertical motion of the string. Clearly

$$F_u = T_2 \sin \theta_2 - T_1 \sin \theta_1.$$

By Newton's second law, this force is equal to the mass of the segment, $\rho \, \Delta x$, times the acceleration which is the second derivative of the displacement with respect to time. That is

$$T_2 \sin \beta - T_1 \sin \alpha = \rho \, \Delta x \frac{\partial^2 u}{\partial t^2}.$$

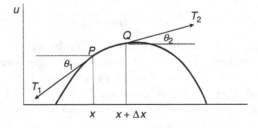

Fig. 5.1. A vibrating string at time t

Dividing this equation by T and using (5.1), we have

$$\frac{T_2 \sin \theta_2}{T_2 \cos \theta_2} - \frac{T_1 \sin \theta_1}{T_1 \cos \theta_1} = \frac{\rho \, \Delta x}{T} \frac{\partial^2 u}{\partial t^2}.$$

Thus

$$\tan \theta_2 - \tan \theta_1 = \frac{\rho \, \Delta x}{T} \frac{\partial^2 u}{\partial t^2}. \tag{5.2}$$

But $\tan \theta_2$ and $\tan \theta_1$ are the slopes of the curve of the string at $x + \Delta x$ and x, respectively,

$$\tan \theta_2 = \left(\frac{\partial u}{\partial x} \right)_{x+\Delta x},$$

$$\tan \theta_1 = \left(\frac{\partial u}{\partial x} \right)_{x}.$$

Hence (5.2) can be written as

$$\left(\frac{\partial u}{\partial x} \right)_{x+\Delta x} - \left(\frac{\partial u}{\partial x} \right)_{x} = \frac{\rho \, \Delta x}{T} \frac{\partial^2 u}{\partial t^2}. \tag{5.3}$$

Recall the definition of a derivative

$$\frac{\mathrm{d}F}{\mathrm{d}x} = \lim_{\Delta x \to 0} \frac{F(x + \Delta x) - F(x)}{\Delta x}.$$

If it is understood that Δx is approaching zero even without the limit sign, then we can write

$$F(x + \Delta x) - F(x) = \frac{\mathrm{d}F}{\mathrm{d}x} \Delta x.$$

Thus it is clear that

$$\left(\frac{\partial u}{\partial x} \right)_{x+\Delta x} - \left(\frac{\partial u}{\partial x} \right)_{x} = \frac{\partial}{\partial x} \left(\frac{\partial u}{\partial x} \right) \Delta x = \frac{\partial^2 u}{\partial x^2} \Delta x.$$

Therefore (5.3) becomes

$$\frac{\partial^2 u}{\partial x^2} = \frac{\rho}{T} \frac{\partial^2 u}{\partial t^2}. \tag{5.4}$$

This is the so called one-dimensional wave equation. We see that it is linear, homogeneous, and of second-order.

If the string is fixed at both ends, we have two boundary conditions

$$\text{B.C.:} \quad u(0, t) = 0; \quad u(L, t) = 0.$$

Furthermore, if the string is initially displaced into a position $u = f(x)$ and released at rest from that position, then we have the following initial conditions:

$$\text{I.C.}:\ u(x,0) = f(x);\ u_t(x,0) = 0,$$

where $u_t(x,0)$ denotes the first partial derivative of $u(x,t)$ with respect to t and then evaluated at $t = 0$

$$u_t(x,0) = \left.\frac{\partial u(x,t)}{\partial t}\right|_{t=0}.$$

The first condition says that the initial shape of the string is $f(x)$, the second condition simply says that at $t = 0$, the velocity everywhere in the string is zero.

Of course, it is possible that the string also has initial velocity. In that case, the initial conditions become

$$u(x,0) = f(x);\ u_t(x,0) = g(x).$$

5.1.2 Separation of Variables

To describe the motion of the string, one must solve the differential equation and the solution must satisfy the boundary and initial conditions. Specifically let us find a formula for the transverse displacement $u(x,t)$ of the stretched string which satisfies (5.4). For simplicity of writing, let us first define

$$a^2 = \frac{T}{\rho}. \tag{5.5}$$

It turns out a has a physical meaning which will become clear later.

A powerful and classical method of solving linear boundary value problems in partial differential equations is the method of separation of variables which reduces a partial differential equation into ordinary differential equations. Although not all problems can be solved by this method and there are other methods, generally separation of variables is the first method one should try.

Let us solve the mathematical problem consisting of the following:

$$\text{D.E.}:\ \frac{\partial^2 u(x,t)}{\partial x^2} = \frac{1}{a^2}\frac{\partial^2 u(x,t)}{\partial t^2}, \tag{5.6}$$

$$\text{B.C.}:\ u(0,t) = 0;\qquad u(L,t) = 0, \tag{5.7}$$

$$\text{I.C.}:\ u(x,0) = f(x);\qquad u_t(x,0) = 0. \tag{5.8}$$

The assumption of separation of variables is that we can write $u(x,t)$ as

$$u(x,t) = X(x)T(t),$$

where X is a function of x alone and T is a function of t alone. The justification of the assumption of this method is that it works. It follows from this assumption, that:

$$\frac{\partial^2 u}{\partial x^2} = \left(\frac{d^2}{dx^2}X(x)\right)T(t) = X''(x)T(t),$$

$$\frac{\partial^2 u}{\partial t^2} = X(x)\left(\frac{d^2}{dt^2}T(t)\right) = X(x)T''(t).$$

Thus (5.6) becomes

$$X''(x)T(t) = \frac{1}{a^2}X(x)T''(t).$$

Dividing both sides of the equation by $X(x)T(t)$

$$\frac{X''(x)T(t)}{X(x)T(t)} = \frac{1}{a^2}\frac{X(x)T''(t)}{X(x)T(t)},$$

we obtain

$$\frac{X''(x)}{X(x)} = \frac{1}{a^2}\frac{T''(t)}{T(t)}.$$

The left-hand side of this equation is a function of x alone, it cannot vary with t. However, it is equal to a function of t which cannot vary with x. This is possible if and only if both sides are equal to the same common constant α. This leads to

$$\frac{X''(x)}{X(x)} = \alpha,$$

$$\frac{1}{a^2}\frac{T''(t)}{T(t)} = \alpha.$$

It follows that:

$$X''(x) = \alpha X(x) \tag{5.9}$$

$$T''(t) = \alpha a^2 T(t). \tag{5.10}$$

The partial differential equation is now decomposed into two ordinary differential equations.

Eigenvalues and Eigenfunctions. If $u(x,t)$ is to satisfy the first boundary condition, then

$$u(0,t) = X(0)T(t) = 0$$

for all t. Since $T(t)$ is changing with t, the only possibility that this can be true is that

$$X(0) = 0.$$

Similarly, the condition $u(L,t) = 0$ leads to

$$X(L) = 0.$$

So far we have not specified the value of the separation constant α, it could be less than zero, equal to zero, or greater than zero. It is easy to show that if $\alpha \geq 0$, there is no solution that can satisfy these boundary conditions.

First, if $\alpha = 0$, the solution of (5.9) is $X(x) = Ax + B$. In this case $X(0) = 0$ requires $B = 0$. Thus, $X(L) = AL$. Since $X(L) = 0$, therefore $A = 0$. This leads to $X(x) = 0$ which is a trivial solution for the case that u is identically equal to zero for all x and t.

When $\alpha > 0$, let us write $\alpha = \mu^2$ with μ being real. Then the solution of $X''(x) = \mu^2 X(x)$ is $X(x) = C \cosh \mu x + D \sinh \mu x$. With $X(0) = 0, C$ must be equal to zero. Thus $X(x) = D \sinh \mu x$. Since $\sinh \mu L \neq 0$, $X(L) = 0$ requires $D = 0$. Again this gives only the trivial solution.

Therefore α must be less than zero. Let us write $\alpha = -\mu^2$, so (5.9) becomes

$$X''(x) = -\mu^2 X(x).$$

The general solution of this equation is

$$X(x) = A \cos \mu x + B \sin \mu x.$$

Thus $X(0) = A$ and the condition $X(0) = 0$ means $A = 0$. Hence we are left with

$$X(x) = B \sin \mu x.$$

To satisfy the condition $X(L) = 0$, μ must be chosen to be

$$\mu = \frac{n\pi}{L}, \quad n = 1, \ 2, \ 3, \ldots.$$

Therefore, for each n, there is a solution $X_n(x)$

$$X_n(x) = B_n \sin \frac{n\pi}{L} x, \quad n = 1, 2, \ldots \tag{5.11}$$

where B_n is an arbitrary constant. The numbers $\alpha = -n^2 \pi^2 / L^2$ for which this problem has nontrivial solutions are called eigenvalues, and the corresponding functions (5.11) are called eigenfunctions.

Solution of the Problem. It is important to keep in mind that α in (5.9) and in (5.10) must be the same. When $\alpha = -n^2 \pi^2 / L^2$, (5.9) is a distinct problem for each different positive integer n. For a fixed integer n, (5.10) becomes

$$T_n''(t) = -\frac{n^2 \pi^2}{L^2} a^2 T_n(t).$$

The solution of this equation is

$$T_n(t) = C_n \cos \frac{n\pi a}{L} t + D_n \sin \frac{n\pi a}{L} t.$$

Thus, each

$$u_n(x, t) = X_n(x) T_n(t)$$

is a solution of the differential equation. An important theorem of the linear homogeneous partial differential equation is the principle of superposition. If u_1 and u_2 are solutions of a linear homogeneous differential equation, then

$$u = c_1 u_1 + c_2 u_2,$$

where c_1 and c_2 are arbitrary constants, is also a solution of that equation. This theorem can be easily proved by showing the equation is satisfied with this combination as the solution.

Therefore the general solution is given by

$$u(x,t) = \sum_{n=1}^{\infty} c_n X_n(x) T_n(t)$$

$$= \sum_{n=1}^{\infty} \left(a_n \cos \frac{n\pi a}{L} t + b_n \sin \frac{n\pi a}{L} t \right) \sin \frac{n\pi}{L} x. \tag{5.12}$$

where we have combined three arbitrary constants $c_n C_n B_n$ into a single constant a_n, and $c_n D_n B_n$ into b_n. Now the coefficients a_n and b_n can be chosen to satisfy the initial conditions.

One of the initial condition is

$$u_t(x,0) = \sum_{n=1}^{\infty} \left[\frac{d}{dt} \left(a_n \cos \frac{n\pi a}{L} t + b_n \sin \frac{n\pi a}{L} t \right) \right]_{t=0} \sin \frac{n\pi}{L} x = 0,$$

which leads to

$$\sum_{n=1}^{\infty} b_n \frac{n\pi a}{L} \sin \frac{n\pi}{L} x = 0.$$

Since $\sin \frac{n\pi}{L} x$ is a complete set in the interval of $0 \leq x \leq L$, all coefficients must be zero. Another way to see that all b_n are zero is the following. This equation is a Fourier sine series of zero. The coefficients are integrals of zero times some sine function. Obviously they are zero. Therefore

$$b_n = 0.$$

Thus we are left with

$$u(x,t) = \sum_{n=1}^{\infty} a_n \sin \frac{n\pi}{L} x \cos \frac{n\pi a}{L} t, \tag{5.13}$$

where the coefficients a_n can be chosen to satisfy the other initial condition.

Since $u(x,0) = f(x)$, it follows from the last equation

$$u(x,0) = \sum_{n=1}^{\infty} a_n \sin \frac{n\pi}{L} x = f(x).$$

This is the half-range Fourier sine series of $f(x)$ between 0 and L. Therefore a_n is given by the Fourier coefficient

$$a_n = \frac{2}{L} \int_0^L f(x) \sin \frac{n\pi x}{L} dx.$$

Thus, the solution of this problem is given by

$$u(x,t) = \sum_{n=1}^\infty \left[\frac{2}{L} \int_0^L f(x') \sin \frac{n\pi x'}{L} dx' \right] \sin \frac{n\pi}{L} x \cos \frac{n\pi a}{L} t. \qquad (5.14)$$

Example 5.1.1. A guitar string of length L is pulled upward at the middle so it reaches height h. What is the subsequent motion of the string if it is released from the rest?

Solution 5.1.1. To find the subsequent motion means to find the displacement of the string as a function of t. That is, we have to solve for $u(x,t)$ from the equation

$$\frac{\partial^2 u(x,t)}{\partial x^2} = \frac{1}{a^2} \frac{\partial^2 u(x,t)}{\partial t^2}.$$

Since the two ends of the guitar string are fixed, so we have to satisfy the boundary conditions

$$u(0,t) = 0, \qquad u(L,t) = 0.$$

Furthermore, it can be readily shown that the initial shape of the string is given by

$$f(x) = \begin{cases} \dfrac{2h}{L} x & \text{for } 0 \le x \le \dfrac{L}{2}, \\ \dfrac{2h}{L}(L-x) & \text{for } \dfrac{L}{2} \le x \le L. \end{cases}$$

Since it is released from rest, so the initial velocity everywhere in the string is zero. This means the derivative of $u(x,t)$ with respect to time evaluated at $t = 0$ is zero. Thus the initial conditions are

$$u(x,0) = f(x), \qquad u_t(x,0) = 0.$$

According to (5.14), $u(x,t)$ is given by

$$u(x,t) = \sum_{n=1}^\infty a_n \sin \frac{n\pi}{L} x \cos \frac{n\pi a}{L} t,$$

where

$$\begin{aligned} a_n &= \frac{2}{L} \int_0^L f(x) \sin \frac{n\pi x}{L} dx \\ &= \frac{2}{L} \int_0^{L/2} \frac{2h}{L} x \sin \frac{n\pi x}{L} dx + \frac{2}{L} \int_{L/2}^L \frac{2h}{L}(L-x) \sin \frac{n\pi x}{L} dx \\ &= \frac{8h}{n^2\pi^2} \sin \frac{n\pi}{2}. \end{aligned}$$

Therefore

$$u(x,t) = \frac{8h}{\pi^2} \sum_{n=1}^{\infty} \frac{1}{n^2} \sin \frac{n\pi}{2} \sin \frac{n\pi}{L} x \cos \frac{n\pi a}{L} t$$

$$= \frac{8h}{\pi^2} \left(\sin \frac{\pi}{L} x \cos \frac{\pi a}{L} t - \frac{1}{3^2} \sin \frac{3\pi}{L} x \cos \frac{3\pi a}{L} t \right.$$

$$\left. + \frac{1}{5^2} \sin \frac{5\pi}{L} x \cos \frac{5\pi a}{2} t - \frac{1}{7^2} \sin \frac{7\pi}{L} x \cos \frac{7\pi a}{L} t + \cdots \right). \quad (5.15)$$

It is interesting to see the time development of the displacements. The shapes of the string at various times are shown in the left-hand side column of Fig. 5.2. The individual components are shown in the right-hand side column of the same figure. The string oscillates up and down as expected. We have shown the positions of the string within half of a cycle. After that it will go back

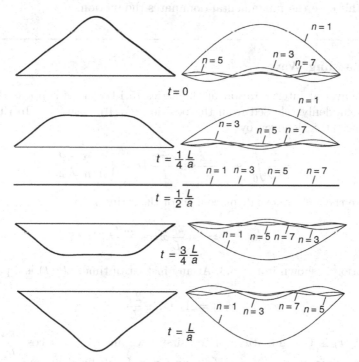

Fig. 5.2. The time development of the displacements of a string after its middle point is pulled up a distance h and released from that position. The left-hand side column is the shape of the string at various times obtained by summing up the first four nonzero terms of the series (5.15). The right-hand side column are the positions of the four individual terms of the series at the correponding times. Although different components oscillate at different frequencies, they sum up to a string going up and down as expected. It is seen that the fundamental (the first term) dominates the motion

to its original position and then repeat the motion. During this time interval of half a cycle, the fundamental (the first term in the series, $\sin \frac{\pi}{L} x \cos \frac{\pi a}{L} t$) also completes half of its cycle. The third harmonic (the second nonzero term, $\sin \frac{3\pi}{L} x \cos \frac{3\pi a}{L} t$) actually completes one and half of its cycle. Various components oscillate at various different frequencies, yet together they sum up to the oscillation shown in the left-hand side column. In fact we have summed up only four nonzero terms, so the lines for the shape of the string are somewhat curved and the corners are somewhat rounded. If we use the computer to plot

$$\frac{8h}{\pi^2} \sum_{n=1}^{N} \frac{1}{n^2} \sin \frac{n\pi}{2} \sin \frac{n\pi}{L} x \cos \frac{n\pi a}{L} t,$$

with $N = 50$, then all the lines in the left-hand side column will be straight and all corners sharply pointed. The amplitudes of higher components are very small, but they do make the sum converge to the exact value. It is seen that in this case the fundamental dominates the motion.

5.1.3 Standing Wave

For the physical interpretation of the series (5.14), let us suppose that the string is suddenly released from the position $u(x, 0) = \sin \frac{2\pi}{L} x$. In this case, the coefficient a_n is given by

$$a_n = \frac{2}{L} \int_0^L \sin \frac{2\pi x}{L} \sin \frac{n\pi x}{L} dx = \begin{cases} 1 & n = 2, \\ 0 & n \neq 2. \end{cases}$$

Therefore the subsequent displacement of the string is

$$u(x, t) = \sin \frac{2\pi}{L} x \cos \frac{2\pi a}{L} t.$$

This motion is shown in Fig. 5.3. At any instant of time, $u(x, t)$ is a pure sine curve

$$u(x, t) = A_2(t) \sin \frac{2\pi}{L} x,$$

where $A_2(t)$ is the amplitude of the sine wave and $A_2(t) = \cos \frac{2\pi a}{L} t$. Note that the points at $x = 0$, $x = L/2$, and $x = L$ are fixed in time. They are called nodes. Between the nodes, the string oscillates up and down. This kind of motion is known as standing wave.

In general (5.13) can be regarded as

$$u(x, t) = \sum_{n=1}^{\infty} a_n u_n(x, t), \tag{5.16}$$

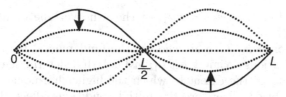

Fig. 5.3. The standing wave of $\sin \frac{2\pi}{L}x \cos \frac{2\pi a}{L}t$

where

$$u_n(x,t) = \sin \frac{n\pi}{L}x \cos \frac{n\pi a}{L}t$$

is known as the nth normal mode. One characteristic of a normal mode is that once the string is vibrating in the standing wave of that mode, it will continue to vibrate in that mode forever. Of course, if there is damping, its amplitude will eventually die down.

The time dependence of each normal mode is given by $\cos \frac{n\pi a t}{L}$ which is a periodic function. The period is defined as the time interval after which the function will return to its original value. Let P_n be the period, so that

$$\cos \frac{n\pi a}{L}(t + P_n) = \cos \frac{n\pi a}{L}t. \tag{5.17}$$

Since

$$\cos \frac{n\pi a}{L}(t + P_n) = \cos \left(\frac{n\pi a}{L}t + \frac{n\pi a}{L}P_n \right)$$

clearly

$$\frac{n\pi a}{L}P_n = 2\pi.$$

Therefore

$$P_n = \frac{2L}{na}.$$

Frequency ν_n is defined as the number of oscillations in one second (the unit of frequency is called Hertz, Hz), that is

$$\nu_n = \frac{1}{P_n} = \frac{na}{2L}.$$

Therefore the series (5.16) represents the motion of a string (of violin or guitar) as a superposition of infinitely many normal modes, each vibrating with a different frequency. The lowest of these frequencies

$$\nu_1 = \frac{a}{2L} = \frac{1}{2L}\sqrt{\frac{T}{\rho}}$$

is called the fundamental frequency. Here we have used the definition of a given in (5.5). The fundamental frequency usually predominates in the sound we hear. The frequency $\nu_n = n\nu_1$, of the nth overtone or harmonic is an integral multiple of ν_1.

Note that once L, T, ρ are chosen, the fundamental frequency is fixed. The initial conditions do not affect ν_1; instead, they determine the coefficients in (5.14) and hence the extent to which the higher harmonics contribute to the sound produced. Therefore the initial conditions affect the overall frequency mixture (known as timbre), rather than the fundamental frequency. For example, if the string of a violin is bowed at some other point than its center, the amplitudes of higher harmonics would have been different than shown in Fig. 5.2. By choosing the point properly any desired harmonic may be emphasized or diminished, a fact well known to musicians.

Once a musical instrument is constructed, the length of string L and the density ρ cannot be changed. Therefore tuning is usually done by changing the tension T.

The spatial dependence of the first few normal modes is shown in Fig. 5.4. The first mode ($n = 1$) is called the fundamental mode, represents a harmonic time dependence of frequency $a/2L$. The second harmonic or first overtone ($n = 2$) vibrates harmonically with frequency a/L, twice as fast as the fundamental mode. Its motion is also shown in Fig. 5.4. Note that, in addition to the two end points, the midpoint of this harmonic is a node. Similarly, the third ($n = 3$) and fourth ($n = 4$) harmonics have two and three nodes, respectively, in addition to the two end points.

In describing the frequency of the oscillation, the angular frequency ω_n (radians per second) in often used

$$\omega_n = 2\pi\nu_n = \frac{\pi n a}{L}.$$

Another quantity associated with wave motion is the wavelength. The wavelength λ_n is defined such that $u_n(x, t)$ will return to its original value if x is increased by λ_n, that is

$$u_n(x + \lambda_n, t_0) = u_n(x, t_0). \tag{5.18}$$

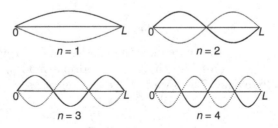

$n = 1$ $n = 2$

$n = 3$ $n = 4$

Fig. 5.4. The first four normal modes of a vibrating string. Each normal mode is a standing wave. The nth normal mode has $n - 1$ nodes, excluding the nodes at the two end points

Since

$$u_n(x + \lambda_n, t_0) = \sin\left(\frac{n\pi}{L}x + \frac{n\pi}{L}\lambda_n\right)\cos\frac{n\pi a}{L}t_0;$$

$$u_n(x, t_0) = \sin\frac{n\pi}{L}x\cos\frac{n\pi a}{L}t_0$$

it is clear that (5.18) will be satisfied if

$$\frac{n\pi}{L}\lambda_n = 2\pi.$$

Therefore

$$\lambda_n = \frac{2L}{n}. \tag{5.19}$$

Thus, for $n = 1$, $L = \frac{1}{2}\lambda$; $n = 2$, $L = \lambda$; $n = 3$, $L = \frac{3}{2}\lambda$, $n = 4$, $L = 2\lambda$. These relations are clearly demonstrated in Fig. 5.4. Therefore the distance between two nodes of a standing wave is half of a wavelength.

Often a quantity known as wave number k_n (number of wavelengths in the interval of 2π)

$$k_n = \frac{2\pi}{\lambda_n} = \frac{n\pi}{L} \tag{5.20}$$

is used to describe the wave form. In this notation, the normal mode $u_n(x, t)$ is written as

$$u_n(x, t) = \sin k_n x \cos \omega_n t. \tag{5.21}$$

A very important relationship between the frequency and the wavelength is

$$\nu_n \lambda_n = \frac{na}{2L}\frac{2L}{n} = a = \sqrt{\frac{T}{\rho}}. \tag{5.22}$$

This relation says that the frequency is proportional to the inverse of the wavelength and the proportionality constant is equal to the square root of tension over density. A standard physics experiment is shown in Fig. 5.5. A string of density ρ and tension T is connected to a vibrator, the frequency of which can be varied. Standing wave patterns will occur for certain discrete values of the frequency. The wavelength of each standing wave can then be measured. After several standing waves of different wavelength are measured,

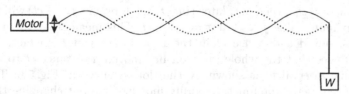

Fig. 5.5. A standing wave experiment to verify the relationship between the frequency and the wavelength

we can plot the frequency against the inverse of the wavelength. The curve is indeed a straight line and the slope of the line is indeed equal to $\sqrt{T/\rho}$.

This is not only a demonstration of a physical principle, but also a demonstration of the power of analysis. We have applied the Newton's law, which relates the force to the acceleration of the particle, to the motion of a string through the use of calculus and concluded that the frequency and the wavelength must satisfy the relation shown in (5.22). This can then be verified in the laboratory.

If the wave is traveling down on an infinite line, one may think that the frequency is the number of wave cycles generated per second and each extends a distance of one wavelength, therefore $\nu_n \lambda_n = a$ is the distance the wave travels in one second. In other words, a is the velocity of a traveling wave. This is indeed the case, as we will clearly see in Sect. 5.1.4.

5.1.4 Traveling Wave

In Sect. 5.1.3, we have shown that each normal mode is a standing wave. Now we wish to show that the same normal mode can also be regarded as a superposition of two traveling waves in the opposite direction.

Using the trigonometric identity

$$\sin a \cos b = \frac{1}{2}\left[\sin(a+b) + \sin(a-b)\right],$$

we can write (5.21) as

$$u_n(x,t) = \frac{1}{2}\sin(k_n x + \omega_n t) + \frac{1}{2}\sin(k_n x - \omega_n t)$$
$$= \frac{1}{2}\sin[k_n(x+at)] + \frac{1}{2}\sin[k_n(x-at)], \qquad (5.23)$$

where we have used the fact

$$\frac{\omega_n}{k_n} = \nu_n \lambda_n = a.$$

Before we discuss the interpretation of (5.23), let us first examine the behavior of the function $f(x-ct)$. In this function, the variables x and t are combined in the particular way of $x-ct$. Suppose at $t=0$, the function $f(x)$ looks like the solid curve in Fig. 5.6. If the maximum value of the function $f(x_m)$ is at $x=x_m$, then at a later time t, the function $f(x-ct)$ will have the same maximum value $f(x_m)$ at $x=x_m+ct$. This means that the maximum point has moved a distance ct in the time interval of t. In fact, it is not difficult to see that the whole function has moved a distance ct to the right in the time interval t, as shown by the dotted curve in Fig. 5.6. Therefore $f(x-ct)$ represents the function bodily moving (without changing the shape of the function) to the right with velocity c. Similarly, $f(x+ct)$ represents the function traveling to the left with velocity c.

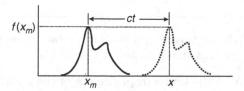

Fig. 5.6. Traveling wave. The *solid curve* shows what the function $f(x - ct)$ might be like at $t = 0$, the *dashed curve* shows what the function is at a later time t

It is now clear that $\sin[k_n(x + at)]$ and $\sin[k_n(x - at)]$ in the normal mode of (5.23) are two sine waves traveling in opposite directions with the same speed a. It is interesting to write (5.13) in terms of traveling waves

$$u(x,t) = \frac{1}{2} \sum_{n=1}^{\infty} a_n \left[\sin k_n(x + at) + \sin k_n(x - at)\right]. \tag{5.24}$$

Since initially at $t = 0$ the string is displaced into the form $f(x)$

$$f(x) = u(x,0) = \sum_{n=1}^{\infty} a_n \sin k_n x \tag{5.25}$$

clearly

$$f(x + at) = \sum_{n=1}^{\infty} a_n \sin k_n(x + at),$$

$$f(x - at) = \sum_{n=1}^{\infty} a_n \sin k_n(x - at).$$

Thus

$$u(x,t) = \frac{1}{2}f(x + at) + \frac{1}{2}f(x - at). \tag{5.26}$$

In other words, when the string is released at $t = 0$ from the displaced position $f(x)$, it will split into two equal parts, one traveling to the right and the other to the left with the same speed a.

However, there is a question about the range over which $f(x)$ is defined. The initial displacement $f(x)$ is defined between 0 and L. But now the argument is $x + at$ or $x - at$. Since t can take any value, the argument certainly exceeds the range between 0 and L. In order to have (5.26) valid for all t, we must extend the argument of the function beyond this range. Since (5.26) is obtained from (5.25) and $\sin k_n x = \sin \frac{n\pi}{L} x$, which is an odd periodic function with period $2L$, the functions in (5.26) must also have this property. So if we denote f^* to be the odd periodic extension of f with period $2L$, then

$$u(x,t) = \frac{1}{2}f^*(x + at) + \frac{1}{2}f^*(x - at) \tag{5.27}$$

is valid for all t.

Example 5.1.2. With the traveling wave interpretation, solve the problem of the previous example of a string pulled at the middle.

Solution 5.1.2. With the initial displacement of the string

$$u(x,0) = f(x) = \begin{cases} \dfrac{2h}{L}x & \text{if } 0 < x < \dfrac{L}{2} \\ \dfrac{2h}{L}(L-x) & \text{if } \dfrac{L}{2} < x < L \end{cases},$$

the subsequent displacement $u(x,t)$ is given by

$$u(x,t) = \frac{1}{2}f^*(x+at) + \frac{1}{2}f^*(x-at).$$

To interpret this expression, first we imagine the function $f(x)$ is antisymmetrically extended from 0 to $-L$, and then periodically extended from $-\infty$ to ∞ with a period $2L$. Then half of this extended function is moving to the right with velocity a and the other half moving to the left with the same velocity as shown in Fig. 5.7. The sum of these two traveling waves in the region $0 \le x \le L$ is the displacement $u(x,t)$ of the string.

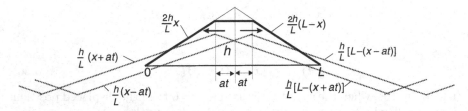

Fig. 5.7. The traveling wave interpretation of the solution of the wave equation with initial and boundary conditions. The displacement $u(x,t)$ is the sum of half of the extended initial function traveling to the left and half traveling to the right with the same velocity a

As a consequence, we see that at any time $t = T$, for $T \le \frac{L}{2a}$, the displacements are

$$u(x,T) = \frac{1}{2}\left\{\frac{2h}{L}(x+aT) + \frac{2h}{L}(x-aT)\right\}$$

$$= \frac{2h}{L}x \quad \text{if } 0 \le x \le \left(\frac{L}{2} - aT\right),$$

$$u(x,T) = \frac{1}{2}\left\{\frac{2h}{L}[L-(x+aT)] + \frac{2h}{L}(x-aT)\right\}$$

$$= \frac{2h}{L}\left(\frac{L}{2}-aT\right) \quad \text{if} \quad \left(\frac{L}{2}-aT\right) \le x \le \left(\frac{L}{2}+aT\right),$$

$$u(x,T) = \frac{1}{2}\left\{\frac{2h}{L}[L-(x+aT)] + \frac{2h}{L}[L-(x-aT)]\right\}$$

$$= \frac{2h}{L}(L-x) \quad \text{if} \quad \left(\frac{L}{2}+aT\right) \le x \le L.$$

These results are shown as the thick line in Fig. 5.7.

The displacements $u(x,t)$ as a function of time are shown in Fig. 5.8. In the left-hand side column, the positions of the string are shown at various times t. Each case is a superposition of two traveling waves, one to the left and one to the right, shown in the right-hand side column. Both of them are traveling with the same speed a. The sum of these two traveling waves describes the exact up and down motion of the string. It is interesting to compare Fig. 5.8 with Fig. 5.2. They describe the same motion but with two different interpretations.

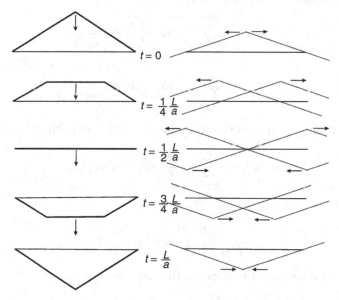

Fig. 5.8. The graph of the solution of the vibrating string with initial displacement $u(x,0)$ shown on the top of the left-hand side column. At various time t, the string will assume such positions as indicated in the left-hand side column. The positions are obtained as the superposition of a wave traveling to the right and a wave traveling to the left shown in the right-hand side column

Problems with Initial Velocity. Let us consider the case that the string is initially at rest but with a initial velocity of $g(x)$. The displacements of the string is the solution of the following problem:

$$\text{D.E.:} \quad \frac{\partial^2 u(x,t)}{\partial x^2} = \frac{1}{a^2}\frac{\partial^2 u(x,t)}{\partial t^2},$$

$$\text{B.C.:} \quad u(0,t) = 0; \ u(L,t) = 0,$$

$$\text{I.C.:} \quad u(x,0) = 0; \ u_t(x,0) = g(x).$$

With separation of variables, we will obtain (5.12) just as before, since the differential equation and the boundary conditions are the same

$$u(x,t) = \sum_{n=1}^{\infty}\left(a_n \cos\frac{n\pi a}{L}t + b_n \sin\frac{n\pi a}{L}t\right)\sin\frac{n\pi}{L}x.$$

The initial condition $u(x,0) = 0$ means that

$$u(x,0) = \sum_{n=1}^{\infty} a_n \sin\frac{n\pi}{L}x = 0.$$

Therefore all a_n must be equal to zero. Thus

$$u(x,t) = \sum_{n=1}^{\infty} b_n \sin\frac{n\pi a}{L}t \sin\frac{n\pi}{L}x$$

and

$$\frac{\partial}{\partial t}u(x,t) = \sum_{n=1}^{\infty} b_n\frac{n\pi a}{L}\cos\frac{n\pi a}{L}t \sin\frac{n\pi}{L}x. \tag{5.28}$$

It follows from the other initial condition $u_t(x,0) = g(x)$ that:

$$u_t(x,0) = \sum_{n=1}^{\infty} b_n\frac{n\pi a}{L}\sin\frac{n\pi}{L}x = g(x). \tag{5.29}$$

This is a Fourier sine series, therefore

$$b_n\frac{n\pi a}{L} = \frac{2}{L}\int_0^L g(x)\sin\frac{n\pi}{L}x\,dx.$$

Thus the solution $u(x,t)$ is given by the infinite series

$$u(x,t) = \sum_{n=1}^{\infty}\left[\frac{2}{n\pi a}\int_0^L g(x')\sin\frac{n\pi}{L}x'\,dx'\right]\sin\frac{n\pi a}{L}t \sin\frac{n\pi}{L}x.$$

This solution is expressed in terms of a summation of infinite standing waves. We can also express it in terms of the sum of two traveling waves. With trigonometric identity

$$\sin a \, \cos b = \frac{1}{2} \left[\sin(a + b) + \sin(a - b) \right],$$

we can write (5.28) as

$$\frac{\partial}{\partial t} u(x, t) = \sum_{n=1}^{\infty} b_n \frac{n\pi a}{L} \cos \frac{n\pi a}{L} t \, \sin \frac{n\pi}{L} x$$

$$= \frac{1}{2} \sum_{n=1}^{\infty} b_n \frac{n\pi a}{L} \sin \frac{n\pi}{L} (x + at) + \frac{1}{2} \sum_{n=1}^{\infty} b_n \frac{n\pi a}{L} \sin \frac{n\pi}{L} (x - at).$$

Using (5.29,) we can write this expression as

$$\frac{\partial}{\partial t} u(x, t) = \frac{1}{2} g^* (x + at) + \frac{1}{2} g^* (x - at),$$

where g^* is the odd periodic extension of g with period of $2L$, for the same reason as f^* is the odd periodic extension of f with a period of $2L$.

An integration of $\frac{\partial}{\partial t} u(x, t)$ will yield $u(x, t)$. The constant of integration is determined by the initial condition $u(x, 0) = 0$. This condition is satisfied by the following integral:

$$u(x, t) = \int_0^t \frac{\partial u(x, t')}{\partial t'} dt' = \frac{1}{2} \int_0^t g^* (x + at') \, dt' + \frac{1}{2} \int_0^t g^* (x - at') \, dt'.$$

With a change of variable

$$\tau = x + at', \qquad dt' = \frac{1}{a} d\tau,$$

the first integral on the right-hand side can be written as

$$\frac{1}{2} \int_0^t g^* (x + at') \, dt' = \frac{1}{2a} \int_x^{x+at} g^*(\tau) d\tau,$$

since at $t' = 0$, $\tau = x$ and at $t' = t$, $\tau = x + at$.

Similarly, the second integral can be written as

$$\frac{1}{2} \int_0^t g^* (x - at') \, dt' = -\frac{1}{2a} \int_x^{x-at} g^*(\tau) d\tau.$$

It follows that:

$$u(x, t) = \frac{1}{2a} \int_x^{x+at} g^*(\tau) d\tau - \frac{1}{2a} \int_x^{x-at} g^*(\tau) d\tau$$

$$= \frac{1}{2a} \int_{x-at}^{x+at} g^*(\tau) d\tau. \tag{5.30}$$

This is the solution for the case that the string has no initial displacement but is given an initial velocity $g(x)$.

Superposition of Solutions. If the string is given both an initial displacement and an initial velocity,

$$u(x,0) = f(x), \quad u_t(x,0) = g(x), \tag{5.31}$$

then the subsequent displacements can be written as the superposition of (5.27) and (5.30,) namely

$$u(x,t) = \frac{1}{2}\left[f^*(x-at) + f^*(x+at)\right] + \frac{1}{2a}\int_{x-at}^{x+at} g^*(\tau)\mathrm{d}\tau. \tag{5.32}$$

Note that both terms satisfy the homogeneous differential equation and the boundary conditions, while their sum clearly satisfies the initial conditions of (5.31).

In general the solution of a linear problem containing more than one nonhomogeneous conditions can be written as a sum of the solutions of problems each of which contains only one nonhomogeneous condition. The resolution of the original problem in this way, although not necessary, often simplifies the process of solving the problem.

5.1.5 Nonhomogeneous Wave Equations

Vibrating String with External Force. If there is an external force acting on the stretched string, then there will be an extra term in the governing differential equation. For example, if the weight of the string is not negligible, then in the derivation of (5.4), we must add to the equation the downward gravitational force, $-\rho\Delta x g$, where g is the constant gravitational acceleration. As a consequence, (5.6) becomes

$$\frac{\partial^2 u(x,t)}{\partial x^2} - \frac{g}{a^2} = \frac{1}{a^2}\frac{\partial^2 u(x,t)}{\partial t^2}. \tag{5.33}$$

Let us solve this equation with the same boundary and initial conditions as the previous problem:

$$u(0,t) = 0, \quad u(L,t) = 0,$$

$$u(x,0) = f(x), \quad u_t(x,0) = 0.$$

Since (5.33) is an nonhomogeneous equation, a straightforward application of separation of variables will not work. However, the following device will reduce this nonhomogeneous partial differential equation into a homogeneous partial differential equation plus an ordinary differential equation which we can solve. Let

$$u(x,t) = U(x,t) + \phi(x),$$

then

$$\frac{\partial^2 u(x,t)}{\partial x^2} = \frac{\partial^2 U(x,t)}{\partial x^2} + \frac{d^2 \phi(x)}{dx^2},$$

$$\frac{\partial^2 u(x,t)}{\partial t^2} = \frac{\partial^2 U(x,t)}{\partial t^2},$$

so the problem becomes

D.E.: $\quad \dfrac{\partial^2 U(x,t)}{\partial x^2} + \dfrac{d^2 \phi(x)}{dx^2} - \dfrac{g}{a^2} = \dfrac{1}{a^2} \dfrac{\partial^2 U(x,t)}{\partial t^2},$

B.C.: $\quad u(0,t) = U(0,t) + \phi(0) = 0, \quad u(L,t) = U(L,t) + \phi(L) = 0,$

I.C.: $\quad u(x,0) = U(x,0) + \phi(x) = f(x), \quad u_t(x,0) = U_t(x,0) = 0.$

Now we require

$$\frac{d^2 \phi(x)}{dx^2} - \frac{g}{a^2} = 0,$$

$$\phi(0) = 0, \quad \phi(L) = 0.$$

This is a second-order ordinary differential equation with two boundary conditions, which can be readily solved to give

$$\phi(x) = \frac{g}{2a^2} \left(x^2 - Lx \right).$$

With $\phi(x)$ so chosen, what we are left with are the differential equation and boundary and initial conditions for $U(x,t)$

$$\frac{\partial^2 U(x,t)}{\partial x^2} = \frac{1}{a^2} \frac{\partial^2 U(x,t)}{\partial t^2},$$

$$U(0,t) = 0, \quad U(L,t) = 0,$$

$$U(x,0) = f(x) - \phi(x), \quad U_t(x,0) = 0.$$

Note that other than the modification of one of the initial conditions, this is the same equation we solved before. Therefore we can write down its solutions immediately,

$$U(x,t) = \sum_{n=1}^{\infty} b_n \cos \frac{n\pi a}{L} t \, \sin \frac{n\pi}{L} x,$$

$$b_n = \frac{2}{L} \int_0^L \left[f(x') - \frac{g}{2a^2} \left(x'^2 - Lx' \right) \right] \sin \frac{n\pi}{L} x' \, dx'.$$

It follows that the displacements of the string, including the effect of its own weight, are given by:

$$u(x,t) = \frac{g}{2a^2}\left(x^2 - Lx\right)$$

$$+ \sum_{n=1}^{\infty}\left\{\frac{2}{L}\int_0^L\left[f(x) - \frac{g}{2a^2}(x^2 - Lx)\right]\sin\frac{n\pi}{L}x\,dx\right\}\cos\frac{n\pi a}{L}t\,\sin\frac{n\pi}{L}x.$$

Forced Vibration and Resonance. Now suppose that the string fixed at both ends is influenced by a periodic external force per unit length $F(t) = F_1\cos\omega t$. In this case, the string will satisfy the nonhomogeneous partial differential equation

$$a^2\frac{\partial^2 u(x,t)}{\partial x^2} + F_0\cos\omega t = \frac{\partial^2 u(x,t)}{\partial t^2}, \tag{5.34}$$

where $F_0 = F_1/\rho$. The boundary conditions remain the same

$$u(0,t) = 0, \quad u(L,t) = 0.$$

If the string is initially at rest in equilibrium when the external force begins to act, then the displacement $u(x,t)$ will also have to satisfy the initial conditions

$$u(x,0) = 0, \quad u_t(x,0) = 0.$$

Since the external force is purely sinusoidal, it is relatively easy to find a solution to satisfy the differential equation and the boundary conditions. Just like solving ordinary differential equation, we know that the particular solution will have to oscillate with $\cos\omega t$. Therefore, let us take the trial solution

$$v(x,t) = X(x)\cos\omega t.$$

Replace $u(x,t)$ with $v(x,t)$ in (5.34), we have

$$a^2 X''(x)\cos\omega t + F_0\cos\omega t = -\omega^2 X(x)\cos\omega t$$

or

$$X''(x) = -\frac{\omega^2}{a^2}X(t) - \frac{F_0}{a^2},$$

which yields the solution

$$X(x) = A\cos\frac{\omega x}{a} + B\sin\frac{\omega x}{a} - \frac{F_0}{\omega^2}.$$

The boundary conditions require

$$X(0) = X(L) = 0.$$

Thus

$$X(0) = A - \frac{F_0}{\omega^2} = 0, \quad A = \frac{F_0}{\omega^2}.$$

Furthermore

$$X(L) = \frac{F_0}{\omega^2} \cos \frac{\omega L}{a} + B \sin \frac{\omega L}{a} - \frac{F_0}{\omega^2} = 0$$

or

$$B = \frac{F_0}{\omega^2} \frac{\left(1 - \cos \frac{\omega L}{a}\right)}{\sin \frac{\omega L}{a}},$$

except for $\omega = n\pi a/L$ with even n, in that case $B = 0$. Therefore, in general

$$X(x) = \frac{F_0}{\omega^2} \cos \frac{\omega x}{a} + \frac{F_0}{\omega^2} \frac{\left(1 - \cos \frac{\omega L}{a}\right)}{\sin \frac{\omega L}{a}} \sin \frac{\omega x}{a} - \frac{F_0}{\omega^2}, \qquad (5.35)$$

$$v(x,t) = X(x) \cos \omega t.$$

But this solution does not satisfy the initial conditions. Therefore we resort to the method of splitting the solution into two parts

$$u(x,t) = v(x,t) + U(x,t).$$

In terms of v and U, the original equation and the boundary of initial conditions become

$$a^2 \frac{\partial^2 v(x,t)}{\partial x^2} + a^2 \frac{\partial^2 U(x,t)}{\partial x^2} + F_0 \cos \omega t = \frac{\partial^2 v(x,t)}{\partial t^2} + \frac{\partial^2 U(x,t)}{\partial t^2},$$

$$u(0,t) = v(0,t) + U(0,t) = 0,$$
$$u(L,t) = v(L,t) + U(L,t) = 0,$$
$$u(x,0) = v(x,0) + U(x,0) = 0,$$
$$u_t(x,0) = v_t(x,0) + U_t(x,0) = 0.$$

Since

$$a^2 \frac{\partial^2 v(x,t)}{\partial x^2} + F_0 \cos \omega t = \frac{\partial^2 v(x,t)}{\partial t^2},$$
$$v(0,t) = 0, \qquad v(L,t) = 0,$$

and

$$v(x,0) = X(x), \qquad v_t(x,0) = -\omega X(x) \sin 0 = 0.$$

Hence

$$a^2 \frac{\partial^2 U(x,t)}{\partial x^2} = \frac{\partial^2 U(x,t)}{\partial t^2},$$
$$U(0,t) = 0, \quad U(L,t) = 0,$$
$$U(x,0) = -X(x), \quad U_t(x,0) = 0.$$

This is the homogeneous differential equation we solved before

$$U(x,t) = \sum_{n=1}^{\infty} \left(-\frac{2}{L} \int_0^L X(x') \sin \frac{n\pi}{L} x' \, dx' \right) \sin \frac{n\pi}{L} x \cos \frac{n\pi a}{L} t. \qquad (5.36)$$

Therefore the solution $u(x,t)$ is given by

$$u(x,t) = X(x) \cos \omega t + U(x,t)$$

with $X(x)$ given by (5.35) and $U(x,t)$ given by (5.36).

This solution is valid for any ω. However, if ω approach $\omega_m = \frac{m\pi a}{L}$ with an odd integer m, then $X(x)$ in (5.35) approaches ∞, thus resonance occurs. But if $\omega = \frac{m\pi a}{L}$ with an even integer m, then

$$X(x) = \frac{F_0}{\omega^2} \cos \frac{m\pi x}{L} - \frac{F_0}{\omega^2}$$

and resonance does not occur in this case.

5.1.6 D'Alembert's Solution of Wave Equations

Using the separation of variables, we have solved the vibrating string problem by first finding the eigenvalues and eigenfunctions dictated by the boundary conditions. In the next step, we used the initial conditions to determine the constants in the Fourier series of the solution. Now we will introduce a method of doing just the opposite. We will first solve the initial values problem, and then find the solution to satisfy the boundary conditions.

Let us solve the following pure initial values problem:

$$\text{D.E.:} \quad \frac{\partial^2 u(x,t)}{\partial x^2} = \frac{1}{a^2} \frac{\partial^2 u(x,t)}{\partial t^2}$$

$$\text{I.C.:} \quad u(x,0) = f(x), \; u_t(x,0) = g(x),$$

for $0 < t < \infty$ and $-\infty < x < \infty$. It turns out the general solution of this equation can be found by a change of variables:

$$\zeta = x + at$$
$$\eta = x - at.$$

According to the chain rule

$$\frac{\partial}{\partial x} = \frac{\partial \zeta}{\partial x} \frac{\partial}{\partial \zeta} + \frac{\partial \eta}{\partial x} \frac{\partial}{\partial \eta} = \frac{\partial}{\partial \zeta} + \frac{\partial}{\partial \eta},$$

$$\frac{\partial}{\partial t} = \frac{\partial \zeta}{\partial t} \frac{\partial}{\partial \zeta} + \frac{\partial \eta}{\partial t} \frac{\partial}{\partial \eta} = a\frac{\partial}{\partial \zeta} - a\frac{\partial}{\partial \eta},$$

so the differential equation becomes

$$\left(\frac{\partial}{\partial\zeta}+\frac{\partial}{\partial\eta}\right)\left(\frac{\partial}{\partial\zeta}+\frac{\partial}{\partial\eta}\right)u=\frac{1}{a^2}\left(a\frac{\partial}{\partial\zeta}-a\frac{\partial}{\partial\eta}\right)\left(a\frac{\partial}{\partial\zeta}-a\frac{\partial}{\partial\eta}\right)u$$

or

$$\left(\frac{\partial^2}{\partial\zeta^2}+2\frac{\partial^2}{\partial\zeta\partial\eta}+\frac{\partial^2}{\partial\eta^2}\right)u=\left(\frac{\partial^2}{\partial\zeta^2}-2\frac{\partial^2}{\partial\zeta\partial\eta}+\frac{\partial^2}{\partial\eta^2}\right)u.$$

Clearly

$$\frac{\partial^2}{\partial\zeta\partial\eta}u=0.$$

This new equation can be solved easily by two straightforward integrations. Integration with respect to ζ gives an arbitrary function $A(\eta)$ of η, that is

$$\frac{\partial}{\partial\eta}u=A(\eta),$$

since

$$\frac{\partial}{\partial\zeta}\left(\frac{\partial}{\partial\eta}u\right)=\frac{\partial}{\partial\zeta}A(\eta)=0.$$

The second integration with respect to η gives

$$u=\int A(\eta)\,d\eta+G(\zeta),$$

where $G(\zeta)$ is an arbitrary function of ζ. Since $A(\eta)$ is arbitrary, we might as well write $F(\eta)$ in place of $\int A(\eta)\,d\eta$. Thus

$$u(\zeta,\eta)=F(\eta)+G(\zeta).$$

Substituting back the original variables, we have

$$u(x,t)=F(x-at)+G(x+at). \tag{5.37}$$

Thus the general solution of the wave equation is a sum of two arbitrary moving waves, each moving in opposite direction with velocity a.

It is easy to see that (5.37) is indeed the solution of the wave equation. We can use the chain rule

$$\frac{\partial u(x,t)}{\partial t}=\frac{dF(x-at)}{d(x-at)}\frac{\partial(x-at)}{\partial t}+\frac{dG(x+at)}{d(x+at)}\frac{\partial(x+at)}{\partial t}$$

$$=-aF'(x-at)+aG'(x+at),$$

$$\frac{\partial^2 u(x,t)}{\partial t^2}=-a\frac{dF'(x-at)}{d(x-at)}\frac{\partial(x-at)}{\partial t}+a\frac{dG'(x+at)}{d(x+at)}\frac{\partial(x+at)}{\partial t}$$

$$=a^2F''(x-at)+a^2G''(x+at).$$

Similarly, we obtain

$$\frac{\partial^2 u(x,t)}{\partial x^2} = F''(x - at) + G''(x + at).$$

It is readily seen that the differential equation

$$\frac{\partial^2 u(x,t)}{\partial x^2} = \frac{1}{a^2}\frac{\partial^2 u(x,t)}{\partial t^2}$$

is satisfied.

Now if we impose the initial conditions

$$u(x,0) = f(x),$$
$$u_t(x,0) = g(x),$$

we have

$$u(x,0) = F(x) + G(x) = f(x), \tag{5.38}$$

$$u_t(x,0) = -aF'(x) + aG'(x) = g(x). \tag{5.39}$$

Integrating (5.39) from any fixed point, say 0, to x gives

$$-aF(x) + aG(x) + aF(0) - aG(0) = \int_0^x g(x')dx'$$

or

$$-F(x) + G(x) = \frac{1}{a}\int_0^x g(x')dx' - F(0) + G(0). \tag{5.40}$$

Solving for $F(x)$ and $G(x)$ from (5.38) and (5.40), we get

$$F(x) = \frac{1}{2}f(x) - \frac{1}{2a}\int_0^x g(x')dx' + \frac{1}{2}\left[F(0) - G(0)\right], \tag{5.41}$$

$$G(x) = \frac{1}{2}f(x) + \frac{1}{2a}\int_0^x g(x')dx' - \frac{1}{2}\left[F(0) - G(0)\right]. \tag{5.42}$$

If we replace the argument x by $x - at$ on both sides of (5.41), we can write

$$F(x - at) = \frac{1}{2}f(x - at) - \frac{1}{2a}\int_0^{x-at} g(x')dx' + \frac{1}{2}\left[F(0) - G(0)\right].$$

Similarly, replacing the argument x by $x + at$ on both sides of (5.42), we have

$$G(x + at) = \frac{1}{2}f(x + at) + \frac{1}{2a}\int_0^{x+at} g(x')dx' - \frac{1}{2}\left[F(0) - G(0)\right].$$

Thus

$$u(x,t) = F(x - at) + G(x + at) = \frac{1}{2}\left[f(x - at) + f(x + at)\right]$$

$$+\frac{1}{2a}\int_0^{x+at} g(x')\mathrm{d}x' - \frac{1}{2a}\int_0^{x-at} g(x')\mathrm{d}x'.$$

Reversing the upper and lower limits of the last integral and combining it with the preceding one, we obtain

$$u(x,t) = \frac{1}{2}\left[f(x - at) + f(x + at)\right] + \frac{1}{2a}\int_{x-at}^{x+at} g(x')\mathrm{d}x'. \tag{5.43}$$

This is the solution for $-\infty < x < \infty$ without any boundary condition.

Now suppose that the string is of finite length from 0 to L, and both ends are fixed, so we have the following boundary conditions:

$$u(0,t) = 0, \qquad u(L,t) = 0.$$

In this case, $f(x)$ and $g(x)$ are, of course, defined only in the range of $0 \leq x \leq L$. We want to find the solution that satisfy these additional conditions.

First, if $u(0,t) = 0$, then according to (5.43)

$$u(0,t) = \frac{1}{2}\left[f(-at) + f(at)\right] + \frac{1}{2a}\int_{-at}^{+at} g(x')\mathrm{d}x' = 0.$$

Since $f(x)$ and $g(x)$ are two unrelated functions, to satisfy this condition we must have

$$f(-at) + f(at) = 0, \tag{5.44}$$

$$\int_{-at}^{+at} g(x')\mathrm{d}x' = 0. \tag{5.45}$$

It is clear from (5.44) that

$$f(x) = -f(-x).$$

Therefore $f(x)$ must be an odd function. In other words, $f(x)$ should be antisymmetrically extended to negative x.

The integral in (5.45) can be written as

$$\int_{-at}^0 g(x')\mathrm{d}x' + \int_0^{at} g(x)\mathrm{d}x = 0.$$

With a change of variable $x' = -x$ in the first integral, we can write it as

$$\int_{-at}^0 g(x')\mathrm{d}x' = \int_0^{at} g(-x)\mathrm{d}x.$$

For

$$\int_0^{at} g(-x)\mathrm{d}x + \int_0^{at} g(x)\mathrm{d}x = 0,$$

we must have

$$g(-x) = -g(x).$$

Therefore, $g(x)$ also is an odd function.

Similarly, to satisfy the other boundary condition

$$u(L,t) = \frac{1}{2}\left[f(L-at) + f(L+at)\right] + \frac{1}{2a}\int_{L-at}^{L+at} g(x')\mathrm{d}x' = 0,$$

we require

$$f(L-at) + f(L+at) = 0, \tag{5.46}$$

$$\int_{L-at}^{L+at} g(x')\mathrm{d}x' = 0. \tag{5.47}$$

Thus

$$f(L-ct) = -f(L+ct).$$

Since $f(x)$ is an odd function, so

$$f(L-ct) = -f(-L+ct).$$

Therefore

$$f(-L+ct) = f(L+ct).$$

It follows that:

$$f(-L+ct+2L) = f(L+ct) = f(-L+ct)$$

This shows that $f(x)$ should be a periodic function of period $2L$.

Now (5.47) can be written as

$$\int_{L-at}^{L} g(x')\mathrm{d}x' + \int_{L}^{L+at} g(x')\mathrm{d}x' = 0.$$

With a change of variable, $x' = L - x$ in the first integral, it becomes

$$\int_{L-at}^{L} g(x')\mathrm{d}x' = \int_0^{ct} g(L-x)\mathrm{d}x.$$

Change the variable x' to $L + x$ in the second integral, we can write

$$\int_{L}^{L+at} g(x')\mathrm{d}x' = \int_0^{ct} g\left(L+x\right)\mathrm{d}x.$$

So for

$$\int_0^{ct} g(L-x)\mathrm{d}x + \int_0^{ct} g(L+x)\mathrm{d}x = 0,$$

we must have

$$g(L-x) = -g(L+x).$$

Since $g(L-x)$ is an odd function, so

$$g(L-x) = -g(-L+x).$$

Thus

$$g(-L+x) = g(L+x).$$

It follows that:

$$g(-L+x+2L) = g(L+x) = g(-L+x)$$

Therefore $g(x)$ should also be a periodic function of period $2L$.

Thus, if we define $f^*(x)$ and $g^*(x)$ as the odd periodic functions of period $2L$, whose definitions in $0 \le x \le L$ are, respectively, $f(x)$ and $g(x)$, then the solution of the problem is given by

$$u(x,t) = \frac{1}{2}[f^*(x-at) + f^*(x+at)] + \frac{1}{2a}\int_{x-at}^{x+at} g^*(\tau)\mathrm{d}\tau, \qquad (5.48)$$

which is exactly the same as (5.32).

This is known as D'Alembert's solution. The French mathematician Jean le Rond D'Alembert (1717–1783) first discovered it around 1750s. This method is very elegant but unfortunately it is limited to the solution of this type of equations. The method of separation of variable which, as we have seen, can also give the same solution, is more general. We shall use it to solve many other types of partial differential equations.

Example 5.1.3. A string of length L with tension T and density ρ, fixed at both ends, is initially given the displacement of a small triangular pulse at $x = L/4$, shown on the top line of Fig. 5.9, and is released from rest. Determine its subsequent motion.

Solution 5.1.3. Let the initial triangular pulse be $f(x)$, the subsequent displacement is simply given by

$$u(x,t) = \frac{1}{2}[f^*(x-at) + f^*(x+at)]$$

where f^* is equal to $f(x)$ in the range of $0 \le x \le L$, outside this range f^* is the antisymmetrical periodic extension of $f(x)$ with a period of $2L$. This extension is shown in Fig. 5.9 as the dotted lines. The actual displacements of

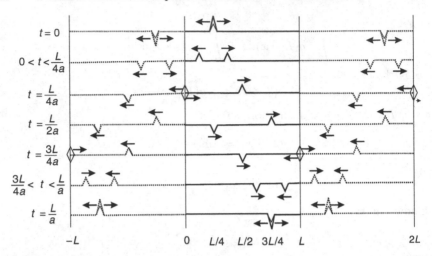

Fig. 5.9. Time development of an initial triangular pulse at $x = L/4$ on a string of length L, fixed at both ends

the string are shown in Fig. 5.9 as the thick dark lines in the "physical" space between 0 and L. However, after $t > L/4a$, what happens in the physical space is the result of some dotted pulse moving from the "mathematical" space into the "physical" space. In other words, the finite string of length L can be regarded as if it is only a segment of an infinitely long line. The pulses are moving on this infinite line. What happens in the segment between 0 and L is the displacements of the real string that we can see. Other parts of this infinitely long line are only mathematical constructs which are used to predict what will happen in the real string.

Soon after the pulse is released, it splits into two equal parts and moving in the opposite direction with the same velocity a which is equal to $\sqrt{T/\rho}$. This motion is shown in the second line of Fig. 5.9. At $t = L/4a$, the left pulse reaches the end point at $x = 0$. It will gradually disappear as shown in the third line of the figure. Then an identical pulse will reappear but upside down. In the time interval $L/4a < t < 3L/4a$, there are two pulses $L/2$ distance apart, one up and one down, both moving to the right as shown in the fourth line. At $t = 3L/4a$, the right pulse reaches the end point at $x = L$ and gradually disappears. This is shown in the fifth line. Soon after an identical pulse reappears upside down, and moves to the left. In the time interval $3L/4a < t < L/a$, two pulses, both upside down, are moving toward each other. This is shown in the sixth line. The end of the first half cycle is at $t = L/a$. At that time, the original pulse appears upside down at $x = 3L/4$. This is shown in the last line of Fig. 5.9. After that the motion will repeat itself in reverse direction until $t = 2L/a$. That is the end of the first cycle and the string will return to its original shape. These are a well-known facts which can be easily checked.

Example 5.1.4. When a piano wire is struck by a narrow hammer at x_0, a localized velocity is imparted to that point. At that moment, the wire is still at rest, but it will start to vibrate thereafter. Describe the motion assuming $a = \sqrt{T/\rho}$ and $0 < x_0 < L/4$.

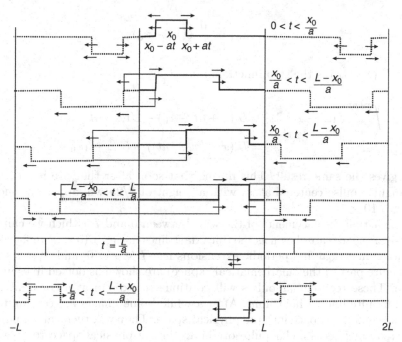

Fig. 5.10. The motion of wire of length L after a localized velocity is imparted at x_0

Solution 5.1.4. To find the displacements of the wire, we have to solve the boundary value problem of the vibrating string with the following initial conditions

$$u(x,0) = 0 \quad \text{and} \quad u_t(x,0) = \delta(x - x_0),$$

where $\delta(x - x_0)$ is a delta function. According to (5.48),

$$u(x,t) = \frac{1}{2a} \int_{x-at}^{x+at} \delta^*(x' - x_0)\, dx',$$

where $\delta^*(x - x_0)$ is the odd periodic function of period $2L$, whose definition between 0 and L is the delta function $\delta(x - x_0)$. By the definition of delta function

$$\int_{x-at}^{x+at} \delta(x' - x_0)\, dx' = \begin{cases} 1 & \text{if } x - at < x_0 < x + at \\ 0 & \text{otherwise} \end{cases}.$$

By adding at to both sides, we see that the condition $x - at < x_0$ is equivalent to $x < x_0 + at$. Similarly, $x_0 < x + at$ is equivalent to $x_0 - at < x$. That is, between $x_0 - at$ and $x_0 + at$, the integral is equal to 1, outside this range, it is equal to zero.

This result can also be obtained by using

$$\delta\left(x - x_0\right) = \frac{\mathrm{d}}{\mathrm{d}x} U\left(x - x_0\right),$$

where $U\left(x - x_0\right)$ is the step function, and

$$\int_{x-at}^{x+at} \delta\left(x' - x_0\right) \mathrm{d}x' = U\left(x + at - x_0\right) - U(x - at - x_0)$$
$$= U\left(x - \left(x_0 - at\right)\right) - U\left(x - \left(x_0 + at\right)\right).$$

This gives the same result. This means that soon after the wire is struck, a rectangular pulse centered at x_0 with a height of $1/2a$ will appear as shown in Fig. 5.10.

The actual displacements of the wire between 0 and L (which we call the "physical" space) are shown as the thick dark lines in Fig. 5.10. The images due to antisymmetrical and periodic extensions in $-L < x < 0$ and $0 < x < 2L$ (which are part of the "mathematical" space) are shown as dotted lines in the figure. These rectangular pulses will continue to expand at a constant rate on an infinite line without limit. After a while, some images are coming from the mathematical space into the physical space. They will then cancel part of the expanding rectangular pulse originally in the physical space to give the actual displacements of the wire. As a result, the motion of the wire appears as follows.

First the width of the rectangular pulse will expand at a constant rate as shown in the first line of Fig. 5.10. At $t = x_0/a$, the left edge of this rectangular pulse reaches the end point at $x = 0$. Immediately it bounces back and moves toward the right. In the time interval $x_0/a < t < (L - x_0)/a$, a rectangular pulse of constant width of $2x_0$ is moving to the right with velocity of a. This motion is shown in the second and the third line of Fig. 5.10. At $t = (L-x_0)/a$, the right edge of the rectangular pulse reaches the end point at $x = L$ and bounces back. The width of the rectangular pulse starts to shrink as shown in the fourth line of Fig. 5.10. At $t = L/a$, the pulse disappears. What happened is that the two negative rectangular pulses have moved from the mathematical space so far into the physical space so that they touch each other. As a consequence, they completely cancel the positive rectangular pulse. This is shown in line five. Soon after that, the two negative rectangular pulses overlap and over compensate the positive rectangular pulse, as a result another reactangular pulse reappears up side down. This is shown in the last line of Fig. 5.10. After that the motion will repeat itself in reverse order. The end of the first cycle is at $t = 2L/a$. At that moment, the pulse will disappear, but will soon reappear to repeat the motion.

5.2 Two-Dimensional Wave Equations

5.2.1 The Governing Equation of a Vibrating Membrane

A vibrating membrane such as a drumhead is a two-dimensional version of a vibrating string. We assume that the membrane is stretched uniformly under a tension T per unit length. That is, at each point of the membrane the tension per unit length along any straight line through that point, independent of the orientation of the line, is perpendicular to that line with a magnitude T.

Let us consider the vibration of such a membrane; we shall suppose that the density ρ defined as the mass per unit area is a constant. If its equilibrium position is taken as the xy plane, then we are concerned with displacements $z(x, y, t)$ perpendicular to this plane. Consider a small rectangular element of sides $\Delta x, \Delta y$ shown in Fig. 5.11. We proceed as before. We assume the weight of the element is negligible compared with the tensile force. Applying Newton's second law to the element $\Delta x \, \Delta y$, we have

$$T_2 \, \Delta y \sin \theta_2 - T_1 \, \Delta y \sin \theta_1 + T_4 \, \Delta x \sin \theta_4 - T_3 \, \Delta x \sin \theta_3 = \rho \, \Delta x \, \Delta y \frac{\partial^2 z}{\partial t^2}. \quad (5.49)$$

Now there is no horizontal motion either in x- or in y-directions, therefore

$$T_2 \cos \theta_2 = T_1 \cos \theta_1, \qquad T_4 \cos \theta_4 = T_3 \cos \theta_3. \quad (5.50)$$

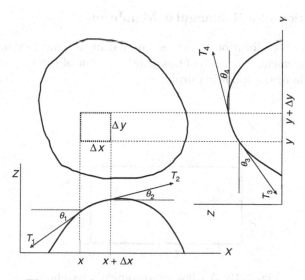

Fig. 5.11. A membrane under uniform tension

We assume the slopes $\frac{\partial z}{\partial x}$ and $\frac{\partial z}{\partial y}$ are uniformly small over the domain, so these horizontal components in (5.50) can be regarded as the tension T of membrane. Since the tension is uniform, so

$$T_2 \cos \theta_2 = T_1 \cos \theta_1 = T_4 \cos \theta_4 = T_3 \cos \theta_3 = T.$$

Dividing both sides of (5.49) by the appropriate expression of T, we obtain

$$\Delta y \tan \theta_2 - \Delta y \tan \theta_1 + \Delta x \tan \theta_4 - \Delta x \tan \theta_3 = \frac{1}{T} \rho \, \Delta x \, \Delta y \frac{\partial^2 z}{\partial t^2}$$

or

$$\Delta y \left(\left. \frac{\partial z}{\partial x} \right|_{x+\Delta x} - \left. \frac{\partial z}{\partial x} \right|_x \right) + \Delta x \left(\left. \frac{\partial z}{\partial y} \right|_{y+\Delta y} - \left. \frac{\partial z}{\partial y} \right|_y \right) = \frac{1}{T} \rho \, \Delta x \, \Delta y \frac{\partial^2 z}{\partial t^2}.$$

In the limit of $\Delta x \to 0$, $\Delta y \to 0$, the last equation can be written as

$$\Delta y \frac{\partial^2 z}{\partial x^2} \Delta x + \Delta x \frac{\partial^2 z}{\partial y^2} \Delta y = \frac{1}{T} \rho \, \Delta x \, \Delta y \frac{\partial^2 z}{\partial t^2}.$$

It follows that:

$$\frac{\partial^2 z}{\partial x^2} + \frac{\partial^2 z}{\partial y^2} = \frac{1}{v^2} \frac{\partial^2 z}{\partial t^2},$$

where

$$v = \sqrt{\frac{T}{\rho}}.$$

5.2.2 Vibration of a Rectangular Membrane

Let us consider the vibration of the rectangular membrane shown in Fig. 5.12.
The displacements $z\,(x, y, t)$ of the membrane out of the xy plane are given by the solution of the following problem:

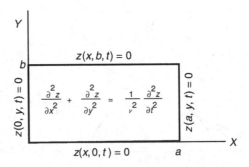

Fig. 5.12. A vibrating rectangular membrane

D.E.: $\dfrac{\partial^2 z}{\partial x^2} + \dfrac{\partial^2 z}{\partial y^2} = \dfrac{1}{v^2}\dfrac{\partial^2 z}{\partial t^2},$

B.C.: $z(0,y,t) = 0, \quad z(a,y,t) = 0,$

$z(x,0,t) = 0, \quad z(x,b,t) = 0,$

I.C.: $z(x,y,0) = f(x,y), \quad z_t(x,y,0) = g(x,y).$

We will again use the method of separation of variables,

$$z(x,y,t) = X(x)Y(y)T(t).$$

The differential equation can be written as

$$X''(x)Y(y)T(t) + X(x)Y''(y)T(t) = \dfrac{1}{v^2}X(x)Y(y)T''(t).$$

Dividing by $X(x)Y(y)T(t)$, we have

$$\dfrac{X''(x)}{X(x)} + \dfrac{Y''(y)}{Y(y)} = \dfrac{1}{v^2}\dfrac{T''(t)}{T(t)}.$$

The left-hand side is a function of x and y, and the right-hand side is a function of t. Since x, y, t are independent variables, both sides must be equal to the same constant

$$\dfrac{1}{v^2}\dfrac{T''(t)}{T(t)} = \lambda, \qquad (5.51)$$

$$\dfrac{X''(x)}{X(x)} + \dfrac{Y''(y)}{Y(y)} = \lambda.$$

We can separate the x and y dependence by writing

$$\dfrac{X''(x)}{X(x)} = \lambda - \dfrac{Y''(y)}{Y(y)}.$$

The left-hand side is a function of x alone and the right-hand side is a function of y alone, so it follows that both sides must be equal to the same constant. The constant must be a negative number for the same reason as the separation constant in the vibrating string problem is negative. Otherwise the boundary conditions in x would not be satisfied. Therefore we write

$$\lambda - \dfrac{Y''(y)}{Y(y)} = -\alpha^2,$$

$$\dfrac{X''(x)}{X(x)} = -\alpha^2.$$

Thus,

$$X''(x) = -\alpha^2 X(x), \qquad (5.52)$$

$$Y''(y) = \left(\lambda + \alpha^2\right) Y(y).$$

Since λ is a constant yet to be determined, so we can combine it with α^2 as another constant. Again to satisfy the boundary conditions in y, that constant must be a negative number, therefore we write

$$\lambda + \alpha^2 = -\beta^2 \tag{5.53}$$

and

$$Y''(y) = -\beta^2 Y(y). \tag{5.54}$$

The boundary conditions in terms of $X(x)$ and $Y(y)$ are

$$X(0) = X(a) = 0; \qquad Y(0) = Y(b) = 0.$$

The solution to (5.52) and (5.54), together with these boundary conditions are

$$X(x) = \sin \alpha x, \qquad \alpha = \frac{n\pi}{a}, \qquad n = 1, 2, 3, \ldots$$
$$Y(y) = \sin \beta y, \qquad \beta = \frac{m\pi}{b}, \qquad m = 1, 2, 3, \ldots.$$

To emphasize the fact that for each integer n, there is a separate eigenfunction solution, we write

$$X_n(x) = \sin \frac{n\pi}{a} x.$$

Similarly, for each m, there is a separate $Y_m(y)$,

$$Y_m(y) = \sin \frac{m\pi}{b} y.$$

It follows from (5.53) that for each pair of n and m, there is a constant λ

$$\lambda_{nm} = -\left[\left(\frac{n\pi}{a}\right)^2 + \left(\frac{m\pi}{b}\right)^2\right].$$

Clearly, λ_{nm} depends on two integers n and m. For each λ_{nm}, there is a time-dependent equation as seen from (5.51)

$$T''_{nm}(t) = \lambda_{nm} v^2 T_{nm}(t).$$

Therefore

$$T_{nm}(t) = a_{nm} \cos \omega_{nm} t + b_{nm} \sin \omega_{nm} t,$$

where

$$\omega_{nm} = \sqrt{-\lambda_{nm} v^2} = \left[\left(\frac{n}{a}\right)^2 + \left(\frac{m}{b}\right)^2\right]^{1/2} \pi v.$$

Thus, for each pair of n and m, we have a solution

$$z_{nm}(x, y, t) = X_n(x) Y_m(t) T_{nm}(t).$$

We can regard this as the (n, m) normal mode. The complete solution to problem of vibrating rectangular membrane can be expressed as a superposition of these normal modes

$$z(x, y, t) = \sum_{n=1}^{\infty} \sum_{m=1}^{\infty} z_{nm}(x, y, t)$$

$$= \sum_{n=1}^{\infty} \sum_{m=1}^{\infty} (a_{nm} \cos \omega_{nm} t + b_{nm} \sin \omega_{nm} t) \sin \frac{n\pi x}{a} \sin \frac{m\pi y}{b}. \quad (5.55)$$

The coefficients a_{nm} and b_{nm} are determined by the initial conditions. Using the condition, at $t = 0$, $z(x, y, 0) = f(x, y)$, we have

$$z(x, y, 0) = \sum_{n=1}^{\infty} \sum_{m=1}^{\infty} a_{nm} \sin \frac{n\pi x}{a} \sin \frac{m\pi y}{b} = f(x, y).$$

This is known as a *double Fourier series*. We will assume that $f(x, y)$ can indeed be expressed in such a series. If we define $R_m(x)$ as

$$R_m(x) = \sum_{n=1}^{\infty} a_{nm} \sin \frac{n\pi x}{a}, \quad (5.56)$$

then

$$f(x, y) = \sum_{m=1}^{\infty} R_m(x) \sin \frac{m\pi y}{b}.$$

For a fixed x, this is a half-range Fourier sine expansion of $f(x, y)$ on $0 \le y \le b$. Therefore

$$R_m(x) = \frac{2}{b} \int_0^b f(x, y) \sin \frac{m\pi y}{b} dy. \quad (5.57)$$

By definition, $R_m(x)$ is also given by (5.56), which is a half-range Fourier sine expansion of $R_m(x)$ on $0 \le x \le a$. So

$$a_{nm} = \frac{2}{a} \int_0^a R_m(x) \sin \frac{n\pi x}{a} dx.$$

Putting $R_m(x)$ of (5.57) into this expression, we obtain

$$a_{nm} = \frac{4}{ab} \int_0^a \int_0^b f(x, y) \sin \frac{n\pi x}{a} \sin \frac{m\pi y}{b} dx \, dy. \quad (5.58)$$

This is the generalized Euler coefficients for the double Fourier series.

To determine b_{nm}, we differentiate (5.55) termwise with respect to t, using the condition $z_t(x, y, 0) = g(x, y)$, we obtain

$$\left. \frac{\partial z}{\partial t} \right|_{t=0} = \sum_{n=1}^{\infty} \sum_{m=1}^{\infty} \omega_{nm} b_{nm} \sin \frac{n\pi x}{a} \sin \frac{m\pi y}{b} = g(x, y).$$

Proceeding as before, we obtain

$$b_{nm} = \frac{1}{\omega_{nm}} \frac{4}{ab} \int_0^a \int_0^b g(x,y) \sin \frac{n\pi x}{a} \sin \frac{m\pi y}{b} dx\, dy. \qquad (5.59)$$

If the initial conditions are

$$u(x,y,0) = f(x,y), \qquad u_t(x,y,0) = g(x,y) = 0,$$

then $b_{nm} = 0$ and

$$z(x,y,t) = \sum_{n=1}^{\infty} \sum_{m=1}^{\infty} a_{nm} \cos \omega_{nm} t \sin \frac{n\pi x}{a} \sin \frac{m\pi y}{b}$$

with a_{nm} given by (5.58).

In general, since

$$a_{nm} \cos \omega_{nm} t + b_{nm} \sin \omega_{nm} t = c_{nm} \cos(\omega_{nm} t + \delta_{nm}),$$

we can write the (n,m) normal mode as

$$z_{nm}(x,y,t) = c_{nm} \cos(\omega_{nm} t + \delta_{nm}) \sin \frac{n\pi x}{a} \sin \frac{m\pi y}{b}. \qquad (5.60)$$

Its frequency is

$$\nu_{nm} = \frac{\omega_{nm}}{2\pi} = \left[\frac{n^2}{a^2} + \frac{m^2}{b^2} \right]^{1/2} \frac{\pi v}{2\pi} = \left[\left(\frac{n^2}{a^2} + \frac{m^2}{b^2} \right) \frac{T}{4\rho} \right]^{1/2}. \qquad (5.61)$$

The fundamental vibration is the $(1,1)$ mode, for which the frequency is

$$\nu_{1,1} = \left[\left(\frac{1}{a^2} + \frac{1}{b^2} \right) \frac{T}{4\rho} \right]^{1/2}.$$

The overtones in (5.61) are not related in any simple numerical way to the fundamental frequency, not like the vibrating string where the overtones (harmonics) are all simple integer multiples of the fundamental. For some reason, our ears find the sound more pleasing if the overtones are simply related to the fundamental. Therefore, the sound of a vibrating rectangular membrane is much less "musical" to the ear than that of a vibrating string.

According to (5.61,) the frequency of vibration depends on two integers m and n. As a result, it may happen that there are several different modes having the same frequency. For example, if $a = 3b$, then $(3,3)$ and $(9,1)$ modes have the same frequency. When two or more modes have the same frequency, we call it degenerate. Any combination of these degenerate modes gives another vibration with the same frequency.

In the (m,n) mode of (5.60) there are nodal lines at $x = 0, \frac{a}{n}, \frac{2a}{n}, \ldots, a$, and $y = 0, \frac{b}{m}, \frac{2b}{m}, \ldots, b$. On opposite sides of any nodal line the displacement has opposite sign. A few normal modes are shown in Fig. 5.13, in which the shaded and the unshaded parts are moving in opposite directions.

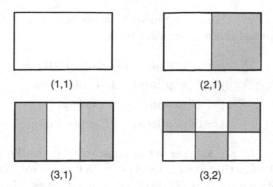

Fig. 5.13. Nodal lines and displcements of the normal modes z_{11}, z_{21}, z_{31}, z_{32} of a rectangular membrane

5.3 Three-Dimensional Wave Equations

A great many physical quantities satisfy the three-dimensional wave equation

$$\frac{\partial^2 u}{\partial x^2} + \frac{\partial^2 u}{\partial y^2} + \frac{\partial^2 u}{\partial z^2} = \frac{1}{c^2}\frac{\partial^2 u}{\partial t^2}.$$

For example, in electrodynamics we learn that the electric field $\mathbf{E}$, the magnatic field $\mathbf{B}$, the scalar potential φ, and the vector potential $\mathbf{A}$, all together 10 quantities, satisfy this equation.

Following the separation of variables:

$$u\,(x,y,z,t) = X(x)Y(y)Z(z)T(t),$$

we obtain four ordinary differential equations

$$\frac{X''(x)}{X(x)} = -l^2, \quad \frac{Y''(y)}{Y(y)} = -m^2, \quad \frac{Z''(z)}{Z(z)} = -n^2,$$

$$\frac{1}{c^2}\frac{T''(t)}{T(t)} = -\alpha^2,$$

where l, m, n, α are separation constants and they must satisfy the relation

$$l^2 + m^2 + n^2 = \alpha^2.$$

The general solutions of these equations can be easily found to be

$$X(x) = A\cos lx + B\sin lx,$$
$$Y(y) = C\cos my + D\sin my,$$
$$Z(z) = E\cos nz + F\sin nz,$$
$$T(t) = G\cos c\alpha t + H\sin c\alpha t, \tag{5.62}$$

where $A, B, \ldots, H$ are constants. Since $e^{ilx} = \cos lx + i\sin lx$, this set of solution can be expressed in an alternative form

$$
\begin{aligned}
X(x) &= A' \exp(ilx) + B' \exp(-ilx),\\
Y(y) &= C' \exp(imy) + D' \exp(-imy),\\
Z(z) &= E' \exp(inz) + F' \exp(-inz),\\
T(t) &= G' \exp(icat) + H' \exp(-icat),
\end{aligned}
\tag{5.63}
$$

where A', $B', \ldots, H'$ are another set of constants. It can be readily verified by direct substitution that the expressions in (5.63) are solutions of the wave equation. Therefore we can use (5.63) and assume that we always refer to the real part, or we can just use (5.63) as it stands, without reference to its real or imaginary parts.

It is also possible that one (or two) of l^2, m^2, n^2 is negative. For example if

$$-l^2 + m^2 + n^2 = \alpha^2,$$

then

$$
\begin{aligned}
X(x) &= A'' \cosh lx + B'' \sinh lx,\\
Y(y) &= C'' \cos my + D'' \sin my,\\
Z(z) &= E'' \cos nz + F'' \sin nz,\\
T(t) &= G'' \cos cat + H'' \sin cat,
\end{aligned}
\tag{5.64}
$$

where A'', $B'', \ldots, H''$ are still another set of constants.

Depending on the geometrical properties of any specific problem, one set of solutions is usually more convenient than others. Furthermore, the boundary conditions of the problem may restrict l, m, n to a set of allowed discrete values.

5.3.1 Plane Wave

Let us take the following set of solutions of the separated equations

$$X(x) = e^{ilx}, \quad Y(y) = e^{imy}, \quad Z(z) = e^{inz},$$
$$T(t) = e^{-icat}.$$

This gives a particular solution of the wave equation

$$u(x,y,z,t) = e^{i(lx+my+nz-cat)}.$$

This expression has a physical interpretation. To make it clear, we define a "wave vector" $\mathbf{k}$

$$\mathbf{k} = l\widehat{\mathbf{i}} + m\widehat{\mathbf{j}} + n\widehat{\mathbf{k}},$$

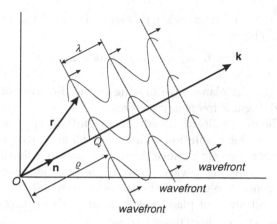

Fig. 5.14. A plane wave propagating in the direction of the vector **k**

where $\widehat{\mathbf{i}}, \widehat{\mathbf{j}}, \widehat{\mathbf{k}}$ are three unit vectors along the coordinate axes. Let **r** be a position vector from origin O to a general point (x, y, z) on a plane perpendicular to **k**, as shown in Fig. 5.14. Since

$$\mathbf{r} = x\widehat{\mathbf{i}} + y\widehat{\mathbf{j}} + z\widehat{\mathbf{k}},$$

so

$$\mathbf{k} \cdot \mathbf{r} = lx + my + nz.$$

Furthermore,

$$\mathbf{k} \cdot \mathbf{k} = k^2 = l^2 + m^2 + n^2 = \alpha^2.$$

Thus $u(x, y, z, t)$ can be written as

$$u(x, y, z, t) = e^{i(\mathbf{k} \cdot \mathbf{r} - ckt)},$$

where we use the fact that $\alpha = k$. This represents a plane wave in three dimension moving in the direction of **k**. A plane wave is one in which the disturbance is constant over all points of a plane perpendicular to the direction of propagation. Such a plane is often called a wavefront.

Let **n** be a unit vector in the direction of **k**, then

$$\mathbf{k} \cdot \mathbf{r} = k\mathbf{n} \cdot \mathbf{r} = k\varrho$$

where ϱ is the perpendicular distance from the origin O measured along the vector **n** to the point Q at which this line meets the wavefront, as shown in Fig. 5.14. Thus

$$e^{i(\mathbf{k} \cdot \mathbf{r} - ckt)} = e^{i(k\varrho - ckt)} = e^{ik(\varrho - ct)}.$$

If **k** is in the x-direction, this expression is just $e^{ik(x - ct)}$, which we recognize as a one-dimensional wave moving with velocity c. Furthermore, $k = 2\pi/\lambda$

and $kc = \omega$, where λ is the wavelength and ω, the angular frequency of this sinusoidal wave. Therefore

$$e^{i(lx+my+nz-c\alpha t)} = e^{i(\mathbf{k}\cdot\mathbf{r}-\omega t)}$$

represents a sinusoidal plane wave moving in the direction of $\mathbf{k}$, with wave length $2\pi/k$ and angular frequency $\omega = kc$.

Since $\mathbf{k}$ can be in any direction with any magnitude, the three-dimensional wave equation can have solutions which are plane waves moving in any direction with any wavelength. Since the wave equation is linear, we may have simultaneously as many plane waves, traveling in as many different directions. Thus the most general solution of the three-dimensional wave equation is a superposition of all kinds of plane waves in all kinds of directions, which is just a Fourier integral in three dimensions.

5.3.2 Particle Wave in a Rectangular Box

A free particle (a particle without force acting on it) is described in quantum mechanics by a somewhat different wave equation, known as Schrödinger equation

$$-\frac{h^2}{8\pi^2 M}\nabla^2\Psi = i\frac{h}{2\pi}\frac{\partial}{\partial t}\Psi,$$

where M is the mass of the particle and h is the Planck's constant. While a discussion of quantum mechanics is outside the scope of this book, we can take it as a mathematical problem.

Using the separation of variables, we assume that

$$\Psi(x,y,z,t) = X(x)Y(y)Z(z)T(t),$$

so the equation becomes

$$-\frac{h^2}{8\pi^2 M}\left(\frac{X''}{X} + \frac{Y''}{Y} + \frac{Z''}{Z}\right) = i\frac{h}{2\pi}\frac{T'}{T}. \tag{5.65}$$

Both sides of this equation must be equal to a constant. Let

$$i\frac{h}{2\pi}\frac{T'}{T} = E.$$

So

$$T(t) = e^{(2\pi E/ih)t}.$$

If we identify $T(t)$ as the time-dependent part of the wavefunction and write

$$e^{(2\pi E/ih)t} = e^{-i\omega t},$$

then

$$E = h\omega/2\pi = h\nu,$$

which is recognized as the energy of the particle, since according to the Planck's rule that energy is equal to h times the frequency.

The separated ordinary differential equations in x, y, z are

$$\frac{X''}{X} = -l^2, \quad \frac{Y''}{Y} = -m^2, \quad \frac{Z''}{Z} = -n^2, \tag{5.66}$$

where l, m, n are separation constants. Because of (5.65), they must satisfy the relation

$$E = \frac{h^2}{8\pi^2 M} \left(l^2 + m^2 + n^2 \right). \tag{5.67}$$

Suppose that the particle is confined in a rectangular box of length a in x-direction, b in y-direction and c in z-direction. The fact that the wavefunction Ψ must vanish at the walls means Ψ must satisfy the following boundary conditions:

$$\Psi(0, y, z, t) = \Psi(a, y, z, t) = 0,$$
$$\Psi(x, 0, z, t) = \Psi(x, b, z, t) = 0,$$
$$\Psi(x, y, 0, t) = \Psi(x, y, c, t) = 0.$$

In order to satisfy these boundary conditions, the space part of the wavefunction must take the form

$$\Psi(x, y, z, t) = \sin \frac{n_1 \pi}{a} x \sin \frac{n_2 \pi}{b} y \sin \frac{n_3 \pi}{c} y,$$

where n_1, n_2, n_3 independently assume the integer values of 1, 2, 3, ... This means that l, m, n in (5.66) must take the values of

$$l = \frac{n_1 \pi}{a}, \quad m = \frac{n_2 \pi}{b}, \quad n = \frac{n_3 \pi}{c}.$$

It follows from (5.67) that the energy is given by

$$E_{n_1, n_2, n_3} = \frac{h^2}{8M} \left[\left(\frac{n_1}{a} \right)^2 + \left(\frac{n_2}{b} \right)^2 + \left(\frac{n_3}{c} \right)^2 \right].$$

Thus we see that the energy is quantized, by which we mean that the particle in the box cannot have just any energy, it has to be one of these special allowed values corresponding to n_1, n_2, n_3, each assuming one of the integer values of 1, 2, 3, ... This is in sharp contrast to the classical case.

Discrete energies exhibited in experiments were one of the main reasons that the quantum mechanics was developed. It is interesting to note that the quantization of energy is the consequence of the boundary conditions on the solutions of the Schrödinger equation.

5.4 Equation of Heat Conduction

To obtain the equation governing the flow of heat, we use the following experimental facts:

- Heat flows in the direction of decreasing temperature.
- The rate at which heat flows through an area is proportional to the area and to the temperature gradient normal to the area. The proportionality constant is called thermal conductivity k.
- The quantity of heat gained or lost by the body when its temperature changes is proportional to the mass of the body and to the temperature change. The proportionality constant is called the specific heat c.

The constants k, c, and density ρ (mass per unit volume) of most materials are listed in the handbooks of chemistry and physics.

Let the temperature be $u(x, y, z, t)$. The quantity of heat ΔQ required to produce a temperature change Δu in the small box with a mass Δm shown in Fig. 5.15 is

$$\Delta Q = c\,\Delta m\,\Delta u = c\rho\,\Delta x\,\Delta y\,\Delta z\,\Delta u. \tag{5.68}$$

The rate at which heat flows across the surface ABCD into the box is

$$\frac{\Delta Q_1}{\Delta t} = -k\left[\frac{\partial u}{\partial y}\right]_y \Delta x\,\Delta z.$$

Note that a positive $\dfrac{\partial u}{\partial y}$ means the temperature is increasing in the positive y-direction and the heat is flowing in the negative y-direction, so the heat is acturally flowing out of the box, thus the negative sign in the equation.

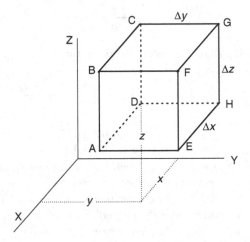

Fig. 5.15. The heat energy increased per unit time in this small element of mass is equal to the heat flux flowing into this element through its six surfaces

The subscript y in $\left[\frac{\partial u}{\partial y}\right]_y$ signifies the gradient is evaluated at the surface perpendicular to y-axis and is at a distance y units from the origin. Hence the heat flowing into the box through ABCD in the time interval Δt is

$$\Delta Q_1 = -k \left[\frac{\partial u}{\partial y}\right]_y \Delta x \, \Delta z \, \Delta t.$$

Similarly the heat flowing into the box through the surface EFGH is

$$\Delta Q_2 = k \left[\frac{\partial u}{\partial y}\right]_{y+\Delta y} \Delta x \, \Delta z \, \Delta t.$$

It follows that:

$$\Delta Q_2 + \Delta Q_1 = k \left[\frac{\partial u}{\partial y}\right]_{y+\Delta y} \Delta x \, \Delta z \, \Delta t - k \left[\frac{\partial u}{\partial y}\right]_y \Delta x \, \Delta z \, \Delta t$$

$$= k \left\{ \left[\frac{\partial u}{\partial y}\right]_{y+\Delta y} - \left[\frac{\partial u}{\partial y}\right]_y \right\} \Delta x \, \Delta z \, \Delta t$$

$$= k \frac{\partial^2 u}{\partial y^2} \Delta y \, \Delta x \, \Delta z \, \Delta t.$$

It can be shown in the same way that the heat flowing into the box through the top surface BFGC and bottom surface AEHD in the time interval Δt is given by

$$\Delta Q_3 + \Delta Q_4 = k \frac{\partial^2 u}{\partial z^2} \Delta z \, \Delta x \, \Delta y \, \Delta t$$

and through the front and back surfaces is

$$\Delta Q_5 + \Delta Q_6 = k \frac{\partial^2 u}{\partial x^2} \Delta x \, \Delta y \, \Delta z \, \Delta t.$$

Thus the total amount heat flowing into the box through its six surfaces must be

$$\Delta Q = \Delta Q_1 + \Delta Q_2 + \Delta Q_3 + \Delta Q_4 + \Delta Q_5 + \Delta Q_6$$

$$= k \left(\frac{\partial^2 u}{\partial y^2} + \frac{\partial^2 u}{\partial z^2} + \frac{\partial^2 u}{\partial x^2} \right) \Delta x \, \Delta y \, \Delta z \, \Delta t. \tag{5.69}$$

This heat is responsible for raising the temperature of the box, therefore ΔQ in (5.69) must be the same ΔQ in (5.68). Thus

$$k \left(\frac{\partial^2 u}{\partial y^2} + \frac{\partial^2 u}{\partial z^2} + \frac{\partial^2 u}{\partial x^2} \right) \Delta x \, \Delta y \, \Delta z \, \Delta t = c\rho \Delta x \, \Delta y \, \Delta z \, \Delta u$$

or

$$\frac{k}{c\rho}\left[\frac{\partial^2 u}{\partial y^2} + \frac{\partial^2 u}{\partial z^2} + \frac{\partial^2 u}{\partial x^2}\right] = \frac{\Delta u}{\Delta t}.$$

In the limit of $\Delta t \to 0$, this equation becomes

$$\frac{\partial^2 u}{\partial x^2} + \frac{\partial^2 u}{\partial y^2} + \frac{\partial^2 u}{\partial z^2} = \frac{1}{\alpha^2}\frac{\partial u}{\partial t}, \qquad (5.70)$$

where

$$\alpha^2 = \frac{k}{c\rho}$$

is known as the thermal diffusivity.

This is the equation of heat conduction. It is interesting to note that nowhere in the derivation of the equation was any use made of boundary conditions. The flow of heat in a body is described by the same equation whether the surface is maintained at a constant temperature, insulated against heat loss, or allowed to cool freely by conduction to the surrounding medium. In general, as we shall soon see, the role of boundary conditions is to determine the form of those solutions of a partial differential equation which are relevant to a particular problem.

This equation differs from the one-dimensional wave equation in that the time derivative is only first order, whereas in the wave equation it is second order.

This equation is also called diffusion equation, because it not only governs the diffusion of heat, it also governs the diffusion of material, such as the diffusion of pollutants in water or the diffusion of drugs in the liver.

5.5 One-Dimensional Diffusion Equations

A number of situations involve only one coordinate. For example, if the lateral surface of a long slender rod of length L in the x-direction is insulated and no heat is generated in the rod, then the temperature distribution in the rod is governed by the one-dimensional heat equation

$$\frac{\partial^2 u}{\partial x^2} = \frac{1}{\alpha^2}\frac{\partial u}{\partial t}.$$

This is because the insulation prevents heat flux in the radial direction, hence the temperature will depend on x coordinate only.

This one-dimensional equation also describes the temperature distribution of a large two-dimensional slab, infinite in y- and z-directions and bounded by planes at $x = 0$ and $x = L$. If the initial and boundary conditions are known, the temperature distribution $u(x, t)$ inside the slab can be found.

In the following subsections, we will solve this one-dimensional problem with several kinds of boundary conditions.

5.5.1 Temperature Distributions with Specified Values at the Boundaries

Both Ends at the Same Temperature. A long rod is subjected to an initial temperature distribution along its axis; the rod is insulated on the lateral surfaces, and the ends of the rod are kept at the same constant temperature. As long as both ends are at the same temperature, we can assume that they are at $0°$. If they are not at $0°$, a simple change of scale can make them equal to $0°$ in the new scale. Let the length of the rod be L and the initial temperature distribution be $f(x)$. To find the temperature $u(x,t)$ anywhere in the rod at a later time is to solve the following problem:

$$\text{D.E.:} \quad \frac{\partial^2 u(x,t)}{\partial x^2} = \frac{1}{\alpha^2} \frac{\partial u(x,t)}{\partial t},$$

$$\text{B.C.:} \quad u(0,t) = 0, \quad u(L,t) = 0,$$

$$\text{I.C.:} \quad u(x,0) = f(x).$$

Following the procedure of separation of variables:

$$u(x,t) = X(x)T(t),$$

the differential equation becomes

$$X''(x)T(t) = \frac{1}{\alpha^2} X(x)T'(t).$$

Dividing both sides by $X(x)T(t)$

$$\frac{X''(x)T(t)}{X(x)T(t)} = \frac{1}{\alpha^2} \frac{X(x)T'(t)}{X(x)T(t)},$$

we have

$$\frac{X''(x)}{X(x)} = \frac{1}{\alpha^2} \frac{T'(t)}{T(t)}.$$

This equation can hold if and only if both sides equal to the same constant

$$\frac{X''(x)}{X(x)} = -\mu^2,$$

$$\frac{1}{\alpha^2} \frac{T'(t)}{T(t)} = -\mu^2.$$

The general solution of $X''(x) = -\mu^2 X(x)$ is

$$X(x) = A \cos \mu x + B \sin \mu x.$$

Since the boundary conditions require

$$X(0) = 0, \quad X(L) = 0,$$

it follows that:

$$A = 0,$$
$$\mu = \frac{n\pi}{L},$$

where n is an integer. For each n, the solution of the space part is

$$X_n(x) = \sin\left(\frac{n\pi}{L}x\right).$$

Corresponding to this n, the equation for $T(t)$ becomes

$$T'(t) = -\left(\alpha\frac{n\pi}{L}\right)^2 T(t).$$

Therefore the time-dependent part is given by

$$T_n(t) = \exp\left[-\left(\frac{\alpha n\pi}{L}\right)^2 t\right].$$

Hence for each integer n, there is a solution $X_n(x)T_n(t)$. The general solution is the linear combination of these individual solutions

$$u(x,t) = \sum_{n=1}^{\infty} c_n \sin\left(\frac{n\pi}{L}x\right) \exp\left[-\left(\frac{\alpha n\pi}{L}\right)^2 t\right].$$

The initial condition requires

$$u(x,0) = \sum_{n=1}^{\infty} c_n \sin\left(\frac{n\pi}{L}x\right) = f(x).$$

This is a Fourier sine series, the coefficient c_n is given by

$$c_n = \frac{2}{L}\int_0^L f(x)\sin\left(\frac{n\pi}{L}x\right)dx.$$

Thus the complete solution of this problem is

$$u(x,t) = \sum_{n=1}^{\infty}\left\{\frac{2}{L}\int_0^L f(x')\sin\left(\frac{n\pi}{L}x'\right)dx'\right\}$$
$$\times \sin\left(\frac{n\pi}{L}x\right)\exp\left[-\left(\frac{\alpha n\pi}{L}\right)^2 t\right].$$

This solution certainly makes sense. No matter what the initial temperature is, as $t \to \infty$ the whole rod will eventually settle to $0°$ as its two ends.

Two Ends at Different Temperatures. A more realistic problem is that the two ends of the rod are at different temperatures. In that case our problem is changed to

$$\text{D.E.:} \quad \frac{\partial^2 u(x,t)}{\partial x^2} = \frac{1}{\alpha^2} \frac{\partial u(x,t)}{\partial t},$$

$$\text{B.C.:} \quad u(0,t) = 0, \quad u(L,t) = K,$$

$$\text{I.C.:} \quad u(x,0) = f(x).$$

One way to solve the problem is to transform it into the problem we just solved. This can be done by splitting the dependent variable $u(x,t)$ in the following way:

$$u(x,t) = v(x,t) - \psi(x).$$

So it follows:

$$\frac{\partial^2 u(x,t)}{\partial x^2} = \frac{\partial^2 v(x,t)}{\partial x^2} - \psi''(x),$$

$$\frac{\partial u(x,t)}{\partial t} = \frac{\partial v(x,t)}{\partial t}.$$

Now, if we require

$$\psi''(x) = 0, \tag{5.71}$$

$$\psi(0) = 0, \quad \psi(L) = -K \tag{5.72}$$

then

$$\frac{\partial^2 v(x,t)}{\partial x^2} = \frac{1}{\alpha^2} \frac{\partial v(x,t)}{\partial t}$$

$$v(0,t) = 0, \quad v(L,t) = 0$$

$$v(x,0) = f(x) + \psi(x).$$

Clearly we can solve for $v(x,t)$ with the same method of last problem. If we can find $\psi(x)$, then $u(x,t)$ can be obtained.

It follows from (5.71) that:

$$\psi(x) = a + bx.$$

The conditions (5.72) requires

$$a = 0 \quad \text{and} \quad b = -\frac{K}{L}.$$

Therefore

$$\psi(x) = -\frac{K}{L}x.$$

Thus

$$u(x,t) = v(x,t) - \psi(x)$$

$$= \frac{K}{L}x + \sum_{n=1}^{\infty} b_n \sin\left(\frac{n\pi}{L}x\right) \exp\left[-\left(\frac{\alpha n\pi}{L}\right)^2 t\right],$$

where

$$b_n = \frac{2}{L} \int_0^L \left[f(x) - \frac{K}{L}x\right] \sin\frac{n\pi}{L}x\, dx.$$

Again, this result makes sense, as $t \to \infty$, the temperature in the rod will increase linearly from 0 to K.

5.5.2 Problems Involving Insulated Boundaries

Both Ends Insulated. The flux of heat across the faces at $x = 0$ and $x = L$ is proportional to $\dfrac{\partial u}{\partial x}$ at those faces. To set $\dfrac{\partial u}{\partial x} = 0$ is to ensure there is no heat transfer. Thus, if both ends are insulated, to find the temperature distribution $u(x,t)$, we have to solve the following problem:

D.E.: $$\frac{\partial^2 u(x,t)}{\partial x^2} = \frac{1}{\alpha^2}\frac{\partial u(x,t)}{\partial t},$$

B.C.: $$\left[\frac{\partial u(x,t)}{\partial x}\right]_{x=0} = 0, \quad \left[\frac{\partial u(x,t)}{\partial x}\right]_{x=L} = 0.$$

I.C.: $$u(x,0) = f(x).$$

Separation of variable now leads to

$$X''(x) = -\mu^2 X(x),$$

$$X'(0) = X'(L) = 0.$$

It can be easily shown that

$$X(x) = \cos\mu x$$

$$\mu = \frac{n\pi}{L},$$

where n is an integer starting from $n = 0$.

The corresponding $T(t)$ function is still

$$T(t) = \exp(-\alpha^2\mu^2 t).$$

Thus the general solution can be written as

$$u(x,t) = \frac{1}{2}c_0 + \sum_{n=1}^{\infty} c_n \cos\frac{n\pi}{L}x \exp\left[-\left(\frac{\alpha n\pi}{L}\right)^2 t\right]. \tag{5.73}$$

The initial condition gives us the Fourier cosine series

$$u(x,0) = \frac{1}{2}c_0 + \sum_{n=1}^{\infty} c_n \cos \frac{n\pi}{L}x = f(x).$$

Thus

$$c_n = \frac{2}{L} \int_0^L f(x) \cos \frac{n\pi}{L}x \, dx, \quad n = 0, 1, 2, \dots . \tag{5.74}$$

Therefore (5.73) with c_n given by (5.74) is our final solution.

One End at a Constant Temperature and the Other End Insulated. In this case, the problem becomes

D.E.: $\quad \dfrac{\partial^2 u(x,t)}{\partial x^2} = \dfrac{1}{\alpha^2}\dfrac{\partial u(x,t)}{\partial t},$

B.C.: $\quad u(0,t) = 0, \quad \left[\dfrac{\partial u(x,t)}{\partial x}\right]_{x=L} = 0,$

I.C.: $\quad u(x,0) = f(x).$

Upon separation of variables $u(x,t) = X(x)T(t)$, we find

$$X''(x) = -\mu^2 X(x), \tag{5.75}$$
$$X(0) = 0, \quad X'(L) = 0$$

and

$$T'(t) = -\alpha^2 \mu^2 T(t). \tag{5.76}$$

From (5.75) we have

$$X(x) = A \cos \mu x + B \sin \mu x.$$

The condition $X(0) = 0$ requires $A = 0$. Therefore we are left with

$$X(x) = B \sin \mu x$$

so

$$X'(x) = \mu B \cos \mu x.$$

The other boundary condition $X'(L) = 0$ becomes

$$X'(L) = \mu B \cos \mu L = 0.$$

This condition requires μL to be a half integer times π, that is

$$\mu = \left(\frac{2n-1}{2}\right)\frac{\pi}{L}, \quad n = 1, 2, 3, \dots .$$

Thus for each n, the solution of the space part is

$$X_n(x) = \sin\left(\frac{2n-1}{2}\right)\frac{\pi}{L}x$$

and the solution of the corresponding equation for $T(t)$ is

$$T_n(t) = \exp\left[-\left(\frac{\alpha(2n-1)\pi}{2L}\right)^2 t\right].$$

Therefore the general solution is a linear combination of $X_n(x)T_n(t)$

$$u(x,t) = \sum_{n=1}^{\infty} c_n \sin\frac{(2n-1)\pi}{2L}x \exp\left[-\left(\frac{\alpha(2n-1)\pi}{2L}\right)^2 t\right]. \qquad (5.77)$$

From the initial condition

$$u(x,0) = \sum_{n=1}^{\infty} c_n \sin\frac{(2n-1)\pi}{2L}x = f(x),$$

the coefficients c_n can be easily determined to be

$$c_n = \frac{2}{L}\int_0^L f(x)\sin\frac{(2n-1)\pi}{2L}x\,dx, \quad n = 1,2,3,\ldots.$$

Therefore (5.77) with c_n given by the above expression is the solution of our problem.

5.5.3 Heat Exchange at the Boundary

If heat exchange takes place, then according to Newton's law of cooling, the temperature function satisfies the relation

$$hu(x,t) + \frac{\partial}{\partial x}u(x,t) = 0,$$

where the constant h is an appropriate heat transfer coefficient.

Suppose we want to know the temperature $u(x,t)$ of a slab which initially has a uniform temperature u_0. The face of the slab at $x = 0$ is held at temperature 0, at the face $x = L$, heat exchange takes place so that

$$u_x(L,t) = -hu(L,t).$$

To find $u(x,t)$, we have to solve the following problem:

D.E.: $\dfrac{\partial^2 u}{\partial x^2} = \dfrac{1}{a^2}\dfrac{\partial u}{\partial t}$,

B.C.: $u(0,t) = 0$, $u_x(L,t) = -hu(L,t)$,

I.C.: $u(x,0) = u_0$.

Again, we assume the variables can be separated,

$$u(x,t) = X(x)T(t),$$

so

$$\frac{X''}{X} = \frac{1}{a^2}\frac{T'}{T} = -\lambda^2$$

and

$$T' = -a^2\lambda^2 T, \quad X'' = -\lambda^2 X.$$

The solution for X is

$$X(x) = A\cos\lambda x + B\sin\lambda x.$$

The boundary condition $u(0,t) = 0$ means $X(0) = 0$. Thus

$$X(0) = A = 0.$$

The other boundary condition at $x = L$, $u_x(L,t) = -hu(L,t)$, becomes

$$X'(L)T(t) = -hX(x)T(t)$$

or

$$X'(L) = -hX(x).$$

Thus

$$B\lambda\cos\lambda L = -hB\sin\lambda L$$

or

$$\tan\lambda L = -\frac{\lambda}{h}. \tag{5.78}$$

The values of λ that satisfy this equation are the eigenvalues of this problem. Let $\lambda L = \beta$, so

$$\tan\beta = -\frac{\beta}{hL}. \tag{5.79}$$

The solution of this equation are the points of intersection of the graphs of $y = \tan x$ and $y = -x/hL$, as indicated in Fig. 5.16. It is apparent from the figure that there is an infinite sequence of positive roots $\beta_1, \beta_2, \beta_3, \ldots$. The eigenvalues of (5.78), are given by

$$\lambda_n = \frac{\beta_n}{L}$$

for $n = 1, 2, 3, \ldots$. In other words,

$$\tan\lambda_n L = -\frac{\lambda_n}{h}. \tag{5.80}$$

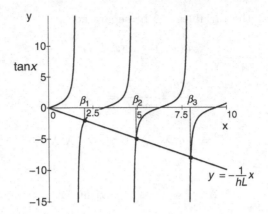

Fig. 5.16. The solutions of $\tan \beta = -\beta/hL$ are the intersections of $y = \tan x$ and $y = -x/hL$

Equation (5.79) appears frequently in many applications, its solutions for various values of hL are tabulated in Table 4.19 of Abramowitz and Stegun, *Handbook of Mathematical Functions* (Dover, New York, 1965).

The eigenfunction $X_n(x)$ associate with the eigenvalue λ_n is

$$X_n(x) = \sin \lambda_n x.$$

According to Sturm–Liouville theory, these eigenfunctions are orthogonal. This can be explicitly shown. Let

$$I_{nm} = \int_0^L \sin \lambda_n x \, \sin \lambda_m x \, \mathrm{d}x.$$

By trigonometrical identity or by changing it into exponential form, one can show that the integral is equal to

$$I_{nm} = \frac{1}{2} \left[\frac{\sin (\lambda_n - \lambda_m) L}{(\lambda_n - \lambda_m)} - \frac{\sin (\lambda_n + \lambda_m) L}{(\lambda_n + \lambda_m)} \right]. \tag{5.81}$$

If $\lambda_n \neq \lambda_m$, then

$$I_{nm} = \frac{1}{(\lambda_n^2 - \lambda_m^2)} \left[\lambda_m \sin \lambda_n L \, \cos \lambda_m L - \lambda_n \sin \lambda_m L \, \cos \lambda_n L \right].$$

Using (5.80,) which can be written as

$$h \sin \lambda_i L = -\lambda_i \cos \lambda_i L, \tag{5.82}$$

We see that

$$I_{nm} = \frac{1}{(\lambda_n^2 - \lambda_m^2)} \left[-h \sin \lambda_n L \, \sin \lambda_m L + h \sin \lambda_m L \, \sin \lambda_n L \right] = 0.$$

Therefore $\{\sin \lambda_n x\}$ is an orthogonal set.

For $\lambda_n = \lambda_m$, we can either use L'Hospital's rule on the first term of (5.81) or integrate directly, the result is

$$I_{nn} = \frac{1}{2}\left[L - \frac{1}{2\lambda_n}\sin 2\lambda_n L\right] = \frac{1}{2}\left[L - \frac{1}{\lambda_n}\sin \lambda_n L \cos \lambda_n L\right].$$

Again by (5.82,)

$$I_{nn} = \frac{1}{2}\left[L + \frac{1}{h}\cos^2 \lambda_n L\right] = \frac{1}{2h}\left(Lh + \cos^2 \lambda_n L\right).$$

Corresponding to each λ_n, the solution to the equation for $T_n(t)$ is

$$T_n(t) = e^{-\lambda_n^2 a^2 t}.$$

So the general solution for $u(x,t)$ can be expressed as

$$u(x,t) = \sum_{n=1}^{\infty} c_n e^{-\lambda_n^2 a^2 t} \sin \lambda_n x.$$

The coefficients c_n can be determined by the initial condition, $u(x,0) = u_0$,

$$\sum_{n=1}^{\infty} c_n \sin \lambda_n x = u_0.$$

Multiplying both sides by $\sin \lambda_m x$ and integrating from 0 to L,

$$\sum_{n=1}^{\infty} c_n \int_0^L \sin \lambda_n x \sin \lambda_m x \, dx = \int_0^L u_0 \sin \lambda_m x \, dx,$$

we have

$$\sum_{n=1}^{\infty} c_n I_{nm} = c_m I_{mm} = \frac{1}{\lambda_m}(1 - \cos \lambda_m L)u_0.$$

Thus

$$c_n = \frac{2hu_0}{Lh + \cos^2 \lambda_n L}\frac{1}{\lambda_n}(1 - \cos \lambda_n L)$$

and

$$u(x,t) = 2hu_0 \sum_{n=1}^{\infty} \frac{1 - \cos \lambda_n L}{\lambda_n (Lh + \cos^2 \lambda_n L)} e^{-\lambda_n^2 a^2 t} \sin \lambda_n x,$$

where λ_1, λ_2, ... are positive roots of

$$\tan \lambda L = -\frac{\lambda}{h}.$$

5.6 Two-Dimensional Diffusion Equations: Heat Transfer in a Rectangular Plate

Suppose that the edges of a rectangular plate are bounded by the line $x = 0$, $x = a$, $y = 0$, and $y = b$. Its faces are insulated so that $\frac{\partial u}{\partial z} = 0$. The edges are held at temperature zero and initially the temperature distribution in the plate is $f(x, y)$. We want to find the expression for $u(x, y, t)$.

Our problem can be formulated as follows:

$$\frac{\partial^2 u(x, y, t)}{\partial x^2} + \frac{\partial^2 u(x, y, t)}{\partial y^2} = \frac{1}{\alpha^2} \frac{\partial u(x, y, t)}{\partial t}$$

$$u(0, y, t) = 0, \quad u(a, y, t) = 0,$$
$$u(x, 0, t) = 0, \quad u(x, b, t) = 0,$$
$$u(x, y, 0) = f(x, y).$$

Again we assume the variables can be separated

$$u(x, y, t) = X(x)Y(y)T(t).$$

Then the differential equation can be written as

$$X''(x)Y(y)T(t) + X(x)Y''(y)T(t) = \frac{1}{\alpha^2} X(x)Y(y)T'(t).$$

Dividing by $X(x)Y(y)T(t)$, we have

$$\frac{X''(x)}{X(x)} + \frac{Y''(y)}{Y(y)} = \frac{1}{\alpha^2} \frac{T'(t)}{T(t)}$$

or

$$\frac{X''(x)}{X(x)} = \frac{1}{\alpha^2} \frac{T'(t)}{T(t)} - \frac{Y''(y)}{Y(y)}.$$

Since the left-hand side is a function of x only, and right-hand side is a function of t and y, they can be equal if and only if both sides equal to the same constant

$$\frac{X''(x)}{X(x)} = -\lambda^2 = \frac{1}{\alpha^2} \frac{T'(t)}{T(t)} - \frac{Y''(y)}{Y(y)}.$$

For the same reason as in the one-dimensional case, we anticipate the constant to be $-\lambda^2$,

$$\frac{X''(x)}{X(x)} = -\lambda^2,$$

$$\frac{1}{\alpha^2} \frac{T'(t)}{T(t)} - \frac{Y''(y)}{Y(y)} = -\lambda^2.$$

The last equation can be written as

$$\frac{Y''(y)}{Y(y)} = \frac{1}{\alpha^2}\frac{T'(t)}{T(t)} + \lambda^2.$$

Again, both sides have to equal to another constant

$$\frac{Y''(y)}{Y(y)} = -\mu^2 = \frac{1}{\alpha^2}\frac{T'(t)}{T(t)} + \lambda^2.$$

It is easy to see that the boundary conditions require

$$X(0) = X(a) = 0, \quad Y(0) = Y(b) = 0.$$

With these conditions, the equation

$$X''(x) = -\lambda^2 X(x)$$

can have solution only if

$$\lambda = \frac{n\pi}{a}, \quad n = 1, 2, 3, \ldots.$$

and corresponding to each n the solution X_n is

$$X_n(x) = \sin\frac{n\pi}{a}x.$$

The equation governing $Y(y)$

$$Y''(y) = -\mu^2 Y(y)$$

has solution only if

$$\mu = \frac{m\pi}{b}, \quad m = 1, 2, 3, \ldots$$

with

$$Y_m(y) = \sin\frac{m\pi}{b}y.$$

The equation for $T(t)$ is

$$T'(t) = -(\lambda^2 + \mu^2)\alpha^2 T(t).$$

For each set of (n, m), the solution of this equation is

$$T_{n,m}(t) = \exp\left\{-\left[\left(\frac{n\pi\alpha}{a}\right)^2 + \left(\frac{m\pi\alpha}{b}\right)^2\right]t\right\}.$$

The linear combination of $X_n(x)Y_m(y)T_{n,m}(t)$ is the most general solution, thus

$$u(x,y,t) = \sum_{n=1}^{\infty} \sum_{m=1}^{\infty} c_{nm} \sin \frac{n\pi}{a} x \sin \frac{m\pi}{b} y \, \exp\left\{ -\left[\left(\frac{n\pi\alpha}{a} \right)^2 + \left(\frac{m\pi\alpha}{b} \right)^2 \right] t \right\}.$$

(5.83)

Applying the initial condition, we obtain

$$u(x,y,0) = f(x,y) = \sum_{n=1}^{\infty} \sum_{m=1}^{\infty} c_{nm} \sin \frac{n\pi}{a} x \sin \frac{m\pi}{b} y.$$

Now if we define

$$g_m(x) = \sum_{n=1}^{\infty} c_{nm} \sin \frac{n\pi}{a} x,$$

(5.84)

we can write

$$f(x,y) = \sum_{m=1}^{\infty} g_m(x) \sin \frac{m\pi}{b} y.$$

We can regard this equation as a Fourier sine series in y for each fixed x, and the coefficient $g_m(x)$ is given by

$$g_m(x) = \frac{2}{b} \int_0^b f(x,y) \sin \frac{m\pi y}{b} \mathrm{d}y.$$

On the other hand (5.84) is a Fourier sine series in x, and the coefficient c_{nm} is given by

$$c_{nm} = \frac{2}{a} \int_0^a g_m(x) \sin \frac{n\pi x}{a} \mathrm{d}x$$

$$= \frac{4}{ab} \int_0^a \int_0^b f(x,y) \sin \frac{m\pi y}{b} \sin \frac{n\pi x}{a} \mathrm{d}y \, \mathrm{d}x.$$

(5.85)

Thus the solution of our problem is (5.83) with coefficient c_{nm} given by (5.85). This problem is another example of the two-dimensional double Fourier series.

5.7 Laplace's Equations

One of the most important partial differential equations in physics is the Laplace's equation

$$\nabla^2 u = 0,$$

named after French mathematician Pierre-Simon Laplace (1740–1827) The theory of the solution of Laplace equation is called potential theory. Solutions of the equation that have second derivatives are called harmonic functions.

Laplace equation may be obtained by setting $\dfrac{\partial u}{\partial t} = 0$ in the heat equation. This describes the steady-state temperature distribution in a solid in which there is no heat source or sink.

Laplace's equation also describes the electrostatic potential in a charge free region. Since the electric field is given by the gradient of a potential

$$\mathbf{E} = -\nabla V$$

and the divergent of $\mathbf{E}$ is equal to zero in a free space, therefore

$$\nabla \cdot \mathbf{E} = -\nabla \cdot \nabla V = -\nabla^2 V = 0,$$

the potential V is the solution of the Laplace equation. Similarly, the potential of the gravitational field in a region containing no matter also satisfies the Laplace equation.

Furthermore, Laplace equation is also very important in hydrodynamics. It applies to the incompressible fluid flow with no source, sink, and vortex. In this case, the velocity is given by the gradient of the velocity potential which satisfies the Laplace equation.

5.7.1 Two-Dimensional Laplace's Equation: Steady-State Temperature in a Rectangular Plate

Suppose that the three edges $x = 0$, $x = a$, and $y = 0$ are maintained at zero temperature

$$u(0, y) = 0, \quad u(a, y) = 0, \quad u(x, 0) = 0, \tag{5.86}$$

and the fourth edge $y = b$ is maintained at a temperature distribution $f(x)$

$$u(x, b) = f(x). \tag{5.87}$$

We want to find the temperature throughout the plate after the steady-state temperature distribution is reached.

To find the answer we must determine the solution of the two-dimensional Laplace equation

$$\frac{\partial^2 u}{\partial x^2} + \frac{\partial^2 u}{\partial y^2} = 0$$

subject to the boundary conditions (5.86) and (5.87).

Again we use the method of separation of variables

$$u(x, y) = X(x)Y(y)$$

to write the equation as

$$X''Y + XY'' = 0.$$

Dividing by $u(x, y)$, we have

$$\frac{X''}{X} = -\frac{Y''}{Y}.$$

It follows that both sides must be equal to a constant

$$\frac{X''}{X} = -\lambda^2, \qquad -\frac{Y''}{Y} = -\lambda^2.$$

Thus

$$X''(x) = -\lambda^2 X(x),$$
$$Y''(y) = \lambda^2 Y(y).$$

Note that we choose a negative constant in order for $X(x)$ to satisfy the boundary conditions. With this choice, $X(x)$ is given by

$$X(x) = A\cos\lambda x + B\sin\lambda x.$$

The boundary conditions $u(0, y) = u(a, y) = 0$ require $X(x)$ to be one of

$$X_n(x) = \sin\lambda_n x,$$

where

$$\lambda_n = \frac{n\pi}{a}, \qquad n = 1, 2, 3, \dots.$$

The solutions for $Y(y)$ can be written in terms of $e^{\lambda y}$ and $e^{-\lambda y}$, or in terms of hyperbolic cosine and sine

$$\cosh\lambda y = \frac{1}{2}\left(e^{\lambda y} + e^{-\lambda y}\right), \qquad \sinh\lambda y = \frac{1}{2}\left(e^{\lambda y} - e^{-\lambda y}\right).$$

In the present problem, it is more convenient to express the solution in the hyperbolic functions, since at $y = 0$, $\cosh\lambda y = 1$, and $\sinh\lambda y = 0$. Thus

$$Y(y) = C\cosh\lambda y + D\sin\lambda y$$

and the boundary condition $u(x, 0) = 0$ requires C to be zero

$$Y(0) = C = 0.$$

Therefore

$$Y_n(y) = \sinh\lambda_n y.$$

A linear combination of $X_n(x)Y_n(y)$ is the general solution

$$u(x, y) = \sum_{n=1}^{\infty} c_n \sinh\frac{n\pi y}{a}\sin\frac{n\pi x}{a}.$$

To satisfy the other boundary condition $u(x, b) = f(x)$, we must have

$$u(x, b) = \sum_{n=1}^{\infty} c_n \sinh\frac{n\pi b}{a}\sin\frac{n\pi x}{a} = f(x).$$

It is clear that $c_n \sinh\frac{n\pi b}{a}$ is the Fourier coefficient of the sine series expansion of $f(x)$. Therefore

$$c_n \sinh \frac{n\pi b}{a} = \frac{2}{L} \int_0^L f(x) \sin \frac{n\pi x}{a} dx.$$

Thus, the steady-state temperature distribution is given by

$$u(x,y) = \sum_{n=1}^{\infty} \left(\frac{2}{L} \int_0^L f(x') \sin \frac{n\pi x'}{a} dx' \right) \left[\sinh \frac{n\pi b}{a} \right]^{-1} \sinh \frac{n\pi y}{a} \sin \frac{n\pi x}{a}.$$

It is clear that the solution of the more general problem where the temperature is prescribed by nonzero functions along all four edges can be obtained by superimposing four solutions analogous to the one obtained here, each corresponding to a problem in which zero temperatures are prescribed along three of the four edges.

5.7.2 Three-Dimensional Laplace's Equation: Steady-State Temperature in a Rectangular Parallelepiped

Suppose that the temperatures of five faces of a rectangular parallelepiped are maintained at zero

$$u(0,y,z) = u(a,y,z) = u(x,0,z) = u(x,b,z) = 0, \qquad (5.88)$$

$$u(x,y,0) = 0$$

and the sixth face is maintained at a prescribed temperature distribution

$$u(x,y,d) = f(x,y).$$

We want to know the steady-state temperature distribution in the interior.
In this case we have to solve the three-dimensional Laplace equation

$$\frac{\partial^2 u}{\partial x^2} + \frac{\partial^2 u}{\partial y^2} + \frac{\partial^2 u}{\partial z^2} = 0$$

with the specified boundary conditions.
With the assumption

$$u(x,y,z) = X(x)Y(y)Z(z),$$

the equation can be written as

$$-\frac{X''}{X} = \frac{Y''}{Y} + \frac{Z''}{Z}.$$

Since both sides must be equal to the same constant, we have

$$-\frac{X''}{X} = \alpha^2$$

$$\frac{Y''}{Y} + \frac{Z''}{Z} = \alpha^2.$$

It follows from the second equation:

$$\frac{Y''}{Y} = \alpha^2 - \frac{Z''}{Z} = -\beta^2.$$

Thus

$$X'' = -\alpha^2 X,$$
$$Y'' = -\beta^2 Y,$$
$$Z'' = (\alpha^2 + \beta^2)Z.$$

The homogeneous boundary conditions (5.88) are satisfied if $X(x)$ and $Y(y)$ are one of the following eigenfunctions

$$X_n(x) = \sin \alpha_n x, \quad \alpha_n = \frac{n\pi}{a}, \quad n = 1, 2, 3, \ldots,$$

$$Y_m(y) = \sin \beta_m y, \quad \beta_m = \frac{m\pi}{b}, \quad m = 1, 2, 3, \ldots$$

The solution for $Z(z)$ corresponding to $X_n(x)$ and $Y_m(y)$ is

$$Z_{nm}(z) = A \cosh \gamma_{nm} z + B \sinh \gamma_{nm} z,$$

where

$$\gamma_{nm} = \left[\left(\frac{n\pi}{a}\right)^2 + \left(\frac{m\pi}{b}\right)^2\right]^{1/2}.$$

The boundary condition $u(x, y, 0) = 0$ requires $Z(0) = 0$

$$Z_{nm}(0) = A = 0.$$

Thus

$$Z_{nm}(z) = \sinh \gamma_{nm} z.$$

Therefore, for each set of (n, m), there is a solution

$$u_{nm}(x, y, z) = X_n(x) Y_m(y) Z_{nm}(z).$$

The general solution is given by the linear combination

$$u(x, y, z) = \sum_{n=1}^{\infty} \sum_{m=1}^{\infty} c_{nm} \sin \alpha_n x \sin \beta_m y \sinh \gamma_{nm} z.$$

To satisfy the boundary condition on the top surface, we have

$$u(x, y, d) = \sum_{n=1}^{\infty} \sum_{m=1}^{\infty} c_{nm} \sin \alpha_n x \sin \beta_m y \sinh \gamma_{nm} d = f(x, y).$$

Clearly $c_{nm} \sinh \gamma_{nm} d$ are the coefficients of the double Fourier series of $f(x, y)$. Therefore

$$c_{nm} \sinh \gamma_{nm} d = \frac{4}{ab} \int_0^a \int_0^b f(x,y) \sin \frac{n\pi x}{a} \sin \frac{m\pi y}{b} dy\, dx.$$

Thus the steady-state temperature distribution inside the parallelepiped is given by

$$u(x,y,z) = \sum_{n=1}^{\infty} \sum_{m=1}^{\infty} a_{nm} \left[\sinh \gamma_{nm} d\right]^{-1} \sin \frac{n\pi x}{a} \sin \frac{m\pi y}{b} \sinh \gamma_{nm} z,$$

where

$$a_{nm} = \frac{4}{ab} \int_0^a \int_0^b f(x,y) \sin \frac{n\pi x}{a} \sin \frac{m\pi y}{b} dy\, dx,$$

$$\gamma_{nm} = \left[\left(\frac{n\pi}{a}\right)^2 + \left(\frac{m\pi}{b}\right)^2\right]^{1/2}.$$

5.8 Helmholtz's Equations

An equivalent approach to the Laplace's equation of Sect. 5.7 consists of first seeking a product in the form of

$$u(x,y,z) = F(x,y)Z(z),$$

where the $Z(z)$ factor is treated in a distinct way since only along a boundary $z = $ constant is a nonhomogeneous condition imposed. The process of separation of variables will give

$$\frac{1}{F(x,y)} \left(\frac{\partial^2 F(x,y)}{\partial x^2} + \frac{\partial^2 F(x,y)}{\partial y^2}\right) = -\frac{\partial^2 Z(z)}{\partial z^2} = -k^2,$$

where k^2 is the separation constant. Thus, $F(x,y)$ is the solution of the following equation:

$$\frac{\partial^2 F(x,y)}{\partial x^2} + \frac{\partial^2 F(x,y)}{\partial y^2} + k^2 F(x,y) = 0.$$

This equation is known as the Helmholtz's equation named after Hermann von Helmholtz (1821–1894) who studied this equation in connection with acoustics.

A great number of important engineering and physics problems can be reduced to solving the Helmholtz equation. As we have seen, if we put a time dependence $\exp(i\omega t)$ in the wave equation, the space part is given by the Helmholtz equation. Similarly, with a time dependence $\exp(-\lambda t)$, the space part of the heat equation is also given by the Helmholtz equation.

So far we have used only cartesian coordinates. The success of solving the problems in this chapter are due to the facts that the Helmholtz equation is separable in cartesian coordinates and the boundaries of straight lines and planes can be easily described by the rectangular coordinates.

Actually the Helmholtz equation is separable in 11 different curvilinear, orthogonal coordinate systems. However, most of the problems in engineering and physics can be formulated in either cartesian, cylindrical, or spherical coordinates. In Chap. 6, we will study the solutions of the Helmholtz equation in the cylindrical and spherical coordinates.

Exercises

1. Show that the solution of

$$\frac{\partial^2 u(x,t)}{\partial x^2} = \frac{1}{v^2} \frac{\partial^2 u(x,t)}{\partial t^2},$$

$$u(0,t) = 0; \quad u(L,t) = 0,$$

$$u(x,0) = f(x); \quad u_t(x,0) = g(x)$$

is given by

$$u(x,t) = \sum_{n=1}^{\infty} \left(A_n \cos \frac{n\pi v}{L} t + B_n \sin \frac{n\pi v}{L} t \right) \sin \frac{n\pi}{L} x,$$

where

$$A_n = \frac{2}{L} \int_0^L f(x) \sin \frac{n\pi x}{L} dx,$$

$$B_n = \frac{2}{n\pi v} \int_0^L g(x) \sin \frac{n\pi x}{L} dx.$$

2. The transverse displacements $u(x,t)$ in a string of length L stretched between the point 0 and L, and initially displaced into a position $u(x,0) = f(x)$ and released from rest is given by the solution of

$$\frac{\partial^2 u(x,t)}{\partial x^2} = \frac{1}{v^2} \frac{\partial^2 u(x,t)}{\partial t^2},$$

$$u(0,t) = 0; \quad u(L,t) = 0,$$

$$u(x,0) = f(x); \quad u_t(x,0) = 0.$$

Show that the solution can be expressed as

$$u(x,t) = \sum_{n=1}^{\infty} \left[A_n \sin \frac{n\pi}{L}(x - vt) + B_n \sin \frac{n\pi}{L}(x + vt) \right].$$

Express A_n and B_n in terms of quantities given in the problem.

3. Show that the motion of every point of the string of the previous problem is periodic in t with a period $2L/v$.

4. If the initial displacement of the previous problem is $f(x) = C \sin \frac{3\pi}{L} x$, find $u(x,t)$. What is the frequency of this vibration?

5. If the initial displacement of the previous problem is $f(x) = A \sin \frac{\pi}{L} x + B \sin \frac{2\pi}{L} x$, find the subsequent displacements $u(x,t)$.

6. Find the solution of the following boundary value problem

$$\text{D.E.:} \quad \frac{\partial^2 u(x,t)}{\partial t^2} = a^2 \frac{\partial^2 u(x,t)}{\partial x^2},$$

$$\text{B.C.:} \quad u(0,t) = 0, \quad u(L,t) = 0,$$

$$\text{I.C.:} \quad u(x,0) = 0, \quad u_t(x,0) = \begin{cases} x & 0 \le x \le \frac{L}{2}, \\ L - x & \frac{L}{2} \le x \le L. \end{cases}$$

Ans. $u(x,t) = \frac{4L^2}{a\pi^3} \sum_{n=1}^{\infty} \frac{1}{n^3} \sin \frac{n\pi}{2} \sin \frac{n\pi x}{L} \sin \frac{n\pi a}{L} t$.

7. Find $u(x,t)$ of the previous problem if the initial velocity is changed to

$$u_t(x,0) = \begin{cases} 0 & 0 \le x < \frac{L}{2} - w, \\ h & \frac{L}{2} - w < x < \frac{L}{2} + w, \\ 0 & \frac{L}{2} + w < x < L. \end{cases}$$

Ans. $u(x,t) = \frac{4Lh}{a\pi^2} \sum_{n=1}^{\infty} \frac{(-1)^{n+1}}{(2n-1)^2} \sin \frac{(2n-1)\pi w}{L} \sin \frac{(2n-1)\pi}{L} x \sin \frac{(2n-1)\pi a}{L} t$.

8. If the vibrating string is subject to viscous damping, the governing equation can be written as

$$\frac{\partial^2 u(x,t)}{\partial t^2} = \frac{\partial^2 u(x,t)}{\partial x^2} - 2h \frac{\partial u(x,t)}{\partial t},$$

where h is a constant. Let the boundary and initial conditions be

$$\text{B.C.:} \quad u(0,t) = 0, \quad u(L,t) = 0,$$

$$\text{I.C.:} \quad u(x,0) = f(x), \quad u_t(x,0) = 0.$$

Show that, assuming $h < \pi/L$, $u(x,t)$ is given by

$$u(x,t) = \sum_{n=1}^{\infty} b_n \, e^{-ht} \left(\cos \frac{k_n}{L} t + \frac{hL}{k_n} \sin \frac{k_n}{L} t \right) \sin \frac{n\pi x}{L},$$

$$\text{where} \quad b_n = \frac{2}{L} \int_0^L f(x) \sin \frac{n\pi x}{L} dx, \quad k_n = \sqrt{(n\pi)^2 - (hL)^2}.$$

9. Show explicitly that the following functions:

(a) $(x + at)^2$,   (b) $2e^{-(x-at)^2}$,   (c) $5\sin[3(x - at)] + (x + at)$.

satisfy the wave equation

$$\frac{\partial^2 u(x,t)}{\partial t^2} = a^2 \frac{\partial^2 u(x,t)}{\partial x^2}.$$

10. What is the solution to the initial value problem

D.E.: $\dfrac{\partial^2 u(x,t)}{\partial t^2} = \dfrac{\partial^2 u(x,t)}{\partial x^2}$ $-\infty < x < \infty,\quad 0 < t < \infty$

I.C.: $\begin{cases} u(x,0) = e^{-x^2} \\ u_t(x,0) = 0 \end{cases}.$

Ans. $u(x,t) = \frac{1}{2}\left[e^{-(x-t)^2} + e^{-(x+t)^2}\right].$

11. What is the solution to the initial value problem

D.E.: $\dfrac{\partial^2 u(x,t)}{\partial t^2} = \dfrac{\partial^2 u(x,t)}{\partial x^2}$ $-\infty < x < \infty,\quad 0 < t < \infty$

I.C.: $\begin{cases} u(x,0) = 0 \\ u_t(x,0) = x\,e^{-x^2} \end{cases}.$

Ans. $u(x,t) = \frac{1}{4}\left[e^{-(x-t)^2} - e^{-(x+t)^2}\right].$

12. A stretched string between 0 and L is set to motion by a sudden blow at x_0. This problem can be modeled by the fact that the string at x_0 is given a certain velocity at $t = 0$. Find the displacements of the string $u(x,t)$ by solving the following boundary problem:

D.E.: $\dfrac{\partial^2 u(x,t)}{\partial x^2} = \dfrac{1}{a^2}\dfrac{\partial^2 u(x,t)}{\partial t^2}$,

B.C.: $u(0,t) = 0,\qquad u(L,t) = 0,$

I.C.: $u(x,0) = 0,\qquad u_t(x,0) = Lv_0\delta(x - x_0).$

Ans. $u(x,t) = \frac{2v_0 L}{\pi a}\sum_{n=1}^{\infty}\frac{1}{n}\sin\frac{n\pi x_0}{L}\sin\frac{n\pi}{L}x\sin\frac{n\pi a}{L}t.$

13. If the external force acting on the stretched string is proportional to the distance from one end, then the displacement $u(x,t)$ is governed by the differential equation

$$a^2\frac{\partial^2 u(x,t)}{\partial x^2} + Ax = \frac{\partial^2 u(x,t)}{\partial t^2}.$$

If the boundary and initial conditions are

$$u(0,t) = 0, \qquad u(L,t) = 0,$$
$$u(x,0) = f(x), \qquad u_t(x,0) = 0,$$

find $u(x,t)$.

Ans. $u(x,t) = \frac{A}{6a^2}x\left(L^2 - x^2\right) + \sum_{n=1}^{\infty} b_n \sin\frac{n\pi}{L}x \cos\frac{n\pi a}{L}t,$

$b_n = \frac{2}{L}\int_0^L \left[f(x) - \frac{A}{6a^2}x\left(L^2 - x^2\right)\right] \sin\frac{n\pi}{L}x\,dx.$

14. Determine the vibrational behavior of a rectangular membrane described by

D.E.: $\quad \dfrac{1}{v^2}\dfrac{\partial^2 z}{\partial t^2} = \dfrac{\partial^2 z}{\partial x^2} + \dfrac{\partial^2 z}{\partial y^2}$

B.C.: $\quad z(0,y,t) = z(a,y,t) = 0,$

$\quad z(x,0,t) = z(x,b,t) = 0,$

if it is initially displaced according to

$$z(x,y,0) = \sin\frac{\pi x}{a} \sin\frac{\pi y}{b},$$

and then released from rest.

Ans. $z(x,y,t) = \sin\frac{\pi x}{a}\sin\frac{\pi y}{b}\cos\left[\left(\frac{1}{a^2}+\frac{1}{b^2}\right)^{1/2} v\pi t\right]$

15. Solve the equation

$$\frac{1}{v^2}\frac{\partial^2 z}{\partial t^2} = \frac{\partial^2 z}{\partial x^2} + \frac{\partial^2 z}{\partial y^2}$$

with the conditions

$$z(0,y,t) = z(a,y,t) = 0,$$
$$z_y(x,0,t) = z_y(x,b,t) = 0,$$
$$z(x,y,0) = f(x,y), \qquad z_t(x,y,0) = 0.$$

Ans. $z(x,y,t) = \sum_{n=1}^{\infty}\sum_{m=0}^{\infty} a_{nm} \cos\omega_{nm}t \sin\frac{n\pi x}{a}\cos\frac{m\pi y}{b}$, where

$\omega_{nm} = v\pi\left(\dfrac{n^2}{a^2}+\dfrac{m^2}{b^2}\right)^{1/2}, \quad a_{nm} = \dfrac{4}{ab}\int_0^a\int_0^b f(x,y)\sin\dfrac{n\pi x}{a}\cos\dfrac{m\pi y}{b}\,dx\,dy.$

16. Find the solution of the following heat conduction problem:

D.E.: $\quad \dfrac{\partial^2 u(x,t)}{\partial x^2} = \dfrac{1}{\alpha^2}\dfrac{\partial u(x,t)}{\partial t}$

B.C.: $\quad u(0,t) = 0, \qquad u(1,t) = 0,$

I.C.: $\quad u(x,0) = \sin(2\pi x) + \dfrac{1}{3}\sin(4\pi x).$

Ans. $u(x,t) = e^{-(2\pi\alpha)^2 t} \sin(2\pi x) + \frac{1}{3} e^{-(4\pi\alpha)^2 t} \sin(4\pi x)$.

17. What would be the solution to the previous problem if the initial condition is changed to
$$u(x,0) = x - x^2, \qquad 0 < x < 1.$$

Ans. $u(x,t) = \frac{8}{\pi^3} \left[e^{-(\pi\alpha)^2 t} \sin(\pi x) + \frac{1}{27} e^{-(3\pi\alpha)^2 t} \sin(3\pi x) + \cdots \right]$.

18. Solve the following nonhomogeneous heat equation:

   D.E.: $\dfrac{1}{\alpha^2} \dfrac{\partial u(x,t)}{\partial t} = \dfrac{\partial^2 u(x,t)}{\partial x^2} + \sin(\pi x)$

   B.C.: $u(0,t) = 0, \qquad u(1,t) = 0,$

   I.C.: $u(x,0) = \sin(2\pi x).$

Ans. $u(x,t) = \frac{1}{\pi^2} \left(1 - e^{-(\pi\alpha)^2 t} \right) \sin(\pi x) + e^{-(2\pi\alpha)^2 t} \sin(2\pi x).$

19. Solve the following problem where $u(x,t)$ is the temperature of a bar of length L with its lateral surface and one end insulated and the other end kept at a constant temperature.

   D.E.: $\dfrac{1}{\alpha^2} \dfrac{\partial u(x,t)}{\partial t} = \dfrac{\partial^2 u(x,t)}{\partial x^2},$

   B.C.: $u_x(0,t) = 0, \; u(L,t) = 1,$

   I.C.: $u(x,0) = 0.$

Ans. $u(x,t) = 1 - \frac{4}{\pi} \sum\limits_{n=0}^{\infty} \frac{(-1)^n}{(2n+1)} e^{-\left[\frac{(2n+1)\alpha\pi}{2L} \right]^2 t} \cos \frac{(2n+1)\pi}{2L} x.$

20. Solve the following problem:

   D.E.: $\dfrac{1}{\alpha^2} \dfrac{\partial u(x,t)}{\partial t} = \dfrac{\partial^2 u(x,t)}{\partial x^2}$

   B.C.: $u(0,t) = 0, \qquad - u_x(1,t) = K,$

   I.C.: $u(x,0) = 0.$

Ans. $u(x,t) = \frac{8K}{\pi^2} \sum\limits_{n=0}^{\infty} \frac{(-1)^n}{(2n+1)^2} e^{-\left[\frac{(2n+1)\alpha\pi}{2} \right]^2 t} \sin \frac{(2n+1)\pi}{2} x - Kx.$

21. Assume the temperature $u(x,t)$ in a bar of thermal diffusivity a^2 and length L that is perfectly insulated, including the ends at $x = 0$ and $x = L$, is initially given by $u(x,0) = f(x)$, the subsequent distribution of the temperature is given by the solution of the following problem:

D.E.: $\dfrac{1}{a^2} \dfrac{\partial u(x,t)}{\partial t} = \dfrac{\partial^2 u(x,t)}{\partial x^2}$

B.C.: $u_x(0,t) = 0, \quad u_x(L,t) = 0,$

I.C.: $u(x,0) = f(x).$

Express $u(x,t)$ in an infinite series.

Ans. $u(x,t) = \displaystyle\sum_{n=0}^{\infty} A_n e^{-\left[\frac{n a \pi}{L}\right]^2 t} \cos \frac{n\pi}{L} x,$ where

$$A_0 = \frac{1}{L} \int_0^L f(x)\,dx, \qquad A_n = \frac{2}{L} \int_0^L f(x) \cos \frac{n\pi}{L} x \, dx, \quad n = 1,2,3,\ldots.$$

22. All four faces of an infinite long rectangular prism with thermal diffusivity a^2 bounded by the planes $x = 0$, $x = a$, $y = 0$, and $y = b$ are kept at temperature zero. If the initial temperature distribution is $f(x,y)$, derive the following formula for the temperature $u(x,y,t)$ in the prism:

$$u(x,y,t) = \sum_{n=1}^{\infty} \sum_{m=1}^{\infty} a_{nm} e^{-\left(\frac{n^2}{a^2} + \frac{m^2}{b^2}\right)\pi^2 a^2 t} \sin \frac{n\pi x}{a} \sin \frac{m\pi y}{b},$$

where

$$a_{nm} = \frac{4}{ab} \int_0^a \int_0^b f(x,y) \sin \frac{n\pi x}{a} \sin \frac{m\pi y}{b}\,dx\,dy.$$

If $f(x,y) = g(x)h(y)$, show that the double Fourier series reduces to product of two series

$$u(x,y,t) = v(x,t)w(y,t)$$

and note that v and w represent temperatures in the slabs $0 \le x \le a$ and $0 \le y \le b$ with their faces at temperature zero and with initial temperatures $g(x)$ and $h(y)$, respectively.

23. Let $u(x,y)$ be the steady-state temperature in a thin plate in the shape of a semi-infinite strip. Let the surface heat transfer takes place at the face so that

$$\frac{\partial^2 u(x,y)}{\partial x^2} + \frac{\partial^2 u(x,y)}{\partial y^2} - bu(x,y) = 0, \quad 0 \le x \le 1; \ y \ge 0.$$

If u is bounded as $y \to \infty$ and satisfies the conditions

$$u(0,y) = 0, \quad u_x(1,y) = -hu(1,y), \quad u(x,0) = 1.$$

Show that

$$u(x,y) = 2h \sum_{n=1}^{\infty} \frac{A_n}{\alpha_n} e^{-\sqrt{b+\alpha_n^2}\,y} \sin \alpha_n x,$$

where

$$A_n = \frac{1 - \cos \alpha_n}{h + \cos^2 \alpha_n}$$

and $\alpha_1, \alpha_2, \alpha_3, \ldots$ are positive roots of the equation

$$\tan \alpha = -\frac{\alpha}{h}.$$

Hint: With $u(x, y) = X(x)Y(y)$, show that

$$X_n(x) = \sin \alpha_n x, \quad Y_n(y) = e^{-\sqrt{b + \alpha_n^2}\, y},$$

$$\int_0^1 \sin^2 \alpha_n x \, dx = \frac{1}{2} - \frac{\sin 2\alpha_n}{4\alpha_n} = 2h \left(h + \cos^2 \alpha_n \right).$$

24. Find the solution of the problem consisting of

$$\frac{\partial u}{\partial t} - a^2 \frac{\partial^2 u}{\partial x^2} = N e^{-kx}$$

$$u(0, t) = 0, \quad u(L, t) = 0,$$

$$u(x, 0) = f(x).$$

Here the term on the right may represent loss of heat due to radioactive decay in the bar.

Ans. $u(x, t) = U(x, t) + \Psi(x)$,

$$U(x, t) = \sum_{n=1}^{\infty} \left[\frac{2}{L} \int_0^L [f(x) - \Psi(x)] \sin \frac{n\pi}{L} x \, dx \right] e^{-(n\pi a/L)^2 t} \sin \frac{n\pi}{L} x$$

$$\Psi(x) = -\frac{N}{a^2 k^2} \left[e^{-kx} + (1 - e^{-kL})x/L - 1 \right].$$

25. Find the electrostatic potential inside an infinitely long rectangular wave guide with conducting walls. The guide measures L by b. One of the sides of length L is held at a constant potential V_0, the other three sides are grounded (That is, $V = 0$).

    Hint: The answer is given by the solution of the problem:

$$\frac{\partial^2 V(x, y)}{\partial x^2} + \frac{\partial^2 V(x, y)}{\partial y^2} = 0,$$

$$V(0, y) = V(L, y) = V(x, 0) = 0,$$

$$V(x, b) = V_0.$$

Ans. $V(x, y) = \frac{4V_0}{\pi} \sum_{n=1}^{\infty} \frac{1}{2n-1} \left[\sinh \frac{(2n-1)\pi b}{L} \right]^{-1} \sinh \frac{(2n-1)\pi}{L} y \sin \frac{(2n-1)\pi}{L} x.$

26. Find the solution of the problem consisting of

$$\frac{\partial^2 u}{\partial x^2} + \frac{\partial^2 u}{\partial y^2} = 0, \quad 0 \le x \le \infty, \; 0 \le y \le b.$$

$$u(x,0) = 0, \quad u(x,b) = 0,$$

$$u(0,y) = u_0, \quad u(\infty, y) = 0.$$

Here $u(x,y)$ is the steady-state temperature distribution in a semi-infinite rectangular plate of width b, the temperature at far end and along its two long sides is fixed at $0°$, and temperature at $x = 0$ is fixed at a constant temperature u_0.

Ans. $u(x,y) = \sum_{n=1} \frac{4u_0}{(2n-1)\pi} e^{-(2n-1)\pi x/b} \sin \frac{2n-1}{b}\pi y.$

27. Determine the solution of the Laplace's equation

$$\frac{\partial^2 u(x,y)}{\partial x^2} + \frac{\partial^2 u(x,y)}{\partial y^2} = 0$$

in the rectangular region $0 \le x \le L$, $0 \le y \le b$, which satisfies the conditions

$$u(0,y) = u(L,y) = 0,$$

$$u(x,0) = f(x), \quad u(x,b) = g(x).$$

Hint: Show that $u_{n,1} = \sinh \frac{n\pi}{L}y \sin \frac{n\pi}{L}x$ satisfies the equation and the boundary conditions: $u(0,y) = u(L,y) = u(x,0) = 0$, and $u_{n,2} = \sinh \frac{n\pi}{L}(b-y) \sin \frac{n\pi}{L}x$ satisfies the equation and the boundary conditions: $u(0,y) = u(L,y) = u(x,b) = 0$. Thus, $u_n = u_{n,1} + u_{n,2}$.

Ans. $u(x,y) = \sum_{n=1} \left[a_n \sinh \frac{n\pi}{L}y \sin \frac{n\pi}{L}x + b_n \sinh \frac{n\pi}{L}(b-y) \sin \frac{n\pi}{L}x \right],$
where

$$a_n = \left[\sinh \frac{n\pi b}{L}\right]^{-1} \frac{2}{L} \int_0^L g(x) \sin \frac{n\pi x}{L} dx,$$

$$b_n = \left[\sinh \frac{n\pi b}{L}\right]^{-1} \frac{2}{L} \int_0^L f(x) \sin \frac{n\pi x}{L} dx.$$

6

Partial Differential Equations with Curved Boundaries

In Chap. 5, we saw that many physical problems can be formulated in terms of partial differential equations. The solutions of these equations have to satisfy certain boundary conditions. Most of these problems involve the Laplacian operator. Depending on the specific problems, either Laplace equation or Helmholtz equation has to be solved.

When the boundaries are mutually perpendicular straight lines or planes, the cartesian coordinate system is very convenient. When the partial differential equation is reduced to ordinary differential equations by the separation of variables, we are able to recognize the solutions of these ordinary differential equations. Eigenvalues and eigenfunctions dictated by the boundary conditions are readily found and the solution of the partial differential equation can be obtained by the Fourier expansion.

However, many physical problems involve boundaries in the shape of circles, cylinders, and spheres. In these cases, it is much simpler to use polar, cylindrical, or spherical coordinates. The partial differential equations expressed in these coordinates are more complicated. For most problems, the radial and angular parts can be separated, but the ordinary differential equations that govern the angular coordinates are different from those for radial coordinates. Some of them are differential equations with variable coefficients which cannot be solved in terms of elementary functions. We have solved most of these ordinary differential equations in Chap. 4 with Bessel and Legendre functions. Generally, the solutions for problems with cylindrical symmetry are expressible in terms of Fourier–Bessel series, and for those with spherical symmetry, in Fourier–Legendre series. But it is the boundary condition that determines how these solutions of ordinary differential equations should be put together to solve a particular problem. Two identical partial differential equations with slightly different boundary conditions can lead to completely different results.

It is not possible to discuss all the different types of partial differential equations encountered in engineering and physics. In this chapter, we will mainly discuss the Laplace's equation $\nabla^2 u = 0$ and the Helmholtz's equation

$\nabla^2 u + ku = 0$, because they occur in many types of practical problems. These two equations look similar, but their solutions are completely different. For example, the radial solution of the Laplace equation in spherical coordinates is given by r raised to negative or positive integer powers, whereas the solutions of the Helmholtz's equation are given by spherical Bessel functions.

Instead of following the usual procedure to classify the partial differential equations in hyperbolic, parabolic, and elliptic types, we will use enough examples to illustrate how similar methods can be used to solve these equations. It is important to know the general characteristics of different kinds of physical problems that can be solved by these methods. Otherwise details may look confusing. A careful examination of the table of contents may help to gain an overall picture of these problems.

6.1 The Laplacian

As we have seen, the Laplacian operator

$$\nabla^2 = \frac{\partial^2}{\partial x^2} + \frac{\partial^2}{\partial y^2} + \frac{\partial^2}{\partial z^2}$$

appears in a many different partial differential equations of mathematical physics. Why should the sum of three second derivatives have so much to do with the law of nature? The answer lies in the fact that the Laplacian of a function is a comparison between the value of the function at a point and the average value of the function at points around it. This is clear in one dimension. If $u''(x) = 0$, then $u(x) = mx + b$. One can readily verify that

$$u(x) = \frac{1}{2}[u(x + \epsilon) + u(x - \epsilon)].$$

That is, the average value of two nearby points is equal to the value at the midpoint. Furthermore, if $u''(x) > o$, then the curve $u(x)$ is concaved upward, and the average value of two nearby points is greater than the value at the midpoint. If $u''(x) < 0$, then the curve is concaved downward, and the average value of two nearby points is less than the value at the midpoint.

The Laplacian may be thought of as the second derivative generalized to higher dimensions. In fact, we will soon show that in two dimensions, if

- $\nabla^2 u(x, y) = 0$, then the average value of u on a small circle is equal to the value of u at the center of the circle. This is what makes the Laplacian so useful, because
- $\nabla^2 u(x, y) > 0$ means the surface is concaved upward, and the value of u at the center of a small circle is less than the average value of u on the circle, and
- $\nabla^2 u(x, y) < 0$ means the surface is concaved downward, and $u(x, y)$ is greater than the average of u at its neighbors.

Similar statements can be made about $\nabla^2 u$ in three dimensions, if we change circle to sphere.

With these principles, we can have an intuitive understanding of some of the basic partial differential equations of physics.

The wave equation:

$$\frac{\partial^2 u}{\partial t^2} = a^2 \nabla^2 u.$$

If we use this equation to describe the vibration of a membrane, then u is the displacement (the height) of the membrane. This equation says the membrane at a point is accelerating upward (pulled up) if the membrane at that point is below the average of its neighbors.

The diffusion equation:

$$\frac{\partial u}{\partial t} = a^2 \nabla^2 u.$$

If we use this equation to describe heat transfer, then u is the temperature. This equation says that the temperature at a point will be increasing (positive rate of change) if the temperature at that point is less than the average temperature on a circle around that point.

Laplace's equation:

$$\nabla^2 u = 0.$$

This equation describes the steady-state of u, in which the rate of change with respect to time is zero. For example, if u is temperature, then this equation says that the temperature will not change if the temperature at a point is equal to the average temperature of the surrounding points.

Poisson's equation:

$$\nabla^2 u(x,y) = -g(x,y).$$

If $g(x,y)$ is positive at a point, and $u(x,y)$ is the temperature at that point, then this equation says that the temperature at that point is greater than its surroundings. In other words, heat is generated at that point, and $g(x,y)$ is the heat source.

Clearly, the meaning of the Laplacian will not change, no matter what coordinate system we choose to express it. Many problems in two and three dimensions are more naturally expressed in polar, cylindrical, or spherical coordinates, either because of the boundary conditions, or because we want to take advantage of the symmetry of the problem. These coordinates are shown in Fig. 6.1. Let us recall the expressions of the Laplacian operator in these systems. We had a detailed discussion of these expressions in Chap. 3 of volume II.

In polar coordinates:

$$\nabla^2 = \frac{1}{\rho}\frac{\partial}{\partial \rho}\left(\rho\frac{\partial}{\partial \rho}\right) + \frac{1}{\rho^2}\frac{\partial^2}{\partial \varphi^2}.$$

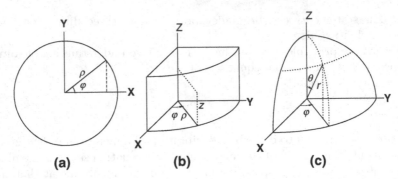

Fig. 6.1. (a) Polar coordinates, (b) cylindrical coordinates, (c) spherical coordinates

In cylindrical coordinates:

$$\nabla^2 = \frac{1}{\rho}\frac{\partial}{\partial \rho}\left(\rho\frac{\partial}{\partial \rho}\right) + \frac{1}{\rho^2}\frac{\partial^2}{\partial \varphi^2} + \frac{\partial^2}{\partial z^2}.$$

In spherical coordinates:

$$\nabla^2 = \frac{1}{r^2}\frac{\partial}{\partial r}\left(r^2\frac{\partial}{\partial r}\right) + \frac{1}{r^2 \sin\theta}\frac{\partial}{\partial \theta}\left(\sin\theta\frac{\partial}{\partial \theta}\right) + \frac{1}{r^2 \sin^2\theta}\frac{\partial^2}{\partial \varphi^2}.$$

6.2 Two-Dimensional Laplace's Equations

6.2.1 Laplace's Equation in Polar Coordinates

Suppose that we need to find the solution of a two-dimensional Laplace's equation that has a prescribed behavior on a circle. Naturally we want to write the equation in the polar coordinates

$$\frac{1}{\rho}\frac{\partial}{\partial \rho}\left(\rho\frac{\partial}{\partial \rho}u(\rho,\varphi)\right) + \frac{1}{\rho^2}\frac{\partial^2}{\partial \varphi^2}u(\rho,\varphi) = 0,$$

so that we can accommodate the boundary condition by examining the solution for $\rho = a$. To solve the equation by separation of variables, we assume $u(\rho,\varphi) = R(\rho)\Phi(\varphi)$ and put it in the above equation

$$\frac{\Phi(\varphi)}{\rho}\frac{\partial}{\partial \rho}\left(\rho\frac{\partial R(\rho)}{\partial \rho}\right) + \frac{R(\rho)}{\rho^2}\frac{\partial^2\Phi(\varphi)}{\partial \varphi^2} = 0.$$

Dividing through by $u(\rho,\varphi) = R(\rho)\Phi(\varphi)$ and multiplying through by ρ^2, we have

$$\frac{\rho}{R(\rho)}\frac{\partial}{\partial \rho}\left(\rho\frac{\partial R(\rho)}{\partial \rho}\right) + \frac{1}{\Phi(\varphi)}\frac{\partial^2\Phi(\varphi)}{\partial \varphi^2} = 0$$

or

$$\frac{1}{R(\rho)} \left(\rho^2 \frac{\partial^2 R(\rho)}{\partial \rho^2} + \rho \frac{\partial R(\rho)}{\partial \rho} \right) = -\frac{1}{\Phi(\varphi)} \frac{\partial^2 \Phi(\varphi)}{\partial \varphi^2}.$$

Since the term on the left-hand side is a function of ρ only, while the term on the right-hand side depends only on φ, both must be equal to the same constant. So we obtain two ordinary differential equations

$$\frac{1}{R(\rho)} \left(\rho^2 \frac{d^2 R(\rho)}{d\rho^2} + \rho \frac{dR(\rho)}{d\rho} \right) = \lambda, \tag{6.1}$$

$$\frac{1}{\Phi(\varphi)} \frac{d^2 \Phi(\varphi)}{d\varphi^2} = -\lambda, \tag{6.2}$$

where λ is the separation constant. The separation constant must be equal to n^2 with n as an integer for the following reason. Since (ρ, φ) and $(\rho, \varphi + 2\pi)$ are the same point, for the description of a real physical system, we must require $\Phi(\varphi)$ to be a periodic function with period 2π. That is, $\Phi(\varphi)$ must satisfy the condition

$$\Phi(\varphi + 2\pi) = \Phi(\varphi).$$

With $\lambda = n^2$, (6.2) becomes

$$\frac{d^2 \Phi_n(\varphi)}{d\varphi^2} = -n^2 \Phi_n(\varphi).$$

Only if n is an integer $(n = 0, 1, 2, \ldots)$ can the solution of this equation be such a periodic function,

$$\Phi_n(\varphi) = A_n \cos n\varphi + B_n \sin n\varphi. \tag{6.3}$$

Note that for $n = 0$, the solution $\Phi_0(\varphi) = \varphi$ is not periodic, therefore not allowed. However, the other solution for $n = 0$, namely $\Phi_0(\varphi) = $ constant satisfies the periodic condition and is included in (6.3).

The radial part of the solution must come from (6.1) with the same n

$$\rho^2 \frac{d^2 R_n(\rho)}{d\rho^2} + \rho \frac{dR_n(\rho)}{d\rho} = n^2 R_n(\rho). \tag{6.4}$$

This is an Euler–Cauchy differential equation. The standard method is to set

$$\rho = e^x, \quad \text{so} \quad x = \ln \rho \quad \text{and} \quad \frac{dx}{d\rho} = \frac{1}{\rho}.$$

It follows that:

$$\frac{dR_n}{d\rho} = \frac{dR_n}{dx} \frac{dx}{d\rho} = \frac{1}{\rho} \frac{dR_n}{dx},$$

$$\frac{d^2 R_n(\rho)}{d\rho^2} = \frac{d}{d\rho} \left(\frac{1}{\rho} \frac{dR_n}{dx} \right) = -\frac{1}{\rho^2} \frac{dR_n}{dx} + \frac{1}{\rho^2} \frac{d^2 R_n}{dx^2},$$

so that (6.4) can be written as

$$\rho^2 \left(\frac{1}{\rho^2} \frac{\mathrm{d}^2 R_n}{\mathrm{d}x^2} - \frac{1}{\rho^2} \frac{\mathrm{d}R_n}{\mathrm{d}x} \right) + \rho \frac{1}{\rho} \frac{\mathrm{d}R_n}{\mathrm{d}x} - n^2 R_n = 0,$$

or

$$\frac{\mathrm{d}^2 R_n}{\mathrm{d}x^2} - n^2 R_n = 0.$$

Clearly

$$R_n(x) = \begin{cases} C_n \mathrm{e}^{nx} + D_n \mathrm{e}^{-nx} & \text{for } n \neq 0, \\ C_0 + D_0 x & \text{for } n = 0. \end{cases}$$

Thus, with $\rho = \mathrm{e}^x$, we have

$$R_n(\rho) = \begin{cases} C_n \rho^n + D_n \rho^{-n} & \text{for } n \neq 0, \\ C_0 + D_0 \ln \rho & \text{for } n = 0. \end{cases}$$

For each integer n, the solution is $u_n(\rho, \varphi) = R_n(\rho) \Phi_n(\varphi)$. A linear combination of them is the general solution

$$u(\rho, \varphi) = \sum_{n=0}^{\infty} e_n u_n(\rho, \varphi).$$

Therefore we can write the solution of the Laplace's equation as

$$u(\rho, \varphi) = a_0 + b_0 \ln \rho + \sum_{n=1}^{\infty} [(a_n \rho^n + b_n \rho^{-n}) \cos n\varphi + (c_n \rho^n + d_n \rho^{-n}) \sin n\varphi],$$

(6.5)

where we have renamed the combinations of constants for convenience.

Whenever we solve the equation in a region containing the origin, we must set b_0, b_n, and d_n to zero, because we are only interested in bounded solutions and $\ln \rho$ and ρ^{-n} go to infinity as ρ approaches zero.

Similarly if the region extends to infinity, then a_0, b_0, a_n, and c_n must be set to zero, unless there is a source at infinity. For example, a uniform electric field extending from $-\infty$ to $+\infty$ must be described by an electric potential that is not vanishing at infinity. (The electric potential satisfies the Laplace's equation, its gradient is the electric field.) In this case a_1 or c_1 may not be zero, but the rest of a_n, c_n, and b_0 must still be all zero.

The following examples will illustrate how these constants are determined by the boundary conditions.

Example 6.2.1. Show that if the temperature $f(\varphi)$ on the boundary of a circular disk of radius r_0 can be expressed as

$$f(\varphi) = \sum_{n=0}^{\infty} (A_n \cos n\varphi + B_n \sin n\varphi),$$

then the steady-state temperature at any point inside the disk is given by

$$u(\rho, \varphi) = \sum_{n=0} (\rho/r_0)^n (A_n \cos n\varphi + B_n \sin n\varphi).$$

Find the steady-state temperature distribution in a circular plate of radius r_0 if the upper semicircular boundary is held at $100°$, and the lower at $0°$.

Solution 6.2.1. The temperature function $u(\rho, \varphi)$ satisfies the Laplace's equation for steady-state distribution. So we have to solve the following boundary value problem:

$$\text{D.E.:} \quad \nabla^2 u(\rho, \varphi) = 0,$$
$$\text{B.C.:} \quad u(r_0, \varphi) = f(\varphi).$$

Since the origin is inside the plate, for the solution to be bounded, we have to set b_n and d_n in (6.5) to zero. What is left in (6.5) can be written as

$$u(\rho, \varphi) = \sum_{n=0} (a_n \rho^n \cos n\varphi + c_n \rho^n \sin n\varphi).$$

Therefore, at $\rho = r_0$, we have

$$u(r_0, \varphi) = \sum_{n=0} (a_n r_0^n \cos n\varphi + c_n r_0^n \sin n\varphi).$$

But it is given that

$$u(r_0, \varphi) = f(\varphi) = \sum_{n=0} (A_n \cos n\varphi + B_n \sin n\varphi).$$

It is clear that

$$a_n = \frac{A_n}{r_0^n}, \quad c_n = \frac{B_n}{r_0^n}.$$

Substituting them into the equation for $u(\rho, \varphi)$, we obtain the required result

$$u(\rho, \varphi) = \sum_{n=0} (\rho/r_0)^n (A_n \cos n\varphi + B_n \sin n\varphi).$$

Now we can expand the given boundary condition

$$u(r_0, \varphi) = f(\varphi) = \begin{cases} 100 & \text{for } 0 < \varphi < \pi, \\ 0 & \text{for } \pi < \varphi < 2\pi \end{cases}$$

into a Fourier series

$$f(\varphi) = \sum_{n=0} (A_n \cos n\varphi + B_n \sin n\varphi),$$

where the coefficients are given by

$$A_0 = \frac{1}{2\pi} \int_0^{2\pi} u(r_0, \varphi) d\varphi = \frac{1}{2\pi} \int_0^{\pi} 100 \, d\varphi = 50,$$

$$A_n = \frac{1}{\pi} \int_0^{2\pi} u(r_0, \varphi) \cos n\varphi \, d\varphi = \frac{1}{\pi} \int_0^{\pi} 100 \cos n\varphi \, d\varphi = 0, \quad n \neq 0,$$

$$B_n = \frac{1}{\pi} \int_0^{2\pi} u(r_0, \varphi) \sin n\varphi \, d\varphi = \frac{100}{\pi} \int_0^{\pi} \sin n\varphi \, d\varphi = \begin{cases} 0 & \text{for } n \text{ even} \\ \frac{200}{n\pi} & \text{for } n \text{ odd} \end{cases}.$$

Thus the temperature distribution in the disk is given by

$$u(\rho, \varphi) = 50 + \frac{200}{\pi} \sum_{n \text{ odd}} \frac{1}{n} \left(\frac{\rho}{r_0} \right)^n \sin n\varphi.$$

It may be noted that $u(\rho, \varphi)$ at $\varphi = 0$ and π, is 50. In general the Fourier series gives the average values at points of discontinuity. In the present case 50 is the average value of 0 and 100.

Example 6.2.2. Determine the steady-state temperature at points of the sector $0 \leq \varphi \leq \theta_0$, $0 \leq \rho \leq r_0$ of a circular plate if the temperature is maintained at zero along the straight edges and at u_0 along the curved edge.

Solution 6.2.2. To find the steady-state temperature $u(\rho, \varphi)$, we have to solve the following boundary value problem:

D.E.: $\nabla^2 u(\rho, \varphi) = 0,$

B.C.: $u(\rho, 0) = 0, \quad u(\rho, \theta_0) = 0, \quad u(r_0, \varphi) = u_0.$

The usual separation assumption $u(\rho, \varphi) = R(\rho)\Phi(\varphi)$ leads to

$$\rho^2 \frac{d^2 R(\rho)}{d\rho^2} + \rho \frac{dR(\rho)}{d\rho} = \lambda R(\rho),$$

$$\frac{d^2 \Phi(\varphi)}{d\varphi^2} = -\lambda \Phi(\varphi).$$

The boundary conditions along the straight edges require that

$$\Phi(0) = 0, \quad \Phi(\theta_0) = 0.$$

The differential equation and the boundary conditions for $\Phi(\varphi)$ actually form a Sturm–Liouville problem. So we know that the eigenvalues λ are positive $(\lambda = \alpha^2)$, and the eigenfunctions are orthogonal. Thus

$$\Phi(\varphi) = A \cos \alpha\varphi + B \sin \alpha\varphi.$$

Since $\Phi(0) = A = 0$, the boundary condition $\Phi(\theta_0) = 0$ requires $\sin\alpha\theta_0 = 0$. Therefore

$$\alpha = \frac{n\pi}{\theta_0},$$

where n is a nonzero integer. Hence the eigenfunctions are

$$\Phi_n(\varphi) = \sin\frac{n\pi}{\theta_0}\varphi, \quad n = 1, 2, 3, \dots .$$

Note that $n = 0$ is not allowed in this case. The corresponding radial equation for $R_n(\rho)$ is

$$\rho^2\frac{\mathrm{d}^2 R_n(\rho)}{\mathrm{d}\rho^2} + \rho\frac{\mathrm{d}R_n(\rho)}{\mathrm{d}\rho} = \left(\frac{n\pi}{\theta_0}\right)^2 R_n(\rho).$$

The bounded solution to this equation is

$$R_n(\rho) = c_n\rho^{n\pi/\theta_0}.$$

Therefore the general solution is

$$u(\rho,\varphi) = \sum_{n=1} c_n\rho^{n\pi/\theta_0}\sin\frac{n\pi}{\theta_0}\varphi.$$

The boundary condition $u(r_0, \varphi) = u_0$ requires c_n to satisfy the relation

$$u_0 = \sum_{n=1} c_n r_0^{n\pi/\theta_0}\sin\frac{n\pi}{\theta_0}\varphi.$$

Since $\left\{\sin\frac{n\pi}{\theta_0}\varphi\right\}$ is a complete orthogonal set in the interval $0 \le \varphi \le \theta_0$,

$$c_n r_0^{n\pi/\theta_0} = \frac{1}{\int_0^{\theta_0}\sin^2\frac{n\pi}{\theta_0}\varphi\,\mathrm{d}\varphi}\int_0^{\theta_0} u_0\sin\frac{n\pi}{\theta_0}\varphi\,\mathrm{d}\varphi.$$

These integrals can be easily evaluated,

$$\int_0^{\theta_0} u_0\sin\frac{n\pi}{\theta_0}\varphi\,\mathrm{d}\varphi = \begin{cases} u_0\frac{2\theta_0}{n\pi} & \text{for } n \text{ odd,} \\ 0 & \text{for } n \text{ even,} \end{cases}$$

$$\int_0^{\theta_0}\sin^2\frac{n\pi}{\theta_0}\varphi\,\mathrm{d}\varphi = \frac{1}{2}\theta_0.$$

Thus the steady-state temperature in the sector is given by

$$u(\rho,\varphi) = \frac{4u_0}{\pi}\sum_{n \text{ odd}}\frac{1}{n}\left(\frac{\rho}{r_0}\right)^{n\pi/\theta_0}\sin\frac{n\pi}{\theta_0}\varphi.$$

Example 6.2.3. Suppose that the temperatures along the inner circle of radius of r_1 and along the outer circle of radius r_2 of an annulus are maintained at $f_1(\varphi)$ and $f_2(\varphi)$, respectively, as shown in Fig. 6.2. Determine the steady-state temperature in the annulus, if

(a) $f_1(\varphi) = 0$, $f_2(\varphi) = \sin\varphi$, $r_1 = 1$, $r_2 = 2$.

(b) $f_1(\varphi) = 1$, $f_2(\varphi) = 10$, $r_1 = 1$, $r_2 = e$.

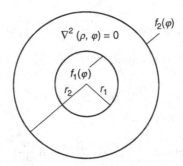

Fig. 6.2. Laplace's equation in an annulus

Solution 6.2.3. The steady-state temperature is determined by the following boundary problem:

D.E.: $\nabla^2 u(\rho, \varphi) = 0$, $r_1 \le \rho \le r_2$,

B.C.: $u(r_1, \varphi) = f_1(\varphi)$, $u(r_2, \varphi) = f_2(\varphi)$.

Since the region of interest neither contains the origin nor extends to infinity, all terms in (6.5) have to be retained. Thus the boundary conditions take the form

$$f_1(\varphi) = a_0 + b_0 \ln r_1 + \sum_{n=1}^{\infty}[(a_n r_1^n + b_n r_1^{-n})\cos n\varphi + (c_n r_1^n + d_n r_1^{-n})\sin n\varphi],$$

$$f_2(\varphi) = a_0 + b_0 \ln r_2 + \sum_{n=1}^{\infty}[(a_n r_2^n + b_n r_2^{-n})\cos n\varphi + (c_n r_2^n + d_n r_2^{-n})\sin n\varphi].$$

According to the theory of Fourier series

$$a_0 + b_0 \ln r_1 = \frac{1}{2\pi}\int_0^{2\pi} f_1(\varphi)d\varphi,$$

$$a_0 + b_0 \ln r_2 = \frac{1}{2\pi}\int_0^{2\pi} f_2(\varphi)d\varphi$$

and for $n \ne 0$

$$a_n r_1^n + b_n r_1^{-n} = \frac{1}{\pi} \int_0^{2\pi} f_1(\varphi) \cos n\varphi \, d\varphi,$$

$$a_n r_2^n + b_n r_2^{-n} = \frac{1}{\pi} \int_0^{2\pi} f_2(\varphi) \cos n\varphi \, d\varphi,$$

$$c_n r_1^n + d_n r_1^{-n} = \frac{1}{\pi} \int_0^{2\pi} f_1(\varphi) \sin n\varphi \, d\varphi,$$

$$c_n r_2^n + d_n r_2^{-n} = \frac{1}{\pi} \int_0^{2\pi} f_2(\varphi) \sin n\varphi \, d\varphi.$$

From these equations, we can solve for the constants a_0, b_0, a_n, b_n, c_n, and d_n.

(a) For $f_1(\varphi) = 0$, $f_2(\varphi) = \sin \varphi$, $r_1 = 1$, $r_2 = 2$, all integrals are equal to zero except

$$\frac{1}{\pi} \int_0^{2\pi} f_2(\varphi) \sin \varphi \, d\varphi = \frac{1}{\pi} \int_0^{2\pi} \sin^2 \varphi \, d\varphi = 1.$$

Therefore all constants are equal to zero except

$$c_1 + d_1 = 0,$$
$$c_1 2 + d_1 \frac{1}{2} = 1.$$

It follows that:

$$c_1 = \frac{2}{3}, \quad d_1 = -\frac{2}{3}.$$

Thus

$$u(\rho, \varphi) = \frac{2}{3} \left(\rho - \frac{1}{\rho} \right) \sin \varphi.$$

(b) For $f_1(\varphi) = 1$, $f_2(\varphi) = 10$, $r_1 = 1$, $r_2 = e$, the only nonvanishing coefficients are

$$a_0 + b_0 \ln r_1 = \frac{1}{2\pi} \int_0^{2\pi} f_1(\varphi) d\varphi = 1,$$

$$a_0 + b_0 \ln r_2 = \frac{1}{2\pi} \int_0^{2\pi} f_2(\varphi) d\varphi = 10.$$

Thus $a_0 = 1$, $b_0 = 10 - a_0 = 9$. Therefore

$$u(\rho, \varphi) = 1 + 9 \ln \rho.$$

Example 6.2.4. Solve the following boundary value problem:

D.E.: $\nabla^2 u(\rho, \varphi) = 0$,

B.C.: $u(\rho \to \infty, \varphi) = E_0 \rho \cos \varphi$, $u(r_0, \varphi) = 0$.

(The solution is the electric potential produced by placing a long grounded conducting cylinder of radius r_0 in a previous uniform electric field $-\mathbf{E}_0$ with the axis of the cylinder perpendicular to the field $\mathbf{E}_0$.)

Solution 6.2.4. Since ρ^{-n} goes to zero as $\rho \to \infty$, asymptotically (6.5) becomes

$$u(\rho \to \infty, \varphi) = a_0 + b_0 \ln \rho + \sum_{n=1}^{\infty} [a_n \rho^n \cos n\varphi + c_n \rho^n \sin n\varphi].$$

Therefore the condition $u(\rho \to \infty, \varphi) = E_0 \rho \cos \varphi$ requires that a_0, b_0, a_n, and c_n to be set to zero, except $a_1 = E_0$. What is left in (6.5) is

$$u(\rho, \varphi) = \sum_{n=1}^{\infty} [b_n \rho^{-n} \cos n\varphi + d_n \rho^{-n} \sin n\varphi] + E_0 \rho \cos \varphi.$$

At $\rho = r_0$, $u(r_0, \varphi) = 0$ becomes

$$(b_1 r_0^{-1} + E_0 r_0) \cos \varphi + d_1 r_0^{-1} \sin \varphi + \sum_{n=2}^{\infty} [b_n r_0^{-n} \cos n\varphi + d_n r_0^{-n} \sin n\varphi] = 0.$$

This requires all coefficients to vanish. That means all b_n and d_n are equal to zero except b_1, and

$$b_1 r_0^{-1} + E_0 r_0 = 0, \quad \text{or} \quad b_1 = -E_0 r_0^2.$$

Therefore the solution of this boundary value problem is given by

$$u(\rho, \varphi) = E_0 \rho \cos \varphi - E_0 r_0^2 \rho^{-1} \cos \varphi = E_0 \left(1 - \frac{r_0^2}{\rho^2} \right) \rho \cos \varphi. \qquad (6.6)$$

6.2.2 Poisson's Integral Formula

Let us return to the solution $u(\rho, \varphi)$ of the Laplace's equation in the interior of a circular disk of radius r_0. If the boundary condition is $u(r_0, \varphi) = f(\varphi)$, then

$$u(\rho, \varphi) = \sum_{n=0}^{\infty} (\rho/r_0)^n (A_n \cos n\varphi + B_n \sin n\varphi),$$

where A_n and B_n are the Fourier coefficients of

$$f(\varphi) = \sum_{n=0} (A_n \cos n\varphi + B_n \sin n\varphi).$$

Substituting the expressions of these Fourier coefficients into $u(\rho, \varphi)$, we have

$$u(\rho, \varphi) = \frac{1}{2\pi} \int_0^{2\pi} f(\varphi') d\varphi' + \sum_{n=1} \left(\frac{\rho}{r_0}\right)^n \left[\frac{1}{\pi} \int_0^{2\pi} f(\varphi') \cos n\varphi' d\varphi'\right] \cos n\varphi$$

$$+ \sum_{n=1} \left(\frac{\rho}{r_0}\right)^n \left[\frac{1}{\pi} \int_0^{2\pi} f(\varphi') \sin n\varphi' d\varphi'\right] \sin n\varphi$$

$$= \frac{1}{2\pi} \int_0^{2\pi} f(\varphi') \left\{1 + 2 \sum_{n=1} \left(\frac{\rho}{r_0}\right)^n [\cos n\varphi' \cos n\varphi + \sin n\varphi' \sin n\varphi]\right\} d\varphi'.$$

Since

$$\cos n\varphi' \cos n\varphi + \sin n\varphi' \sin n\varphi = \cos n(\varphi' - \varphi) = \mathrm{Re}\left[e^{in(\varphi' - \varphi)}\right],$$

where Re stands for real part, so

$$\sum_{n=1} \left(\frac{\rho}{r_0}\right)^n [\cos n\varphi' \cos n\varphi + \sin n\varphi' \sin n\varphi] = \mathrm{Re} \sum_{n=1} \left[\frac{\rho}{r_0} e^{i(\varphi' - \varphi)}\right]^n.$$

Using the facts that

$$\frac{1}{1-z} = 1 + z + z^2 + \cdots = \sum_{n=0}^{\infty} z^n, \quad \text{for} \quad |z| < 1$$

and $\rho/r_0 < 1$, we can write

$$\sum_{n=1} \left[\frac{\rho}{r_0} e^{i(\varphi' - \varphi)}\right]^n = \frac{1}{1 - \frac{\rho}{r_0} e^{i(\varphi' - \varphi)}} - 1 = \frac{\frac{\rho}{r_0} e^{i(\varphi' - \varphi)}}{1 - \frac{\rho}{r_0} e^{i(\varphi' - \varphi)}}$$

$$= \frac{\frac{\rho}{r_0} e^{i(\varphi' - \varphi)}}{1 - \frac{\rho}{r_0} e^{i(\varphi' - \varphi)}} \cdot \frac{1 - \frac{\rho}{r_0} e^{-i(\varphi' - \varphi)}}{1 - \frac{\rho}{r_0} e^{-i(\varphi' - \varphi)}} = \frac{\frac{\rho}{r_0} e^{i(\varphi' - \varphi)} - \left(\frac{\rho}{r_0}\right)^2}{1 - 2\frac{\rho}{r_0} \cos(\varphi' - \varphi) + \left(\frac{\rho}{r_0}\right)^2}$$

and

$$\mathrm{Re} \sum_{n=1} \left[\frac{\rho}{r_0} e^{i(\varphi' - \varphi)}\right]^n = \frac{(\rho/r_0) \cos(\varphi' - \varphi) - (\rho/r_0)^2}{1 - 2(\rho/r_0) \cos(\varphi' - \varphi) + (\rho/r_0)^2}.$$

Thus

$$u(\rho, \varphi) = \frac{1}{2\pi} \int_0^{2\pi} f(\varphi') \left\{1 + 2\frac{(\rho/r_0) \cos(\varphi' - \varphi) - (\rho/r_0)^2}{1 - 2(\rho/r_0) \cos(\varphi' - \varphi) + (\rho/r_0)^2}\right\} d\varphi'$$

$$= \frac{1}{2\pi} \int_0^{2\pi} f(\varphi') \left\{\frac{r_0^2 - \rho^2}{r_0^2 - 2\rho r_0 \cos(\varphi' - \varphi) + \rho^2}\right\} d\varphi'.$$

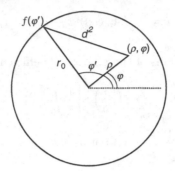

Fig. 6.3. Poisson's integral formula

This remarkable result is known as Poisson integral formula.

According to the law of cosine

$$r_0^2 - 2\rho r_0 \cos(\varphi' - \varphi) + \rho^2 = d^2,$$

where d is the distance between (ρ, φ) and the point (r_0, φ) on the boundary, as shown in Fig. 6.3.

So this formula can be written as

$$u(\rho, \varphi) = \frac{r_0^2 - \rho^2}{2\pi} \int_0^{2\pi} \frac{f(\varphi')}{d^2} d\varphi'.$$

It shows that the value of $u(\rho, \varphi)$ at all points inside a circle of radius r_0 is the weighted sum of the values of the function $f(\varphi')$ on the circumference of the circle, and the weight is given by $(r_0^2 - \rho^2)/d^2$. If we were to increase the values of $f(\varphi')$ on even a small segment of the boundary, this would produce a corresponding increase in the values of u at every interior points. This is in accordance with the fact that the physical system described by the Laplace's equation is in a steady-state. An increase or decrease on the boundary requires the interior region to readjust to bring the system into equilibrium again.

At the center of the circle, $\rho = 0$ and $r_0 = d$, the formula reduces to

$$u(\text{center}) = \frac{1}{2\pi} \int_0^{2\pi} f(\varphi') d\varphi'.$$

In other words, the average value of u on a circle is equal to the value of u at the center of the circle. This is known as the average value property of the Laplace equation.

From this property, we can deduce the fact that the solution of Laplace equation cannot have local maximum or minimum in the interior region. If $u(p)$ were a local maximum at the point P, then there must be a small circle around P, everywhere on it the value of u is smaller than $u(p)$. Obviously the average of the values of u on that circle could not possibly equal to $u(p)$. This violates the average value property of the Laplace equation. Similar argument will also show that there cannot be any local minimum.

6.3 Two-Dimensional Helmholtz's Equations in Polar Coordinates

As we have seen in Chap. 5, when the time dependence part is separated out, the space part of both wave equation and heat equation is given by the Helmholtz equation

$$\nabla^2 u + k^2 u = 0.$$

In polar coordinates, the Helmholtz equation takes the form

$$\frac{1}{\rho}\frac{\partial}{\partial\rho}\left(\rho\frac{\partial}{\partial\rho}u(\rho,\varphi)\right) + \frac{1}{\rho^2}\frac{\partial^2}{\partial\varphi^2}u(\rho,\varphi) + k^2 u(\rho,\varphi) = 0.$$

Proceeding in the same way as we solved the Laplace equation, with the assumption $u(\rho,\varphi) = R(\rho)\,\Phi(\varphi)$, we find that the angular part satisfies the equation

$$\frac{\mathrm{d}^2}{\mathrm{d}\varphi^2}\Phi_n(\varphi) = -n^2\Phi_n(\varphi)$$

and has the familiar solution

$$\Phi_n(\varphi) = A_n \cos n\varphi + B_n \sin n\varphi,$$

where n is an integer. The radial equation

$$\rho^2\frac{\mathrm{d}^2 R_n(\rho)}{\mathrm{d}\rho^2} + \rho\frac{\mathrm{d}R_n(\rho)}{\mathrm{d}\rho} + (k^2\rho^2 - n^2)R_n(\rho) = 0$$

differs from that found in the solution of Laplace equation. The solution of this equation, as we have shown in Chap. 4, is

$$R_n(\rho) = C_n J_n(k\rho) + D_n N_n(k\rho),$$

where $J_n(k\rho)$ and $N_n(k\rho)$ are, respectively, Bessel and Neumann functions of order n. Recall that $N_n(k\rho)$ goes to $-\infty$ as ρ approaches zero. Therefore for solutions that are required to be finite at the origin, we must set $D_n = 0$. Thus, for most applications, the solution of the Helmholtz equation is given by a linear combination of $u_n(\rho,\varphi)$ with

$$u_n(\rho,\varphi) = J_n(k\rho)(A_n \cos n\varphi + B_n \sin n\varphi). \tag{6.7}$$

In what follows, we shall show that not only the two-dimensional wave equation and heat equation reduce to the Helmholtz equation when the time-dependent parts are separated out, but also the three-dimensional Laplace equation in cylindrical coordinates reduce to the Helmholtz equation when the z dependence part is separated out. In that case, the separation constant can be either $+k^2$ or $-k^2$.

6.3.1 Vibration of a Drumhead: Two-Dimensional Wave Equation in Polar Coordinates

Whenever a membrane is a plane and its material is elastic, its vibrations are governed by the wave equation

$$\nabla^2 z = \frac{1}{a^2} \frac{\partial^2}{\partial t^2} z,$$

where z is the displacement (the height of the membrane from the plane). For a circular drumhead, naturally we want to use the polar coordinates.

Let the membrane be stretched over a fixed circular frame of radius c in the plane of $z = 0$. Initially the membrane is deformed in the shape of $z(\rho, \varphi, 0) = f(\rho, \varphi)$ and released from rest in that position. Hence to find the displacements $z(\rho, \varphi, t)$ is to solve the following problem:

D.E.: $\left[\dfrac{\partial^2}{\partial \rho^2} + \dfrac{1}{\rho} \dfrac{\partial}{\partial \rho} + \dfrac{1}{\rho^2} \dfrac{\partial^2}{\partial \varphi^2} \right] z(\rho, \varphi, t) = \dfrac{1}{a^2} \dfrac{\partial^2}{\partial t^2} z(\rho, \varphi, t),$

B.C.: $z(c, \varphi, t) = 0,$

I.C.: $z(\rho, \varphi, 0) = f(\rho, \varphi), \quad \dfrac{\partial z}{\partial t}\bigg|_{t=0} = 0.$

To separate the time-dependent part and the space-dependent part with the assumption $z(\rho, \varphi, t) = u(\rho, \varphi)T(t)$, we find two ordinary differential equations

$$\frac{T''(t)}{a^2 T(t)} = \lambda$$

and

$$\frac{1}{u(\rho, \varphi)} \left[\frac{1}{\rho} \frac{\partial}{\partial \rho} \left(\rho \frac{\partial}{\partial \rho} u(\rho, \varphi) \right) + \frac{1}{\rho^2} \frac{\partial^2}{\partial \varphi^2} u(\rho, \varphi) \right] = \lambda.$$

For the same reason as the vibration of a string, the separation constant must be

$$\lambda = -k^2,$$

so the time-dependent part is governed by

$$T''(t) = -k^2 a^2 T(t) \tag{6.8}$$

and the space-dependent part by the Helmholtz equation

$$\frac{1}{\rho} \frac{\partial}{\partial \rho} \left(\rho \frac{\partial}{\partial \rho} u(\rho, \varphi) \right) + \frac{1}{\rho^2} \frac{\partial^2}{\partial \varphi^2} u(\rho, \varphi) + k^2 u(\rho, \varphi) = 0$$

with (6.7) as its solution

$$u(\rho, \varphi) = J_n(k\rho)(A_n \cos n\varphi + B_n \sin n\varphi).$$

In order to satisfy the boundary condition $z(c, \varphi, t) = 0$, we must have $u(c, \varphi) = 0$. This means that k cannot be any constant, it must satisfy the condition

$$J_n(kc) = 0.$$

Let $k_{nj}c$ be the jth zero of nth order Bessel function $J_n(x)$, then k must be equal to one of the k_{nj}. The same k_{nj} must be used in (6.8), so

$$T(t) = c_1 \cos k_{nj}at + c_2 \sin k_{nj}at.$$

Furthermore, to satisfy the initial condition

$$\left.\frac{\partial}{\partial t} z(\rho, \varphi, t)\right|_{t-0} = u(\rho, \varphi) \left.\frac{dT(t)}{dt}\right|_{t=0} = 0$$

the derivative of $T(t)$ must be zero, therefore c_2 must be set to zero. Thus for each n and j

$$z_{nj}(\rho, \varphi, t) = J_n(k_{nj}\rho)(A_n \cos n\varphi + B_n \sin n\varphi) \cos(ak_{nj}t).$$

The general solution, which is a linear combination of all these terms, is a double sum of all possible n and j

$$z(\rho, \varphi, t) = \sum_{n=0} \sum_{j=1} J_n(k_{nj}\rho)(a_{nj} \cos n\varphi + b_{nj} \sin n\varphi) \cos(ak_{nj}t),$$

where we have consolidated the constants into the coefficients a_{nj} and b_{nj}.

These coefficients can be determined by the other initial condition $z(\rho, \varphi, 0) = f(\rho, \varphi)$

$$z(\rho, \varphi, 0) = \sum_{n=0} \sum_{j=1} J_n(k_{nj}\rho)(a_{nj} \cos n\varphi + b_{nj} \sin n\varphi) = f(\rho, \varphi).$$

Let us first define

$$F_n(\rho) = \sum_{j=1} a_{nj} J_n(k_{nj}\rho), \qquad (6.9)$$

$$G_n(\rho) = \sum_{j=1} b_{nj} J_n(k_{nj}\rho) \qquad (6.10)$$

and express $z(\rho, \varphi, 0)$ in terms of them

$$\sum_{n=0} \{F_n(\rho) \cos n\varphi + G_n(\rho) \sin n\varphi\} = f(\rho, \varphi).$$

Regarding ρ as a parameter, this equation is in the form of a Fourier series, therefore $F_n(\rho)$ and $G_n(\rho)$ are given by

$$F_n(\rho) = \frac{1}{\pi} \int_0^{2\pi} f(\rho, \varphi) \cos n\varphi \, d\varphi, \quad n = 1, 2, \dots .$$

$$= \frac{1}{2\pi} \int_0^{2\pi} f(\rho, \varphi) \cos n\varphi \, d\varphi, \quad n = 0,$$

$$G_n(\rho) = \frac{1}{\pi} \int_0^{2\pi} f(\rho, \varphi) \sin n\varphi \, d\varphi, \quad n = 1, 2, \dots .$$

Putting them back into (6.9) and (6.10), we have

$$\sum_{j=1} a_{nj} J_n(k_{nj}\rho) = \frac{1}{\pi} \int_0^{2\pi} f(\rho, \varphi) \cos n\varphi \, d\varphi, \quad n = 1, 2, \dots,$$

$$\sum_{j=1} a_{nj} J_n(k_{nj}\rho) = \frac{1}{2\pi} \int_0^{2\pi} f(\rho, \varphi) \cos n\varphi \, d\varphi, \quad n = 0,$$

$$\sum_{j=1} b_{nj} J_n(k_{nj}\rho) = \frac{1}{\pi} \int_0^{2\pi} f(\rho, \varphi) \sin n\varphi \, d\varphi, \quad n = 1, 2, \dots .$$

For each fixed n, the series is recognized as a Fourier–Bessel series. The coefficients can be determined by multiplying both sides by $\rho J_n(k_{ni}\rho)$ and integrating from 0 to c

$$\int_0^c \rho J_n(k_{ni}\rho) \sum_{j=1} a_{nj} J_n(k_{nj}\rho) d\rho$$

$$= \int_0^c \rho J_n(k_{ni}\rho) \frac{1}{\pi} \int_0^{2\pi} f(\rho, \varphi) \cos n\varphi \, d\varphi \, d\rho, \quad n = 1, 2, \dots .$$

Because of the orthogonality of the Bessel functions, all terms on the left-hand side are zero except the term with $j = i$

$$a_{ni} \int_0^c \rho J_n^2(k_{ni}\rho) d\rho = \frac{1}{\pi} \int_0^c \int_0^{2\pi} \rho J_n(k_{ni}\rho) f(\rho, \varphi) \cos n\varphi \, d\varphi d\rho.$$

Recall

$$\int_0^c \rho J_n^2(k_{ni}\rho) d\rho = \frac{1}{2} c^2 J_{n+1}^2(k_{ni}c),$$

therefore

$$a_{ni} = \frac{2}{\pi c^2 J_{n+1}^2(k_{ni}c)} \int_0^c \int_0^{2\pi} \rho J_n(k_{ni}\rho) f(\rho, \varphi) \cos n\varphi \, d\varphi d\rho, \quad n = 1, 2, \dots \tag{6.11}$$

For $n = 0$, we have

$$a_{0i} = \frac{2}{2\pi c^2 J_1^2(k_{0i}c)} \int_0^c \int_0^{2\pi} \rho J_0(k_{0i}\rho) f(\rho, \varphi) d\varphi d\rho. \tag{6.12}$$

Similarly,

$$b_{ni} = \frac{2}{\pi c^2 J_{n+1}^2(k_{ni}c)} \int_0^c \int_0^{2\pi} \rho J_n(k_{ni}\rho) f(\rho, \varphi) \sin n\varphi \, d\varphi \, d\rho, \quad n = 1, 2, \ldots .$$

$$(6.13)$$

Thus the solution of problem is given by

$$z(\rho, \varphi, t) = \sum_{n=0} \sum_{i=1} z_{nj} = \sum_{n=0} \sum_{i=1} J_n(k_{ni}\rho)(a_{ni} \cos n\varphi + b_{ni} \sin n\varphi) \cos(ak_{ni}t)$$

$$(6.14)$$

with the coefficients given by (6.11)–(6.13).

Each $z_{nj}(\rho, \varphi, t)$ term is known as a normal mode. Each mode is vibrating in a periodic motion with a frequency of $ak_{nj}/2\pi$. If the initial displacements happen to be in the shape of a particular mode, then the system will vibrate in that mode. For example, if

$$z(\rho, \varphi, 0) = f(\rho, \varphi) = A J_2(k_{21}\rho) \cos 2\varphi,$$

then all the coefficients in (6.14) are equal to zero, except $a_{21} = A$, and

$$z(\rho, \varphi, t) = A J_2(k_{21}\rho) \cos 2\varphi \cos(ak_{21}t).$$

Each mode has nodal lines (lines without motion), which consist of circles and radial lines. Figure 6.4 shows a few of these modes of vibration, shaded parts being displaced in the opposite direction to unshaded parts. If the drumhead is vibrating in one of these normal modes, saw-dust sprinkled on it will collect along the nodal lines, so that you can see them. It is not easy to obtain the pure normal modes with initial conditions. However, if an oscillator vibrating

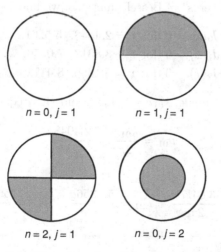

$n = 0, j = 1$ $n = 1, j = 1$

$n = 2, j = 1$ $n = 0, j = 2$

Fig. 6.4. Normal modes of drumhead vibration

in the frequency of a particular mode is placed nearby, the drumhead will eventually be vibrating in that mode, and the nodal lines become visible.

The general motion of a vibrating drumhead is a superposition of all these normal modes. If nodal lines exist at all, they will not usually be of the simple patterns.

Note that if the initial displacement is independent of φ, then $f(\rho, \varphi)$ should be replaced by $f(\rho)$. In that case, it follows from (6.11) and (6.13) that $a_{ni} = 0$ and $b_{ni} = 0$ for all n except for $n = 0$. For $n = 0$, (6.12) gives

$$a_{0i} = \frac{2}{c^2 J_1^2(k_{0i}c)} \int_0^c \rho J_0(k_{0i}\rho) f(\rho) d\rho.$$

In this case the solution is given by

$$z(\rho, \varphi, t) = \sum_{i=1} \left[\frac{2}{c^2 J_1^2(k_{0i}c)} \int_0^c \rho' J_0(k_{0i}\rho') f(\rho) d\rho' \right] J_0(k_{0i}\rho) \cos(ak_{0i}t).$$

$$(6.15)$$

This is what one would expect if both boundary and initial conditions are independent of φ, then there is no reason that the final solution is dependent on φ. Therefore in this case one can start with

$$z = R(\rho)T(t)$$

and obtain the result of (6.15) directly.

Example 6.3.1. Find the frequency of the normal modes shown in Fig. 6.4.

Solution 6.3.1. The frequency of vibration for the nj normal mode is $k_{nj}a/2\pi$. From the "Table of zeros" of Bessel functions, we find

$$J_0(x) = 0 \quad \text{for } x = 2.4048, \ 5.5201, \ \ldots$$
$$J_1(x) = 0 \quad \text{for } x = 3.8317, \ 7.0156, \ \ldots$$
$$J_2(x) = 0 \quad \text{for } x = 5.1356, \ 8.4172, \ \ldots$$

Therefore for the 01 mode, $k_{01}c = 2.4048$, and the frequency of this mode is given by

$$\nu_{01} = \frac{k_{01}a}{2\pi} = \frac{2.4048a}{2\pi c}.$$

Similarly,

$$\nu_{11} = \frac{3.8317a}{2\pi c}, \quad \nu_{21} = \frac{5.1356a}{2\pi c}, \quad \nu_{02} = \frac{5.5201a}{2\pi c}.$$

Example 6.3.2. If the membrane and its frame are moving as a rigid body with unit velocity perpendicular to the membrane and the frame is suddenly brought to rest, then the membrane will start to vibrate. The vibration can be modeled with the following boundary value problem:

$$\text{D.E.:} \quad \nabla^2 z(\rho, t) = \frac{1}{a^2} \frac{\partial^2}{\partial t^2} z(\rho, t),$$

$$\text{B.C.:} \quad z(c, t) = 0,$$

$$\text{I.C.:} \quad z(\rho, 0) = 0, \quad \left.\frac{\partial z}{\partial t}\right|_{t=0} = 1.$$

Find the displacements $z(\rho, t)$.

Solution 6.3.2. With $z(\rho, t) = R(\rho)T(t)$, the time- and space-dependent parts are, respectively, given by

$$T(t) = c_1 \cos kat + c_2 \sin kat$$

and

$$R(\rho) = J_0(k\rho).$$

The boundary condition requires that $J_0(kc) = 0$. Therefore k must be equal to one of k_{0j}, where $k_{0j}c$ is the jth root of $J_0(x) = 0$. The initial condition requires $T(0) = c_1 = 0$. Thus

$$z(\rho, t) = \sum_{j=1}^{\infty} b_j J_0(k_{0j}\rho) \sin k_{0j}at.$$

Now the other initial condition requires that

$$\left.\frac{\partial z}{\partial t}\right|_{t=0} = \sum_{j=1}^{\infty} b_j k_{0j} a J_0(k_{0j}\rho) = 1.$$

Multiplying by $\rho J_0(k_{0j}\rho)$, integrating from 0 to c, we obtain

$$b_j k_{0j} a = \frac{2}{c^2 J_1^2(k_{0j}c)} \int_0^c \rho J_0(k_{0j}\rho) d\rho.$$

Recall

$$\int_0^c x J_0(x) dx = \int_0^c d[x J_1(x)],$$

so

$$\int_0^c \rho J_0(k_{0j}\rho) d\rho = \frac{1}{k_{0j}} c J_1(k_{0j}c).$$

It follows:

$$b_j k_{0j} a = \frac{2}{c^2 J_1^2(k_{0j}c)} \frac{1}{k_{0j}} c J_1(k_{0j}c),$$

$$b_j = \frac{2}{ack_{0j}^2} \frac{1}{J_1(k_{0j}c)}.$$

Therefore

$$z(\rho,t) = \frac{2}{ac} \sum_{j=1} \frac{\sin(k_{0j}at)}{k_{0j}^2 J_1(k_{0j}c)} J_0(k_{0j}\rho)$$

is the displacement, where $k_{0j}c$ are the positive roots of $J_0(x) = 0$.

6.3.2 Heat Conduction in a Disk: Two-Dimensional Diffusion Equation in Polar Coordinates

The solution of the diffusion equation

$$\nabla^2 F = \frac{1}{a^2} \frac{\partial}{\partial t} F$$

in the form of $F(\rho, \varphi, t) = u(\rho, \varphi) T(t)$ is similar to that of the wave equation. In terms of $u(\rho, \varphi)$ and $T(t)$, the equation can be written as

$$\frac{1}{u(\rho,\varphi)} \left[\frac{\partial^2}{\partial \rho^2} + \frac{1}{\rho} \frac{\partial}{\partial \rho} + \frac{1}{\rho^2} \frac{\partial^2}{\partial \varphi^2} \right] u(\rho,\varphi) = \frac{1}{T(t)} \frac{1}{a^2} \frac{\partial}{\partial t} T(t).$$

Again both sides must equal to the same separation constant

$$\frac{1}{u(\rho,\varphi)} \left[\frac{\partial^2}{\partial \rho^2} + \frac{1}{\rho} \frac{\partial}{\partial \rho} + \frac{1}{\rho^2} \frac{\partial^2}{\partial \varphi^2} \right] u(\rho,\varphi) = \lambda,$$

$$\frac{1}{T(t)} \frac{1}{a^2} \frac{\partial}{\partial t} T(t) = \lambda.$$

The separation constant must be negative, $\lambda = -k^2$, just as in the wave equation, but for a slightly different reason. If λ were equal to $+k^2$, then $T(t) = e^{k^2 a^2 t}$, which would make the temperature increase without bound as time increases. This is physically unreasonable. (Formally we say that it violates the principle of conservation of energy.) Not only that, $\lambda = k^2$ would render the space part fail to satisfy the boundary conditions. It is possible to have $\lambda = 0$, indicating that the system has already reached equilibrium. However, we should be able to solve the problem with $\lambda \neq 0$, and show that the solution reduces to that of $\lambda = 0$ as $t \to \infty$.

With $\lambda = -k^2$, $T(t) = e^{-k^2 a^2 t}$ and the space part is given by the Helmholtz equation. While this equation is identical to the one obtained in solving the wave equation, it is the boundary conditions that can make a difference.

In the following examples, we will consider the heat conduction in a coin. Its circumference can either be kept at a constant temperature, or be insulated.

Example 6.3.3. Find the temperature in the disk of radius c, the flat faces of which are insulated. Initially the disk is at a prescribed temperature $f(\rho)$. Its outer edge is kept at $100°$ for all times.

Solution 6.3.3. Since neither the boundary condition nor the initial condition has angular dependence, so we know that the temperature distribution in the disk is independent of the angle. Thus the temperature in the disk $F(\rho,t)$ is the solution of the following problem:

$$\text{D.E.:} \quad \nabla^2 F = \frac{1}{a^2}\frac{\partial}{\partial t}F,$$

$$\text{B.C.:} \quad F(c,t) = 100,$$

$$\text{I.C.:} \quad F(\rho,0) = f(\rho).$$

The problem is easier to solve if we make a change of variable. Let

$$u(\rho,t) = F(\rho,t) - 100.$$

The differential equation that governs $u(\rho,t)$ remains the same

$$\nabla^2 u(\rho,t) = \frac{1}{a^2}\frac{\partial}{\partial t}u(\rho,t)$$

but the boundary condition and the initial condition are changed to

$$u(c,t) = 0, \quad u(\rho,0) = f(\rho) - 100.$$

The solution is given by

$$u(\rho,t) = J_0(k\rho)e^{-k^2 a^2 t}.$$

In order to satisfy the boundary condition $u(c,t) = 0$, k must be equal to one of k_{0j}, where $k_{0j}c$ is the jth root of $J_0(x) = 0$. Therefore the general solution is the following linear combination:

$$u(\rho,t) = \sum_{j=1}^{\infty} c_j J_0(k_{0j}\rho)e^{-k_{0j}^2 a^2 t}.$$

Since $u(\rho,0) = f(\rho) - 100$, so

$$f(\rho) - 100 = u(\rho,0) = \sum_{j=1}^{\infty} c_j J_0(k_{0j}\rho).$$

It follows that:

$$c_j = \frac{2}{c^2 J_1^2(k_{0j}c)} \int_0^c (f(\rho) - 100) J_0(k_{0j}\rho)\rho \, d\rho.$$

Therefore the temperature distribution in the disk is given by

$$F(\rho, t) = 100 + \sum_{j=1} A_j J_0(k_{0j}\rho) e^{-k_{0j}^2 a^2 t},$$

$$A_j = \frac{2}{c^2 J_1^2(k_{0j}c)} \int_0^c (f(\rho') - 100) J_0(k_{0j}\rho')\rho' \, d\rho'.$$

Clearly, as $t \to \infty$, the temperature everywhere in the disk will settle down to $100°$, as it should be, regardless of the initial temperature.

Example 6.3.4. Replace the boundary condition of the previous problem with the condition that the edge of the disk is thermally insulated. Find the temperature distribution $u(\rho, t)$ in the disk if $u(\rho, 0) = f(\rho)$.

Solution 6.3.4. An insulated edge means that there is no heat flowing in and out of the disk. Since the heat flux is proportional to the gradient of the temperature, an insulated edge corresponds to the boundary condition

$$\frac{\partial}{\partial \rho} u(\rho, t)\bigg|_{\rho=c} = 0.$$

The solution of the diffusion equation is still given by

$$u(\rho, t) = J_0(k\rho) e^{-k^2 a^2 t}.$$

Now the boundary condition requires that

$$\frac{d}{d\rho} J_0(k\rho)\bigg|_{\rho=c} = 0.$$

Recall

$$\frac{d}{dx} J_0(x) = -J_1(x).$$

This means

$$\frac{d}{d\rho} J_0(k\rho)\bigg|_{\rho=c} = -kJ_1(kc) = 0.$$

Thus k must be equal to one of k_{1j}, where $k_{1j}c$ is the jth root of $J_1(x) = 0$. Therefore the general solution is

$$u(\rho, t) = \sum_{j=0} c_j J_0(k_{1j}\rho) e^{-k_{1j}^2 a^2 t}.$$

Note that in this case $k_{10} = 0$ is also an eigenvalue, since $J_1(0) = 0$. Furthermore, since $J_0(0) = 1$, we can write this expansion as

$$u(\rho, t) = c_0 + \sum_{j=1}^{\infty} c_j J_0(k_{1j}\rho)e^{-k_{1j}^2 a^2 t}.$$

With the initial condition $u(\rho, 0) = f(\rho)$, we have

$$f(\rho) = c_0 + \sum_{j=1}^{\infty} c_j J_0(k_{1j}\rho). \tag{6.16}$$

The coefficient c_j can be determined with the help of the orthogonality relations of the Bessel functions

$$\int_0^c \rho J_0(k_{1j}\rho) J_0(k_{1i}\rho) d\rho = \delta_{ij}\beta_{0j}^2,$$

where β_{0j}^2 is given by

$$\beta_{0j}^2 = \frac{1}{2}c^2 J_0^2(k_{1j}c),$$

which follows from (4.55) with $n = 0$ and $A = 0$. Furthermore

$$\int_0^c \rho J_0(k_{1j}\rho) d\rho = \frac{c}{k_{1j}} J_1(k_{1j}c) = 0,$$

since $k_{1j}c$ is one of the roots of $J_1(x) = 0$. Multiplying (6.16) by ρ and integrating from 0 to c, we have

$$\int_0^c \rho f(\rho) d\rho = \int_0^c \rho c_0 d\rho = \frac{1}{2}c^2 c_0$$

or

$$c_0 = \frac{2}{c^2} \int_0^c \rho f(\rho) \, d\rho.$$

Multiplying (6.16) by $\rho J_0(k_{1i}\rho)$ and integrating from 0 to c, we have

$$c_i = \frac{1}{\beta_{0i}^2} \int_0^c \rho J_0(k_{1i}\rho) f(\rho) \, d\rho.$$

Thus the temperature distribution in the disk is given by

$$u(\rho, t) = \frac{2}{c^2} \int_0^c \rho f(\rho) d\rho + \sum_{j=1}^{\infty} B_j J_0(k_{1j}\rho)e^{-k_{1j}^2 a^2 t},$$

$$B_j = \frac{2}{c^2 J_0^2(k_{1j}c)} \int_0^c \rho' J_0(k_{1j}\rho') f(\rho') d\rho'.$$

6.3.3 Laplace's Equations in Cylindrical Coordinates

Three-dimensional Laplace equation $\nabla^2 V = 0$ in cylindrical coordinates takes the form

$$\left[\frac{\partial^2}{\partial \rho^2} + \frac{1}{\rho} \frac{\partial}{\partial \rho} + \frac{1}{\rho^2} \frac{\partial^2}{\partial \varphi^2} + \frac{\partial^2}{\partial z^2} \right] V(\rho, \varphi, z) = 0.$$

With $V(\rho, \varphi, z) = u(\rho, \varphi) Z(z)$, this equation can be written as

$$\frac{1}{u(\rho, \varphi)} \left[\frac{\partial^2}{\partial \rho^2} + \frac{1}{\rho} \frac{\partial}{\partial \rho} + \frac{1}{\rho^2} \frac{\partial^2}{\partial \varphi^2} \right] u(\rho, \varphi) = -\frac{1}{Z(z)} \frac{\partial^2}{\partial z^2} Z(z).$$

Again both sides must be equal to the same separation constant,

$$\frac{1}{u(\rho, \varphi)} \left[\frac{\partial^2}{\partial \rho^2} + \frac{1}{\rho} \frac{\partial}{\partial \rho} + \frac{1}{\rho^2} \frac{\partial^2}{\partial \varphi^2} \right] u(\rho, \varphi) = \lambda,$$

$$-\frac{1}{Z(z)} \frac{\partial^2}{\partial z^2} Z(z) = \lambda.$$

It is seen that after the $Z(z)$ is separated out, the remaining equation for $u(\rho, \varphi)$ is again a Helmholtz equation in polar coordinates. In fact, the equation looks exactly the same as that obtained from wave equation or diffusion equation. However, there is a subtle difference. The difference is that in this case, the separation constant λ can be either positive $+k^2$ or negative $-k^2$, depending on the boundary conditions. In the following examples, we will illustrate that under certain boundary conditions we have to use $+k^2$, while under other boundary conditions we need to use $-k^2$. In general, both $+k^2$ and $-k^2$ are needed.

Example 6.3.5. Consider a solid circular cylinder of radius c and length L. The temperatures at the bases of the cylinder are maintained at $0°$ while the temperature of its lateral surface is specified by a given function $f(z)$, as shown is Fig. 6.5. Find the steady-state distribution of the temperature inside the cylinder.

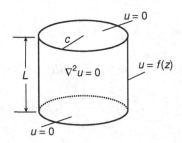

Fig. 6.5. Laplace equation in a circular cylinder. With this set of boundary conditions, the solution is given by a series of products of trigonometric and modified Bessel functions

Solution 6.3.5. The steady-state temperature u satisfies the Laplace equation $\nabla^2 u = 0$. Furthermore, since the boundary conditions do not depend on angle, the solution will also not be a function of φ. Therefore our task is to solve the following boundary value problem:

D.E.: $\nabla^2 u(\rho, z) = 0,$

B.C.: $u(\rho, 0) = u(\rho, L) = 0, \quad u(c, z) = f(z).$

With $u(\rho, z) = R(\rho)Z(z)$, we have

$$\left[\frac{\partial^2}{\partial \rho^2} + \frac{1}{\rho}\frac{\partial}{\partial \rho} \right] R(\rho) = \lambda R(\rho)$$

and

$$\frac{d^2}{dz^2} Z(z) = -\lambda Z(z).$$

The boundary condition $u(\rho, 0) = R(\rho)Z(0) = 0$ translates into $Z(0) = 0$. Similarly $Z(L) = 0$. The Z equation and its boundary conditions remind us of the problem of one-dimensional heat conduction with both ends at $0°$. These boundary conditions can be satisfied if the solutions are cosine and sine functions. With the separation constant λ chosen as $+k^2$,

$$Z(z) = c_1 \cos kz + c_2 \sin kz.$$

The corresponding R equation can be written as

$$\left(\rho^2 \frac{\partial^2}{\partial \rho^2} + \rho \frac{\partial}{\partial \rho} - k^2\rho^2 \right) R(\rho) = 0.$$

As we have seen in Chap. 4, this is the modified Bessel function of order 0. Its solution is

$$R(\rho) = c_3 I_0(k\rho) + c_4 K_0(k\rho),$$

where I_0 and K_0 are modified Bessel functions of first and second kind of order 0. Since K_0 diverges as $\rho \to 0$, to keep the temperature finite on the axis of the cylinder, c_4 must be set to zero. The condition $Z(0) = 0$ requires $c_1 = 0$. The condition $Z(L) = 0$ requires k to be one of k_n, where $k_n = n\pi/L$ with n as an integer ($n = 1, 2, \ldots$). Thus the general solution is in the form of

$$u(\rho, z) = \sum_{n=1}^{\infty} A_n I_0 \left(\frac{n\pi}{L}\rho \right) \sin \left(\frac{n\pi}{L} z \right).$$

The coefficients A_n are determined by the other boundary condition $u(c, z) = f(z)$,

$$f(z) = \sum_{n=1}^{\infty} A_n I_0 \left(\frac{n\pi}{L}c\right) \sin\left(\frac{n\pi}{L}z\right).$$

This is a Fourier sine series. It follows that:

$$A_n I_0 \left(\frac{n\pi}{L}c\right) = \frac{2}{L} \int_0^L f(z) \sin\left(\frac{n\pi}{L}z\right) dz.$$

Therefore the temperature distribution is given by

$$u(\rho, z) = \sum_{n=1}^{\infty} \left[\frac{1}{I_0\left(\frac{n\pi}{L}c\right)} \frac{2}{L} \int_0^L f(z') \sin\left(\frac{n\pi}{L}z'\right) dz'\right] I_0\left(\frac{n\pi}{L}\rho\right) \sin\left(\frac{n\pi}{L}z\right).$$

Example 6.3.6. Solve the previous problem with the boundary conditions replaced by the conditions that both the curved and bottom surfaces are kept at $0°$ and the top surface is specified by a given function $g(\rho)$, as shown in Fig. 6.6.

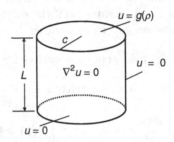

Fig. 6.6. Laplace equation in a circular cylinder. With this set of boundary conditions, the solution is given by a series of products of hyperbolic and Bessel functions

Solution 6.3.6. This time the problem we need to solve is

D.E.: $\nabla^2 u(\rho, z) = 0,$

B.C.: $u(c, z) = 0, \quad u(\rho, 0) = 0, \quad u(\rho, L) = g(\rho).$

Again with $u(\rho, z) = R(\rho)Z(z)$ we have

$$\left[\frac{\partial^2}{\partial\rho^2} + \frac{1}{\rho}\frac{\partial}{\partial\rho}\right] R(\rho) = \lambda R(\rho)$$

and

$$\frac{d^2}{dz^2} Z(z) = -\lambda Z(z).$$

The boundary condition $u(c, z) = R(c)Z(z) = 0$ requires $R(c) = 0$.

Now if $\lambda = +k^2$, as we have seen in the last example

$$R(\rho) = I_0(k\rho).$$

Since I_0 is a monotonically increasing function and is never equal to zero, this choice cannot possibly satisfy the boundary condition.

If $\lambda = 0$, the R equation becomes a Euler–Cauchy differential equation, its solution is

$$R(\rho) = c_1 + c_2 \ln \rho.$$

To keep the temperature finite on the axis of the cylinder, we must set $c_2 = 0$. To make $R(c) = 0$, c_1 must also be zero. Therefore we will not get a solution for $\lambda = 0$.

Finally, if $\lambda = -k^2$, the R equation becomes

$$\left(\rho^2 \frac{\partial^2}{\partial \rho^2} + \rho \frac{\partial}{\partial \rho} + k^2 \rho^2 \right) R(\rho) = 0,$$

and the solution that is finite on the axis of the cylinder is given by the Bessel function of zeroth-order

$$R(\rho) = J_0(k\rho).$$

To satisfy the boundary condition $R(c) = 0$, k must be equal to k_{0j}, where $k_{0j}c$ is the jth root of $J_0(x) = 0$.

The corresponding Z equation is

$$\frac{\mathrm{d}^2}{\mathrm{d}z^2} Z(z) = k_{0j}^2 Z(z).$$

The solution of this equation is

$$Z(z) = c_3 \cosh k_{0j} z + c_4 \sinh k_{0j} z.$$

The condition $u(\rho, 0) = R(\rho)Z(0) = 0$ requires $Z(0) = 0$, therefore c_3 must be zero. Thus the general solution is in the form of

$$u(\rho, z) = \sum_{j=1} A_j J_0(k_{0j}\rho) \sinh k_{0j} z.$$

The coefficients A_j can be determined by the condition $u(\rho, L) = g(\rho)$,

$$g(\rho) = \sum_{j=1} A_j J_0(k_{0j}\rho) \sinh k_{0j} L.$$

It follows that:

$$A_j \sinh k_{0j} L = \frac{2}{c^2 J_1^2(k_{0j}c)} \int_0^c g(\rho) J_0(k_{0j}\rho)\rho\, d\rho.$$

Therefore the temperature distribution in this case is given by

$$u(\rho, z) = \sum_{j=1} \left[\frac{2}{c^2 J_1^2(k_{0j}c) \sinh k_{0j} L} \int_0^c g(\rho') J_0(k_{0j}\rho')\rho'\, d\rho' \right] J_0(k_{0j}\rho) \sinh k_{0j} z.$$

Example 6.3.7. Find the steady-state temperature $u(\rho, z)$ in the cylinder of the previous example, subject to the boundary conditions: $u(\rho, 0) = 0$, $u(c, z) = T_1$, $u(\rho, L) = T_2$, where T_1 and T_2 are two constants.

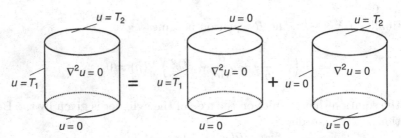

Fig. 6.7. Principle of superposition. The problem on the left-hand side can be broken apart as the sum of the two problems on the right-hand side

Solution 6.3.7. This problem can be solved by using the principle of superposition and the results of the previous examples. The problem is shown on the left-hand side of Fig. 6.7. We break it into two parts on the right-hand side of Fig. 6.7. Suppose that we have solved the following two problems:

D.E.: $\nabla^2 u_1(\rho, z) = 0,$

B.C.: $u_1(\rho, 0) = 0, \quad u_1(\rho, L) = 0, \quad u_1(c, z) = T_1$

and

D.E.: $\nabla^2 u_2(\rho, z) = 0,$

B.C.: $u_2(\rho, 0) = 0, \quad u_2(\rho, L) = T_2, \quad u_2(c, z) = 0.$

Clearly the required solution is

$$u(\rho, z) = u_1(\rho, z) + u_2(\rho, z),$$

since

$$\nabla^2 u(\rho, z) = \nabla^2 u_1(\rho, z) + \nabla^2 u_2(\rho, z),$$
$$u(\rho, 0) = u_1(\rho, 0) + u_2(\rho, 0) = 0,$$
$$u(\rho, L) = u_1(\rho, L) + u_2(\rho, L) = T_2,$$
$$u(c, z) = u_1(c, z) + u_2(c, z) = T_1.$$

From the results of previous examples

$$u_1(\rho, z) = \sum_{n=1}^{\infty} \left[\frac{1}{I_0\left(\frac{n\pi}{L}c\right)} \frac{2}{L} \int_0^L T_1 \sin \frac{n\pi}{L} z' dz' \right] I_0\left(\frac{n\pi}{L}\rho\right) \sin\left(\frac{n\pi}{L}z\right)$$

and

$$u_2(\rho, z) = \sum_{j=1}^{\infty} \left[\frac{2}{c^2 J_1^2(k_{0j}c) \sinh k_{0j}L} \int_0^c T_2 J_0(k_{0j}\rho')\rho' d\rho' \right] J_0(k_{0j}\rho) \sinh k_{0j}z.$$

Since

$$\frac{2}{L} \int_0^L T_1 \sin \frac{n\pi}{L} z' dz' = \begin{cases} \frac{4T_1}{n\pi} & n = \text{odd} \\ 0 & n = \text{even} \end{cases}$$

and

$$\int_0^c T_2 J_0(k_{0j}\rho')\rho' d\rho' = T_2 \frac{c}{k_{0j}} J_1(k_{0j}c),$$

it follows that the required solution is given by:

$$u(\rho, z) = \frac{4T_1}{\pi} \sum_{n=\text{odd}} \frac{1}{n} \frac{1}{I_0\left(\frac{n\pi}{L}c\right)} I_0\left(\frac{n\pi}{L}\rho\right) \sin\left(\frac{n\pi}{L}z\right)$$

$$+ \frac{2T_2}{c} \sum_{j=1}^{\infty} \frac{1}{k_{0j} J_1(k_{0j}c) \sinh k_{0j}L} J_0(k_{0j}\rho) \sinh k_{0j}z.$$

6.3.4 Helmholtz's Equations in Cylindrical Coordinates

Generalizing the above method for Helmholtz equation in cylindrical coordinates is straightforward. However, it is important to appreciate the distinction
between solving Laplace equation and solving Helmholtz equation as an eigenvalue problem. For Helmholtz equation, a homogeneous boundary condition
can be imposed (the function can vanish on all boundary surfaces), whereas
the only solution of Laplace equation satisfying the homogeneous boundary
condition is the trivial solution that the function is identically zero everywhere. This is best illustrated by an example.

Example 6.3.8. **Heat Conduction in a Finite Cylinder.** Consider a solid
circular cylinder of radius c and length L, its surfaces are kept at $0°$ all the
time. Initially the cylinder has a uniform temperature T_0, as shown in Fig. 6.8.
Find the temperature distribution inside the cylinder as a function of time.

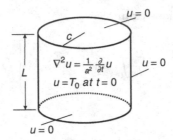

Fig. 6.8. Diffusion equation in a cylinder. Homogeneous boundary conditions can be imposed for a Helmholtz equation

Solution 6.3.8. Let u be the temperature. It satisfies the diffusion equation, which in cylindrical coordinates take the form

$$\left[\frac{\partial^2}{\partial\rho^2} + \frac{1}{\rho}\frac{\partial}{\partial\rho} + \frac{\partial^2}{\partial z^2}\right] u(\rho, z, t) = \frac{1}{a^2}\frac{\partial}{\partial t}u(\rho, z, t),$$

where we have taken the advantage that this problem has no angular dependence because of the axial symmetry. The boundary and initial conditions are

$$u(c, z, t) = u(\rho, 0, t) = u(\rho, L, t) = 0,$$
$$u(\rho, z, 0) = T_0.$$

With $u(\rho, z, t) = R(\rho)Z(z)T(t)$, this equation is first separated into two equations

$$\frac{1}{R(\rho)Z(z)}\left[\frac{\partial^2}{\partial\rho^2} + \frac{1}{\rho}\frac{\partial}{\partial\rho} + \frac{\partial^2}{\partial z^2}\right] R(\rho)Z(z) = -k^2,$$

$$\frac{1}{T(t)}\frac{1}{a^2}\frac{\mathrm{d}}{\mathrm{d}t}T(t) = -k^2.$$

The separation constant has to be a negative number $-k^2$, because the temperature should be in the form of $\exp(-k^2a^2t)$. The differential equation of the space part is a Helmholtz equation in cylindrical coordinates without angular dependence. It can be written as:

$$\frac{1}{R(\rho)}\left[\frac{\partial^2}{\partial\rho^2} + \frac{1}{\rho}\frac{\partial}{\partial\rho}\right] R(\rho) + \frac{1}{Z(z)}\frac{\partial^2}{\partial z^2}Z(z) = -k^2.$$

We can set

$$\frac{1}{R(\rho)}\left[\frac{\partial^2}{\partial\rho^2} + \frac{1}{\rho}\frac{\partial}{\partial\rho}\right] R(\rho) = -k_1^2,$$

$$\frac{1}{Z(z)}\frac{\partial^2}{\partial z^2}Z(z) = -k_2^2,$$

so that $k^2 = k_1^2 + k_2^2$. To satisfy the boundary conditions $Z(0) = Z(L) = 0$, k_2 has to be $k_2 = n\pi/L$, so that

$$Z(z) = \sin\left(\frac{n\pi}{L}z\right), \quad n = 1, 2, \ldots .$$

The solution for $R(\rho)$ is the zeroth order Bessel function

$$R(\rho) = J_0(k_1\rho).$$

In order to satisfy the condition $R(c) = 0$, k_1 must be one of the k_{0j} where $k_{0j}c$ is the jth root of $J_0(x) = 0$. Thus

$$k^2 = k_{0j}^2 + \left(\frac{n\pi}{L}\right)^2$$

and

$$u(\rho, z, t) = \sum_{n=1}^{\infty}\sum_{j=1}^{\infty} a_{nj} J_0(k_{0j}\rho) \sin\left(\frac{n\pi}{L}z\right) e^{-[k_{0j}^2 + (n\pi/L)^2]a^2 t}.$$

The initial condition is such that

$$u(\rho, z, 0) = \sum_{n=1}^{\infty}\sum_{j=1}^{\infty} a_{nj} J_0(k_{0j}\rho) \sin\left(\frac{n\pi}{L}z\right) = T_0.$$

It follows that:

$$a_{nj} = \frac{4}{c^2 J_1^2(k_{0j}c)L} \int_0^c \int_0^L T_0 \rho J_0(k_{0j}\rho) \sin(\frac{n\pi}{L}z) dz d\rho.$$

Since

$$\int_0^L \sin\left(\frac{n\pi}{L}z\right) dz = \begin{cases} \frac{2L}{n\pi} & n \text{ odd} \\ 0 & n \text{ even} \end{cases},$$

$$\int_0^c \rho J_0(k_{0j}\rho) d\rho = \frac{c}{k_{0j}} J_1(k_{0j}c),$$

therefore

$$u(\rho, z, t) = \frac{8}{\pi c} \sum_{n=\text{odd}}^{\infty}\sum_{j=1}^{\infty} \frac{1}{n k_{0j} J_1(k_{0j}c)} J_0(k_{0j}\rho) \sin\left(\frac{n\pi}{L}z\right) e^{-[k_{0j}^2 + (n\pi/L)^2]a^2 t}.$$

As expected $u(\rho, z, t) \to 0$ for $t \to \infty$, at all points.

6.4 Three-Dimensional Laplacian in Spherical Coordinates

6.4.1 Laplace's Equations in Spherical Coordinates

The Laplace equation $\nabla^2 F = 0$ written in the spherical coordinates is in the form of

$$\left[\frac{1}{r^2} \frac{\partial}{\partial r} r^2 \frac{\partial}{\partial r} + \frac{1}{r^2 \sin\theta} \frac{\partial}{\partial \theta} \sin\theta \frac{\partial}{\partial \theta} + \frac{1}{r^2 \sin^2\theta} \frac{\partial^2}{\partial \varphi^2} \right] F(r,\theta,\varphi) = 0.$$

With the assumption $F(r,\theta,\varphi) = R(r)\Theta(\theta)\Phi(\varphi)$, we can multiply this equation by r^2/F and obtain

$$\frac{1}{R(r)} \frac{\partial}{\partial r} r^2 \frac{\partial}{\partial r} R(r) + \frac{1}{\Theta(\theta)\Phi(\varphi)} \left[\frac{1}{\sin\theta} \frac{\partial}{\partial \theta} \sin\theta \frac{\partial}{\partial \theta} + \frac{1}{\sin^2\theta} \frac{\partial^2}{\partial \varphi^2} \right] \Theta(\theta)\Phi(\varphi) = 0$$

or

$$\frac{1}{R(r)} \frac{\partial}{\partial r} r^2 \frac{\partial}{\partial r} R(r) = -\frac{1}{\Theta(\theta)\Phi(\varphi)} \left[\frac{1}{\sin\theta} \frac{\partial}{\partial \theta} \sin\theta \frac{\partial}{\partial \theta} + \frac{1}{\sin^2\theta} \frac{\partial^2}{\partial \varphi^2} \right] \Theta(\theta)\Phi(\varphi).$$

The left-hand side is a function or r, and the right-hand side is a function of θ and φ. Since r, θ, φ are independent variables, they can equal to each other, if and only if both sides equal to the same constant. That is

$$\frac{1}{R(r)} \frac{\partial}{\partial r} r^2 \frac{\partial}{\partial r} R(r) = \lambda, \tag{6.17}$$

$$-\frac{1}{\Theta(\theta)\Phi(\varphi)} \left[\frac{1}{\sin\theta} \frac{\partial}{\partial \theta} \sin\theta \frac{\partial}{\partial \theta} + \frac{1}{\sin^2\theta} \frac{\partial^2}{\partial \varphi^2} \right] \Theta(\theta)\Phi(\varphi) = \lambda. \tag{6.18}$$

Multiplying (6.18) by $\sin^2\theta$, we have

$$-\frac{1}{\Theta(\theta)} \sin\theta \frac{\partial}{\partial \theta} \sin\theta \frac{\partial}{\partial \theta} \Theta(\theta) - \frac{1}{\Phi(\varphi)} \frac{\partial^2}{\partial \varphi^2} \Phi(\varphi) = \lambda \sin^2\theta.$$

Rearranging

$$\frac{1}{\Theta(\theta)} \sin\theta \frac{\partial}{\partial \theta} \sin\theta \frac{\partial}{\partial \theta} \Theta(\theta) + \lambda \sin^2\theta = -\frac{1}{\Phi(\varphi)} \frac{\partial^2}{\partial \varphi^2} \Phi(\varphi),$$

we see again that both sides must be equal to another constant

$$\frac{1}{\Theta(\theta)} \sin\theta \frac{\partial}{\partial \theta} \sin\theta \frac{\partial}{\partial \theta} \Theta(\theta) + \lambda \sin^2\theta = \mu, \tag{6.19}$$

$$-\frac{1}{\Phi(\varphi)} \frac{\partial^2}{\partial \varphi^2} \Phi(\varphi) = \mu.$$

The $\Phi(\varphi)$ Part of the Laplace's Equation

The last equation governs $\Phi(\varphi)$

$$\frac{\partial^2}{\partial\varphi^2}\Phi(\varphi) = -\mu\Phi(\varphi).$$

So $\Phi(\varphi)$ is given by

$$\Phi(\varphi) = \exp[\pm i\sqrt{\mu}\varphi]$$

or by its real value equivalent

$$\Phi(\varphi) = c_1 \cos\sqrt{\mu}\varphi + c_2 \sin\sqrt{\mu}\varphi.$$

But when φ is increased by 2π, we come back to the same point. Therefore

$$\Phi(\varphi + 2\pi) = \Phi(\varphi),$$

which means

$$\exp[\pm i\sqrt{\mu}(\varphi + 2\pi)] = \exp[\pm i\sqrt{\mu}\varphi].$$

This requires $\sqrt{\mu}$ to be an integer, $\sqrt{\mu} = m$ $(m = 0, 1, 2, \ldots)$. In other words

$$\mu = m^2$$

and

$$\Phi(\varphi) = \exp[\pm im\varphi].$$

The $\Theta(\theta)$ Part of the Laplace's Equation

With $\mu = m^2$, (6.19) becomes

$$\frac{1}{\Theta(\theta)}\sin\theta\frac{\partial}{\partial\theta}\sin\theta\frac{\partial}{\partial\theta}\Theta(\theta) + \lambda\sin^2\theta = m^2$$

or

$$\frac{1}{\sin\theta}\frac{\partial}{\partial\theta}\sin\theta\frac{\partial}{\partial\theta}\Theta(\theta) + \left[\lambda - \frac{m^2}{\sin^2\theta}\right]\Theta(\theta) = 0. \qquad (6.20)$$

Make a change of variable

$$x = \cos\theta,$$

$$\frac{\partial}{\partial\theta} = \frac{\partial x}{\partial\theta}\frac{\partial}{\partial x} = -\sin\theta\frac{\partial}{\partial x},$$

$$\sin^2\theta = 1 - \cos^2\theta = 1 - x^2,$$

and let $\Theta(\theta) = y(x)$, so (6.20) becomes

$$\frac{d}{dx}\left[(1 - x^2)\frac{d}{dx}y(x)\right] + \left[\lambda - \frac{m^2}{1 - x^2}\right]y(x) = 0. \qquad (6.21)$$

As we have discussed in Chap. 4, the solution of this equation will diverge at $x = \pm 1$, unless λ is equal to $l(l+1)$, where l is an integer ($l = 0, 1, 2, \ldots$). In other words, in order to have a physically acceptable solution, the separation constant λ, as an eigenvalue, must be

$$\lambda = l(l+1).$$

With this λ, the equation

$$\frac{\mathrm{d}}{\mathrm{d}x}\left[(1-x^2)\frac{\mathrm{d}}{\mathrm{d}x}y(x)\right] + \left[l(l+1) - \frac{m^2}{1-x^2}\right]y(x) = 0$$

is known as the associated Legendre equation. An additional requirement for this equation to yield an acceptable solution is that m must be in the range of $-l \le m \le l$. If these requirements are met, the solution of this equation is known as the associated Legendre polynomials

$$y(x) = P_l^{|m|}(x).$$

Associated Legendre polynomials are all related to (derivable from) Legendre polynomial $P_l(x)$, which is the solution with $m = 0$. The equation

$$\frac{\mathrm{d}}{\mathrm{d}x}\left[(1-x^2)\frac{\mathrm{d}}{\mathrm{d}x}y(x)\right] + l(l+1)y(x) = 0$$

is known as the Legendre equation. Its solution is known as the Legendre polynomial $P_l(x)$. Thus the $\Theta(\theta)$ part of the Laplace equation is given by

$$\Theta(\theta) = P_l^{|m|}(\cos\theta).$$

Therefore, together with $\Phi(\varphi) = \mathrm{e}^{\pm im\varphi}$, the angular part of the solution of the Laplace equation in spherical coordinates can be expressed as spherical harmonics $Y_l^m(\theta, \varphi)$.

Radial Part $R(r)$ of the Laplace Equation

The radial part of the Laplace equation is given by (6.17). With $\lambda = l(l+1)$, this equation becomes

$$r^2\frac{\mathrm{d}^2}{\mathrm{d}r^2}R(r) + 2r\frac{\mathrm{d}}{\mathrm{d}r}R(r) = l(l+1)R(r). \tag{6.22}$$

This type of equation is known as Euler–Cauchy equation, and can be solved by a change of variable

$$r = \mathrm{e}^t, \quad t = \ln r, \quad \frac{\mathrm{d}t}{\mathrm{d}r} = \frac{1}{r}.$$

$$\frac{dR}{dr} = \frac{dR}{dt}\frac{dt}{dr} = \frac{1}{r}\frac{dR}{dt},$$

$$\frac{d^2R}{dr^2} = \frac{d}{dr}\left[\frac{1}{r}\frac{dR}{dt}\right] = -\frac{1}{r^2}\frac{dR}{dt} + \frac{1}{r}\frac{d}{dr}\frac{dR}{dt}$$

$$= -\frac{1}{r^2}\frac{dR}{dt} + \frac{1}{r}\frac{d}{dt}\left[\frac{dR}{dt}\right]\frac{dt}{dr} = -\frac{1}{r^2}\frac{dR}{dt} + \frac{1}{r^2}\frac{d^2R}{dt^2}.$$

So

$$r^2\frac{d^2}{dr^2}R + 2r\frac{d}{dr}R = \frac{d^2R}{dt^2} + \frac{dR}{dt}.$$

Thus (6.22) becomes

$$\frac{d^2R}{dt^2} + \frac{dR}{dt} = l(l+1)R.$$

This is a differential equation with constant coefficients. The standard way to solve it is to set

$$R(t) = e^{mt},$$

so

$$\frac{dR}{dt} = me^{mt}, \qquad \frac{d^2R}{dt^2} = m^2 e^{mt}.$$

Thus m can be determined from

$$m^2 + m = l(l+1)$$

or

$$(m - l)(m + l + 1) = 0.$$

Clearly

$$m = \begin{cases} l, \\ -l - 1. \end{cases}$$

Therefore

$$R(r) = \begin{cases} e^{lt} = e^{l\ln r} = r^l, \\ e^{-(l+1)t} = e^{-(l+1)\ln r} = \dfrac{1}{r^{l+1}}. \end{cases}$$

Thus, the general solution of the Laplace equation is given by

$$F(r,\theta) = \sum_{l=0}^{\infty}\sum_{m=-l}^{l}\left(a_{lm}r^l + b_{lm}\frac{1}{r^{l+1}}\right)P_l^{|m|}(\cos\theta)e^{\pm m\varphi}.$$

If the problem has no φ dependence ($m = 0$, a common case), the solution reduces to

$$F(r,\theta) = \sum_{l=0}^{\infty}\left(a_l r^l + b_l\frac{1}{r^{l+1}}\right)P_l(\cos\theta).$$

The coefficients a_{lm} and b_{lm} are to be determined by the boundary conditions.

Example 6.4.1. A hollow conducting sphere of radius c is divided into two halves at the equator by a thin insulating strip. The upper hemisphere is charged to a potential V_0 and the lower hemisphere is kept at zero potential, as shown in Fig. 6.9. Find the potential inside and outside the sphere.

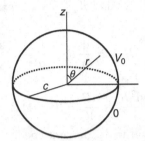

Fig. 6.9. A hollow conducting sphere

Solution 6.4.1. The electrostatic potential V satisfies the Laplace equation and the problem is axially symmetric, therefore the general form of the potential is

$$V(r, \theta) = \sum_{l=0}^{\infty} \left(a_l r^l + b_l \frac{1}{r^{l+1}} \right) P_l(\cos \theta).$$

The boundary conditions are

$$V(c, \theta) = \begin{cases} V_0 & \text{for } 0 \leq \theta \leq \pi/2, \\ 0 & \text{for } \pi/2 \leq \theta \leq \pi. \end{cases}$$

Inside the sphere (for $r < c$) we require the solution to be finite at the origin and so $b_l = 0$ for all l. Imposing the boundary conditions at $r = c$, we have

$$\sum_{l=0}^{\infty} a_l c^l P_l(\cos \theta) = V(c, \theta).$$

The orthogonality of the Legendre polynomials enable us to determine the coefficients a_l. We multiply both sides of this equation by $P_n(\cos \theta) \sin \theta$ and integrate from $\theta = 0$ to π,

$$\sum_{l=0}^{\infty} a_l c^l \int_{\theta=0}^{\pi} P_l(\cos \theta) P_n(\cos \theta) \sin \theta \, d\theta = \int_{\theta=0}^{\pi} V(c, \theta) P_n(\cos \theta) \sin \theta \, d\theta.$$

Since

$$\int_{\theta=0}^{\pi} P_l(\cos \theta) P_n(\cos \theta) \sin \theta \, d\theta = \int_{-1}^{1} P_l(x) P_n(x) dx = \frac{2}{2n+1} \delta_{nl},$$

so

$$a_n c^n \frac{2}{2n+1} = \int_{\theta=0}^{\pi} V(c,\theta) P_n(\cos\theta) \sin\theta\, d\theta = V_0 \int_{\theta=0}^{\pi/2} P_n(\cos\theta) \sin\theta\, d\theta.$$

Changing n to l and $\cos\theta$ to x, we see that a_l is given by

$$a_l = \frac{1}{c^l} \frac{2l+1}{2} V_0 \int_0^1 P_l(x) dx.$$

Therefore

$$V(r,\theta) = \sum_{l=0}^{\infty} \left(\frac{2l+1}{2} V_0 \int_0^1 P_l(x) dx \right) \left(\frac{r}{c} \right)^l P_l(\cos\theta).$$

Since low order Legendre polynomials are relatively simple, the first few coefficients can be easily evaluated

$$a_0 = \frac{1}{2} V_0, \quad a_1 = \frac{3}{4c} V_0, \quad a_2 = 0, \quad a_3 = -\frac{7}{16c^3} V_0, \dots .$$

Thus

$$V(r,\theta) = \frac{1}{2} V_0 + \frac{3}{4} V_0 \frac{r}{c} P_1(\cos\theta) - \frac{7}{16} \left(\frac{r}{c} \right)^3 P_3(\cos\theta) + \cdots .$$

For this problem, the general coefficients a_l can be expressed in a closed expression. Sometimes, this is quite important. For example, if we want to program the computer to sum up the potential to a high degree of accuracy, then it is useful to have an analytic expression. Recall

$$\int_0^1 P_l(x) dx = \frac{1}{2l+1} [P_{l-1}(0) - P_{l+1}(0)]$$

and $P_{l-1}(0) = P_{l+1}(0) = 0$ if l is an even number (so $l-1$ and $l+1$ are odd). Thus $a_l = 0$ for all even l except a_0, which is equal to $V_0/2$. Therefore, $V(r,\theta)$ can be written as

$$V(r,\theta) = \frac{1}{2} V_0 + \sum_{n=0}^{\infty} a_{2n+1} r^{2n+1} P_{2n+1}(\cos\theta),$$

where

$$a_{2n+1} = \frac{1}{c^{2n+1}} \frac{4n+3}{2} V_0 \int_0^1 P_{2n+1}(x) dx.$$

Now, since

$$\int_0^1 P_{2n+1}(x) dx = \frac{1}{4n+3} [P_{2n}(0) - P_{2n+2}(0)]$$

and we have shown in Chap. 4 that

$$P_{2n}(0) = (-1)^n \frac{(2n)!}{2^{2n}(n!)^2},$$

so

$$\int_0^1 P_{2n+1}(x)\mathrm{d}x = \frac{1}{4n+3}\left[(-1)^n \frac{(2n)!}{2^{2n}(n!)^2}\frac{4n+3}{2n+2}\right]. \tag{6.23}$$

Therefore $V(r,\theta)$ inside the sphere is given by

$$V(r,\theta) = \frac{V_0}{2}\left\{1 + \sum_{n=0}^{\infty}\left[(-1)^n\frac{(2n)!}{2^{2n}(n!)^2}\frac{4n+3}{2n+2}\right]\left(\frac{r}{c}\right)^{2n+1}P_{2n+1}(\cos\theta)\right\}.$$

Outside the sphere (for $r > c$), we require the potential to be bounded as $r \to \infty$. So we must set a_l in the general expression to zero for all l. In this case, by imposing the boundary condition at $r = c$, we have

$$\sum_{l=0}^{\infty} b_l \frac{1}{c^{l+1}} P_l(\cos\theta) = V(c,\theta).$$

Following the same argument as before, we find

$$b_l \frac{1}{c^{l+1}} = \frac{2l+1}{2}V_0 \int_0^1 P_l(x)\mathrm{d}x.$$

Therefore $V(r,\theta)$ outside the sphere is given by

$$V(r,\theta) = \frac{V_0}{2}\frac{c}{r}\left\{1 + \sum_{n=0}^{\infty}\left[(-1)^n\frac{(2n)!}{2^{2n}(n!)^2}\frac{4n+3}{2n+2}\right]\left(\frac{c}{r}\right)^{2n+1}P_{2n+1}(\cos\theta)\right\}.$$

It may be noted that the solution for $r < c$ and the solution for $r > c$ are equal at $r = c$.

Example 6.4.2. Electrostatic Potential of a Ring of Charge. Find the electrostatic potential at a general point in space due to a total charge of Q uniformly distributed over a ring of radius c.

Solution 6.4.2. From elementary physics we learnt that the electrostatic potential V at a field point P due to a point charge Q is

$$V = \frac{Q}{d},$$

where d is the distance between the field point P and the charge. Therefore at the point P on the z-axis at a distance z above the plane of the ring (shown in Fig. 6.10), the electrostatic potential is given by

$$V = \frac{Q}{(z^2 + c^2)^{1/2}},$$

since all points of the ring are at the same distance $(z^2 + c^2)^{1/2}$ from the field point P.

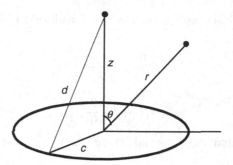

Fig. 6.10. Electrostatic potential of a ring of charge

To calculate the potential at a general point off the z axis would be quite difficult if we are trying to sum up the potential due to the charges on the ring. However, we can find the potential by using the facts that the potential $V(r, \theta)$ satisfies the Laplace equation and it reduces to $V(z, 0)$ when $r = z$ and $\theta = 0$. First if $r > c$, since we require $V(r, \theta)$ to be bounded as $r \rightarrow \infty$, all a_l in the solution of the Laplace equation must be set to zero, and what is left is

$$V(r, \theta) = \sum_{l=0}^{\infty} b_l \frac{1}{r^{l+1}} P_l(\cos \theta).$$

Furthermore, for $r = z$ and $\theta = 0$,

$$V(z, 0) = \sum_{l=0}^{\infty} b_l \frac{1}{z^{l+1}} P_l(1) = \sum_{l=0}^{\infty} b_l \frac{1}{z^{l+1}}, \tag{6.24}$$

since $P_l(1) = 1$. But for $r > c$,

$$V(z, 0) = \frac{Q}{(z^2 + c^2)^{1/2}} = \frac{Q}{z} \left[1 + \left(\frac{c}{z} \right)^2 \right]^{-1/2},$$

and

$$\left[1 + \left(\frac{c}{z} \right)^2 \right]^{-1/2} = 1 - \frac{1}{2} \left(\frac{c}{z} \right)^2 + \frac{3}{8} \left(\frac{c}{z} \right)^4 + \cdots = \sum_{k=0}^{\infty} c_{2k} \left(\frac{c}{z} \right)^{2k},$$

where the binomial coefficients c_{2k} are given by

$$c_{2k} = \frac{\left[-\frac{1}{2} \right] \left[-\frac{1}{2} - 1 \right] \cdots \left[-\frac{1}{2} - (k-1) \right]}{k!} = (-1)^k \frac{1 \cdot 3 \cdot 5 \cdots (2k-1)}{2^k k!}$$

$$= (-1)^k \frac{1 \cdot 3 \cdot 5 \cdots (2k-1)}{2^k k!} \frac{(2k)(2k-2) \cdots 2}{(2k)(2k-2) \cdots 2} = (-1)^k \frac{(2k)!}{2^{2k} (k!)^2}.$$

Therefore

$$V(z, 0) = \frac{Q}{z} \left\{ 1 + \sum_{k=1}^{\infty} \left[(-1)^k \frac{(2k)!}{2^{2k} (k!)^2} \right] \left(\frac{c}{z} \right)^{2k} \right\}.$$

Comparing with (6.24), we see that $b_l = 0$ for all odd l. Thus (6.24) can be written as

$$V(z,0) = \frac{1}{z} \sum_{l=0}^{\infty} \frac{b_{2l}}{z^{2l}}$$

and

$$b_{2l} = Q(-1)^l \frac{(2l)!}{2^{2l}(l!)^2} c^{2l}.$$

Now we conclude that the potential $V(r,\theta)$ for $r > c$ must be

$$V(r,\theta) = \frac{Q}{r} \left\{ 1 + \sum_{l=1}^{\infty} (-1)^l \frac{(2l)!}{2^{2l}(l!)^2} \left(\frac{c}{r}\right)^{2l} P_{2l}(\cos\theta) \right\}.$$

This is so because this function without any arbitrary constant satisfies the Laplace equation and reduces to the correct answer on z-axis. The uniqueness theorem (shown in Chap. 2, vol. II) states that there is only one such function.

The potential for $r < c$ can be found in a similar way. Because of the axial symmetry and the requirement that the potential is bounded at $r = 0$, the solution of the Laplace equation must be in the following form:

$$V(r,\theta) = \sum_{l=0}^{\infty} a_l r^l P_l(\cos\theta).$$

For $r = z$, $\theta = 0$, this solution becomes

$$V(z,0) = \sum_{l=0}^{\infty} a_l z^l P_l(1) = \sum_{l=0}^{\infty} a_l z^l.$$

Comparison of this expression with

$$V(z,0) = \frac{Q}{(z^2 + c^2)^{1/2}} = \frac{Q}{c} \left[1 + \left(\frac{z}{c}\right)^2 \right]^{-1/2}$$

$$= \frac{Q}{c} \left\{ 1 + \sum_{k=1}^{\infty} \left[(-1)^k \frac{(2k)!}{2^{2k}(k!)^2} \right] \left(\frac{z}{c}\right)^{2k} \right\}, \quad \text{for } z < c$$

clearly shows

$$a_0 = \frac{Q}{c}, \quad a_{2l+1} = 0,$$

$$a_{2l} = \frac{Q}{c^{2l+1}} (-1)^l \frac{(2l)!}{2^{2l}(l!)^2} c^{2l}.$$

Therefore the potential for $r < c$ is given by

$$V(r,\theta) = \frac{Q}{c} \left\{ 1 + \sum_{l=1}^{\infty} (-1)^l \frac{(2l)!}{2^{2l}(l!)^2} \left(\frac{r}{c}\right)^{2l} P_{2l}(\cos\theta) \right\}.$$

Example 6.4.3. Conducting sphere in a uniform electric field. A grounded conducting sphere of radius r_0 is placed in a previously uniform electric field $\mathbf{E_0}$. Find the electrostatic potential outside the sphere.

Solution 6.4.3. The electric field $\mathbf{E}$ is related to the potential V by $-\nabla V = \mathbf{E}$. Therefore the uniform field, taken in the direction of z-axis, has an electrostatic potential

$$V = -E_0 z = -E_0 r \cos \theta,$$

since

$$-\nabla(-E_0 z) = kE_0 \frac{\partial}{\partial z} z = \mathbf{E_0}.$$

This potential satisfies the Laplace equation $\nabla^2 V = 0$, as must the potential when the sphere is present. We select the spherical coordinates (r, θ, φ) because of the spherical shape of the conductor. Furthermore, with the positive z-axis chosen to be the direction of the field, this problem is clearly axially symmetric. Therefore the general solution of the Laplace equation is given by

$$V(r, \theta) = \sum_{l=0}^{\infty} \left(a_l r^l + b_l \frac{1}{r^{l+1}} \right) P_l(\cos \theta). \tag{6.25}$$

Since the original unperturbed electrostatic field is kE_0, we require, as one boundary condition

$$V(r \to \infty) = -E_0 r \cos \theta = -E_0 r P_1(\cos \theta).$$

At large r, (6.25) becomes

$$V(r, \theta) = \sum_{l=0}^{\infty} a_l r^l P_l(\cos \theta),$$

since $1/r^{l+1} \to 0$, as $r \to \infty$. The last two equations clearly show that

$$a_1 = -E_0 \quad \text{and} \quad a_l = 0, \ l \neq 1.$$

The other boundary condition requires $V(r_0, \theta) = 0$, because the sphere is grounded. That is

$$V(r_0, \theta) = -E_0 r_0 P_1(\cos \theta) + \sum_{l=0}^{\infty} b_l \frac{1}{r_0^{l+1}} P_l(\cos \theta) = 0.$$

Clearly

$$b_1 = E_0 r_0^3 \quad \text{and} \quad b_l = 0, \ l \neq 1.$$

Thus the potential is given by

$$V(r, \theta) = -E_0 r \cos \theta \left[1 - \frac{r_0^3}{r^3} \right]. \tag{6.26}$$

This solution is similar to (6.6), which describes the potential produced by a conducting cylinder placed in a uniform field. The only difference is that $(r_0/r)^3$ is replaced by $(r_0/\rho)^2$.

Example 6.4.4. Sphere in a uniform stream. For an ideal incompressible fluid, the velocity **v** of the flow is derivable from a velocity potential U, such that $\mathbf{v} = \nabla U$, where U satisfies the Laplace equation $\nabla^2 U = 0$. A sphere of radius r_0 is placed in a uniform stream, find the velocity potential.

Solution 6.4.4. At infinite distance from the sphere, the flow is uniform. Let the velocity v be in the direction of z-axis,

$$U(r \to \infty) = vz = vr\cos\theta = vrP_1(\cos\theta).$$

The velocity normal to the surface of the sphere must be zero

$$\left[\frac{\partial U}{\partial r}\right]_{r=r_0} = 0.$$

Again our solution is given by

$$U(r,\theta) = \sum_{l=0}^{\infty}\left(a_l r^l + b_l \frac{1}{r^{l+1}}\right)P_l(\cos\theta).$$

To satisfy the asymptotic boundary condition

$$U(r \to \infty) = \sum_{l=0}^{\infty} a_l r^l P_l(\cos\theta) = vrP_1(\cos\theta),$$

we must have

$$a_1 = v \quad \text{and} \quad a_l = 0,\ l \neq 1.$$

Thus

$$U(r,\theta) = vrP_1(\cos\theta) + \sum_{l=0}^{\infty} b_l \frac{1}{r^{l+1}}P_l(\cos\theta).$$

Now

$$\frac{\partial U}{\partial r} = vP_1(\cos\theta) - \sum_{l=0}^{\infty} b_l \frac{(l+1)}{r^{l+2}}P_l(\cos\theta),$$

$$\left[\frac{\partial U}{\partial r}\right]_{r=r_0} = vP_1(\cos\theta) - \sum_{l=0}^{\infty} b_l \frac{(l+1)}{r_0^{l+2}}P_l(\cos\theta) = 0.$$

Clearly

$$b_1 = \frac{1}{2}vr_0^3 \quad \text{and} \quad b_l = 0,\ l \neq 1.$$

Therefore

$$U(r,\theta) = v \cos\theta \left[r + \frac{r_0^3}{2r^2} \right]. \tag{6.27}$$

This result is similar to (6.26) but not identical. Note that the tangential component of the electric field is zero at the surface of the sphere, whereas the normal component of the velocity of the flow at the surface of the sphere is zero.

6.4.2 Helmholtz's Equations in Spherical Coordinates

In spherical coordinates, the Helmholtz equation takes the form

$$\left[\frac{1}{r^2}\frac{\partial}{\partial r}r^2\frac{\partial}{\partial r} + \frac{1}{r^2\sin\theta}\frac{\partial}{\partial\theta}\sin\theta\frac{\partial}{\partial\theta} + \frac{1}{r^2\sin^2\theta}\frac{\partial^2}{\partial\varphi^2} + k^2 \right] F(r,\theta,\varphi) = 0.$$

With $F(r,\theta,\varphi) = R(r)\Theta(\theta)\Phi(\varphi)$, this equation can be written as

$$\frac{1}{R}\left[\frac{\partial}{\partial r}r^2\frac{\partial}{\partial r} + k^2r^2 \right] R = -\frac{1}{\Theta\Phi}\left[\frac{1}{\sin\theta}\frac{\partial}{\partial\theta}\sin\theta\frac{\partial}{\partial\theta} + \frac{1}{\sin^2\theta}\frac{\partial^2}{\partial\varphi^2} \right]\Theta\Phi.$$

Both sides of this equation must be equal to the same separation constant. It is seen that the angular part of the solution $\Theta(\theta)\Phi(\varphi)$ is identical to those of Laplace equation. Therefore the separation constant must be $l(l+1)$. Therefore the radial equation takes the form

$$r^2\frac{d^2}{dr^2}R(r) + 2r\frac{d}{dr}R(r) + [k^2r^2 - l(l+1)]R(r) = 0.$$

With a change of variable $x = kr$, this equation becomes the spherical Bessel equation. Therefore the solutions of this equation are spherical Bessel and Neumann functions $j_l(kr)$ and $n_l(kr)$. The following example is an illustration of the use of these solutions.

Example 6.4.5. A particle of mass m is contained in a sphere of radius R. The particle is described by a wavefunction satisfying the Schrödinger wave equation

$$-\frac{\hbar^2}{2m}\nabla^2\psi(r,\theta,z) = E\psi(r,\theta,z)$$

and the conditions that the wavefunction ψ remains finite and goes to zero over the surface of the sphere. Find the lowest permitted energy E.

Solution 6.4.5. This equation can be written in the form of a Helmholtz equation

$$(\nabla^2 + k^2)\psi = 0$$

with

$$k^2 = \frac{2m}{\hbar^2} E.$$

The radial part $R_l(r)$ of the solution $\psi = R_l(r)Y_l^m(\theta, \varphi)$ is given by

$$R_l(r) = a_l j_l(kr) + b_l n_l(kr).$$

Since the function has to remain finite, we set $b_l = 0$ as the spherical Neumann function diverges as $r \to 0$. For the function to go to zero over the surface of the sphere, kR must be one the roots of $j_l(x) = 0$. For the lowest energy, kR must be the first zero of $j_0(x)$. Since

$$j_0(x) = \frac{1}{x} \sin x$$

the zeros are at $x = n\pi$, n an integer. Thus $k = n\pi/R$. Therefore the energies are

$$E_{\min} = \frac{\hbar^2}{2m} k^2 = \frac{\hbar^2}{2m} \left(\frac{n\pi}{R}\right)^2$$

with the minimum energy for $n = 1$,

$$E_{\min} = \frac{\hbar^2}{2m} \left(\frac{\pi}{R}\right)^2.$$

6.4.3 Wave Equations in Spherical Coordinates

In Chap. 5, we have seen that one of the solutions of the wave equation

$$\frac{\partial^2}{\partial x^2} u = \frac{1}{a^2} \frac{\partial^2}{\partial t^2} u$$

is the plane wave

$$u(x, t) = e^{i(kx - kat)} = e^{i(kx - \omega t)},$$

where the wavelength is $2\pi/k$ and the angular frequency is $\omega = ka$. This is a plane wave because its wave front is a plane perpendicular to the x-axis. It moves from left to right with velocity a.

By expressing the wave equation in spherical coordinates, we can study the spherical wave, the wavefront of which is a spherical.

First let us assume that the spherical wave depends only on the radial distance r. In this case, the wave equation can be written as

$$\left[\frac{\partial^2}{\partial r^2} + \frac{2}{r}\frac{\partial}{\partial r}\right] u(r, t) = \frac{1}{a^2}\frac{\partial^2}{\partial t^2} u(r, t).$$

Multiplying both the sides by r, we have

$$\left[r\frac{\partial^2}{\partial r^2} + 2\frac{\partial}{\partial r} \right] u(r,t) = \frac{1}{a^2}\frac{\partial^2}{\partial t^2}[ru(r,t)].$$

Since

$$\left[r\frac{\partial^2}{\partial r^2} + 2\frac{\partial}{\partial r} \right] u(r,t) = \frac{\partial^2}{\partial r^2}[ru(r,t)],$$

we can write the three-dimensional wave equation as

$$\frac{\partial^2}{\partial r^2}[ru(r,t)] = \frac{1}{a^2}\frac{\partial^2}{\partial t^2}[ru(r,t)].$$

Clearly $ru(r,t)$ plays the same role as $u(x,t)$ in the one-dimensional wave equation. So one of its solutions can be written as

$$ru(r,t) = e^{i(kr-\omega t)}$$

or

$$u(r,t) = \frac{1}{r}e^{i(kr-\omega t)}.$$

We can interpret this solution as the outgoing spherical wave with a wave length $2\pi/k$ and a angular frequency ω. Note that its amplitude is decreasing by a factor of $1/r$, as r is increasing. This is what it should be, since the area of the wave front is increasing by a factor of r^2. So the intensity of the wave (energy per unit area which is proportional to $|u|^2$) summed over the wave front is a constant.

In general, spherical waves also depend on θ and φ. The angular dependence is determined by the boundary conditions, or on how it is generated.

The separated solution of the wave equation is the product of a time-dependent part and a space part. The time-dependent part is exactly the same as that of the one- and two-dimensional wave equations, namely

$$T(t) = e^{ikat},\ e^{-ikat}$$

or their real part equivalents. Furthermore, $e^{\pm ikat}$ can be written as $e^{\pm i\omega t}$, where ω is the angular frequency.

The space part is governed by the Helmholtz equation. In spherical coordinates, solutions are given by

$$F_{lm}(r,\theta,\varphi) = [a_{lm}j_l(kr) + b_{lm}n_l(kr)]Y_l^m(\theta,\varphi).$$

If we are looking for an outgoing spherical wave at large r, then we have to combine the spherical Bessel function $j_l(kr)$ and the spherical Neumann function $n_l(kr)$ into the spherical Hankel function $h_l^{(1)}(kr)$ of the first kind. That is, b_{lm} has to be equal to ia_{lm}. Recall as $x \to \infty$

$$h_l^{(1)}(x) \to \frac{1}{x} e^{i[x-(l+1)\pi/2]}.$$

In this way, asymptotically $u(r,\theta,\varphi,t) = F(r,\theta,\varphi)T(t)$ will take the form of

$$u_{lm}(r,\theta,\varphi,t) \to \frac{1}{kr} e^{i(kr\pm\omega t)} Y_l^m(\theta,\varphi).$$

The minus sign $(-\omega t)$ is for a spherical outgoing wave, and the plus sign $(+\omega t)$ is for a spherical incoming wave. Thus, the general outgoing spherical wave is a linear combination of the these components

$$u(r,\theta,\varphi,t) = \sum_{l=0}^{\infty} \sum_{m=-l}^{l} a_{lm} h_l^{(1)}(kr) Y_l^m(\theta,\varphi) e^{-i\omega t}.$$

Example 6.4.6. The acoustic wave $u(\mathbf{r},t)$, satisfying the wave equation

$$\nabla^2 u = \frac{1}{a^2} \frac{\partial^2}{\partial t^2} u$$

is emitted from a split spherical antenna. At $r = r_0$, it satisfies the boundary condition

$$u = \begin{cases} V_0 e^{-i\omega t} & 0 < \theta < \frac{\pi}{2}, \\ -V_0 e^{-i\omega t} & \frac{\pi}{2} < \theta < \pi. \end{cases}$$

Find the outgoing solution of the wave equation.

Solution 6.4.6. First the separation constant k must be equal to

$$k = \frac{\omega}{a}.$$

Then we note that the boundary conditions are axially symmetric, we need only to consider solutions with $m = 0$. Therefore

$$u(r,\theta,t) = \sum_{l=0}^{\infty} a_l h_l^{(1)}(kr) P_l(\cos\theta) e^{-i\omega t}.$$

At $r = r_0$

$$u(r_0,\theta,0) = \sum_{l=0}^{\infty} a_l h_l^{(1)}(kr_0) P_l(\cos\theta) = \begin{cases} V_0 & 0 < \theta < \frac{\pi}{2}, \\ -V_0 & \frac{\pi}{2} < \theta < \pi. \end{cases}$$

This is a Fourier–Legendre series, the coefficients are

$$a_l h_l^{(1)}(kr_0) = \frac{2l+1}{2} V_0 \left[\int_0^{\pi/2} P_l(\cos\theta)\sin\theta\, d\theta - \int_{\pi/2}^{\pi} P_l(\cos\theta)\sin\theta\, d\theta \right]$$

$$= \begin{cases} 0 & l \text{ even}, \\ (2l+1)V_0 \int_0^1 P_l(x)dx & l \text{ odd}. \end{cases}$$

Therefore l has to be odd and can be written as $l = 2n+1$ with $n = 0, 1, 2, \ldots$.
Thus

$$u(r, \theta, t) = \sum_{n=0}^{\infty} a_{2n+1} h_{2n+1}^{(1)}(kr) P_{2n+1}(\cos \theta) e^{-i\omega t}$$

with

$$a_{2n+1} = \frac{4n+3}{h_{2n+1}^{(1)}(kr_0)} V_0 \int_0^1 P_{2n+1}(x)\mathrm{d}x.$$

With (6.23)

$$\int_0^1 P_{2n+1}(x)\mathrm{d}x = \frac{1}{4n+3}\left[(-1)^n \frac{(2n)!}{2^{2n}(n!)^2}\frac{4n+3}{2n+2}\right],$$

the outgoing acoustic wave is therefore given by

$$u(r, \theta, t) = \sum_{n=0}^{\infty} A_n V_0 \frac{1}{h_{2n+1}^{(1)}(kr_0)} h_{2n+1}^{(1)}(kr) P_{2n+1}(\cos \theta) e^{-i\omega t},$$

$$A_n = (-1)^n \frac{(2n)!}{2^{2n}(n!)^2}\frac{4n+3}{2n+2}.$$

6.5 Poisson's Equations

Poisson's equation is a nonhomogeneous partial differential equation. The most familiar one is probably

$$\nabla^2 V(\mathbf{r}) = -\varrho(\mathbf{r}),$$

where $V(\mathbf{r})$ is the electrostatic potential and $\varrho(\mathbf{r})$ is the density of electric charge (see Chap. 2 of vol. II). This is a field equation. Although the potential at the point $\mathbf{r}$ is created by all the charges everywhere, yet $\nabla^2 V(\mathbf{r})$ is miraculously related, through the Poisson equation, to only the charge density at the same point $\mathbf{r}$. The discovery of this kind of field equation is one of the greatest achievements of analysis. At places where there is no charge, this equation reduces to the Laplace equation, $\nabla^2 V(\mathbf{r}) = 0$.

In elementary physics, we learnt that the electrostatic potential due to the charges uniformly distributed over a sphere is the same as if all the charges are concentrated at the center, provided the point where the potential is evaluated is outside the sphere. In the following example, we will illustrate how this problem is solved in the context of theory of partial differential equations.

Example 6.5.1. Find the electrostatic potential V which satisfies the differential equation

$$\nabla^2 V(\mathbf{r}) = \begin{cases} -\varrho_0 & r < r_0, \\ 0 & r > r_0 \end{cases}$$

and the boundary conditions

$$V \to 0, \text{ as } r \to \infty,$$

and that V remains finite everywhere.

Solution 6.5.1. First we observe that the Poisson's equation is a second-order differential equation. The nonhomogeneous term, although discontinuous, is finite everywhere, therefore the solution and its first derivative must be continuous. Furthermore, we note that this problem is spherically symmetric, the solution will not have angular dependence. Thus for $r < r_0$,

$$\nabla^2 V(\mathbf{r}) = \frac{1}{r^2} \frac{d}{dr} \left[r^2 \frac{d}{dr} V(r) \right] = -\varrho_0,$$

which can be written as

$$r^2 \frac{d^2}{dr^2} V(r) + 2r \frac{d}{dr} V(r) = -\varrho_0 r^2.$$

This equation is an Euler–Cauchy equation. With a substitution of $r = \exp(t)$, it can be changed into a nonhomogeneous equation of constant coefficients,

$$\frac{d^2}{dt^2} V(t) + \frac{d}{dt} V(t) = -\varrho_0 e^{2t}.$$

The solution of this equation is

$$V(t) = c_1 + c_2 e^{-t} - \frac{1}{6} \varrho_0 e^{2t}.$$

Therefore

$$V(r) = c_1 + c_2 \frac{1}{r} - \frac{1}{6} \varrho_0 r^2.$$

Because of the requirement that $V(r)$ must be finite at $r = 0$, c_2 must be set to zero. Thus, for $r < r_0$,

$$V(r) = c_1 - \frac{1}{6} \varrho_0 r^2 = V_{\text{in}}(r).$$

For $r > r_0$, the part of the solution of the Laplace equation that has no angular dependence (depends on $P_0(\cos \theta) = 1$) is

$$V(r) = a_0 + b_0 \frac{1}{r}.$$

The coefficient a_0 must be set to zero to make $V(r) \to 0$ as $r \to \infty$. Therefore, for $r > r_0$,

$$V(r) = b_0 \frac{1}{r} = V_{\text{out}}(r).$$

At $r = r_0$,

$$V_{\text{in}}(r_0) = V_{\text{out}}(r_0), \qquad \left. \frac{dV_{\text{in}}(r)}{dr} \right|_{r=r_0} = \left. \frac{dV_{\text{out}}(r)}{dr} \right|_{r=r_0}.$$

Therefore

$$c_1 - \frac{1}{6} \varrho_0 r_0^2 = b_0 \frac{1}{r_0}, \qquad -\frac{1}{3} \varrho_0 r_0 = -b_0 \frac{1}{r_0^2}.$$

Thus

$$V(r) = \begin{cases} \frac{1}{r} \left(\frac{1}{3} \varrho_0 r_0^3 \right) & \text{for } r > r_0, \\ \frac{1}{2} \varrho_0 \left(r_0^2 - \frac{1}{3} r^2 \right) & \text{for } r < r_0. \end{cases}$$

Not surprisingly, these results are in agreement with what we obtain with Gauss' law in elementary physics. For example, the potential produced by a point charge Q at the origin is

$$V(r) = \frac{1}{4\pi} \frac{Q}{r}.$$

Our solution for $r > r_0$ can be written as

$$V(r) = \frac{1}{4\pi} \frac{1}{r} \left(\varrho_0 \frac{4\pi}{3} r_0^3 \right).$$

Since $4\pi r_0^3/3$ is the volume of the sphere, we can see that this result is the same as if all the charges are put at the center.

6.5.1 Poisson's Equation and Green's Function

The Green's function approach to solving ordinary differential equations can be extended to partial differential equations.

The Green's function $G(\mathbf{r}, \mathbf{r}')$ for Poisson's equation is defined as

$$\nabla^2 G(\mathbf{r}, \mathbf{r}') = \delta(\mathbf{r} - \mathbf{r}'), \qquad (6.28)$$

where $\delta(\mathbf{r} - \mathbf{r}')$ is a three-dimensional delta function. In electrostatics, the Green's function is the potential at $\mathbf{r}$ due to a point charge of unit strength at $\mathbf{r}'$. In terms of Green's function, the solution of the Poisson's equation

$$\nabla^2 u(\mathbf{r}) = \varrho(\mathbf{r}) \qquad (6.29)$$

can be expressed as

$$u(\mathbf{r}) = \iiint_V G(\mathbf{r}, \mathbf{r}') \varrho(\mathbf{r}') \mathrm{d}^3 \tau',$$

where the volume integral is over the entire space. It can be easily shown that $u(\mathbf{r})$ so expressed satisfies the Poisson's equation

$$\nabla^2 u(\mathbf{r}) = \nabla^2 \iiint_V G(\mathbf{r}, \mathbf{r}') \varrho(\mathbf{r}') \mathrm{d}^3 \tau' = \iiint_V \nabla^2 G(\mathbf{r}, \mathbf{r}') \varrho(\mathbf{r}') \mathrm{d}^3 \tau'$$

$$= \iiint_V \delta(\mathbf{r} - \mathbf{r}') \varrho(\mathbf{r}') \mathrm{d}^3 \tau' = \varrho(\mathbf{r}).$$

We can bring ∇^2 inside the integral because ∇^2 is operating on $\mathbf{r}$ and the integration is respect to $\mathbf{r}'$. Physically, (6.29) is a statement of principle of superposition, the total potential is the sum of individual potentials created by all the charges everywhere.

Now let us find $G(\mathbf{r}, \mathbf{r}')$ satisfying (6.28) and the asymptotical (boundary) condition $G(\mathbf{r}, \mathbf{r}') \to 0$, as $|\mathbf{r} - \mathbf{r}'| \to \infty$. Since the problem is spherically symmetric about $\mathbf{r}'$, let us consider a sphere with center at $\mathbf{r}'$. Integrating (6.28) over this sphere, by definition of the delta function we obtain

$$\iiint_V \nabla^2 G(\mathbf{r}, \mathbf{r}') \mathrm{d}^3 \tau = \iiint_V \delta(\mathbf{r} - \mathbf{r}') \mathrm{d}^3 \tau = 1. \qquad (6.30)$$

However,

$$\iiint_V \nabla^2 G(\mathbf{r}, \mathbf{r}') \mathrm{d}^3 \tau = \iiint_V \nabla \cdot \nabla G(\mathbf{r}, \mathbf{r}') \mathrm{d}^3 \tau$$

and according to divergence theorem

$$\iiint_V \nabla \cdot \nabla G(\mathbf{r}, \mathbf{r}') \mathrm{d}^3 \tau = \iint_S \nabla G(\mathbf{r}, \mathbf{r}') \cdot \mathbf{n} \, \mathrm{d}S,$$

where S is the surface of the sphere and $\mathbf{n}$ is the normal to the surface. Note that in the surface of the integral, $\mathbf{r}$ is at the surface of the sphere and $\mathbf{r}'$ is at the center of the sphere. Because of the symmetry around $\mathbf{r}'$, we expect that G has the same value everywhere on the surface of the sphere, i.e.,

$$G(\mathbf{r}, \mathbf{r}') = G(|\mathbf{r} - \mathbf{r}'|) = G(r),$$

where r is radius of the sphere. Thus

$$\nabla G(\mathbf{r}, \mathbf{r}') \cdot \mathbf{n} = \nabla G(r) \cdot \hat{\mathbf{r}} = \frac{\partial}{\partial r} G(r)$$

and

$$\iint_S \nabla G(\mathbf{r}, \mathbf{r}') \cdot \mathbf{n} \, \mathrm{d}S = 4\pi r^2 \frac{\partial}{\partial r} G(r).$$

Comparing with (6.30), we have

$$4\pi r^2 \frac{\partial}{\partial r} G(r) = 1 \quad \text{or} \quad \frac{\partial}{\partial r} G(r) = \frac{1}{4\pi r^2}.$$

Integrating this expression, we obtain

$$G(r) = -\frac{1}{4\pi r} + C.$$

Since we require $G(r) \to 0$ as $r \to \infty$, the constant C must be zero. Consequently the Green's function is given by

$$G(\mathbf{r}, \mathbf{r}') = -\frac{1}{4\pi|\mathbf{r} - \mathbf{r}'|}.$$

This Green's function is sometimes called fundamental solution to distinguish it from other Green's functions which satisfy additional boundary conditions. Before we discuss them, we will use this Green's function to solve the problem of the previous example.

Example 6.5.2. Solve the problem of the previous example with the Green's function method.

Solution 6.5.2. Let $u(\mathbf{r})$ be the solution of the Poisson's equation $\nabla^2 u(\mathbf{r}) = \varrho(\mathbf{r})$, so

$$u(\mathbf{r}) = \iiint_V G(\mathbf{r}, \mathbf{r}')\varrho(\mathbf{r}')\mathrm{d}^3\tau'. \tag{6.31}$$

Since

$$\varrho(\mathbf{r}') = \begin{cases} -\varrho_0 & r' < r_0 \\ 0 & r' > r_0 \end{cases}$$

and

$$G(\mathbf{r}, \mathbf{r}') = -\frac{1}{4\pi|\mathbf{r} - \mathbf{r}'|},$$

the problem is clearly spherically symmetric. Therefore $u(\mathbf{r})$ is a function of r only

$$u(r) = 2\pi \int_0^{r_0} \int_0^{\pi} \frac{1}{4\pi|\mathbf{r} - \mathbf{r}'|} \varrho_0 r'^2 \sin\theta \, \mathrm{d}\theta \mathrm{d}r',$$

where r, r', and θ are defined in the following figure:

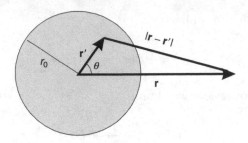

Using the law of cosine and the generating function of Legendre polynomials, we have

$$\frac{1}{|\mathbf{r} - \mathbf{r}'|} = \frac{1}{(r^2 - 2rr' \cos\theta + r'^2)^{1/2}}$$

$$= \begin{cases} \frac{1}{r} \sum_{l=0}^{\infty} \left(\frac{r'}{r}\right)^l P_l(\cos\theta) & \text{for } r > r' \\ \frac{1}{r'} \sum_{l=0}^{\infty} \left(\frac{r}{r'}\right)^l P_l(\cos\theta) & \text{for } r < r' \end{cases}.$$

If $r > r_0$, then clearly $r > r'$, and $u(r)$ can be expressed as

$$u(r) = \frac{1}{2}\varrho_0 \int_0^{r_0} \int_0^\pi \frac{1}{r} \sum_{l=0}^{\infty} \left(\frac{r'}{r}\right)^l P_l(\cos\theta) r'^2 \sin\theta \, d\theta dr'.$$

With $P_0(\cos\theta) = 1$ and the orthogonality of the Legendre polynomials, we can carry out the θ integration

$$\int_0^\pi P_l(\cos\theta) \sin\theta \, d\theta = \int_0^\pi P_0(\cos\theta) P_l(\cos\theta) \sin\theta \, d\theta = \frac{2}{2l+1}\delta_{l0}.$$

Therefore for $r > r_0$,

$$u(r) = \frac{1}{2}\varrho_0 \int_0^{r_0} \frac{1}{r} 2r'^2 dr' = \frac{1}{r}\varrho_0 \frac{1}{3}r_0^3.$$

For $r < r_0$, the r' integration can be divided into two parts, first from 0 to r, then from r to r_0. That is

$$u(r) = u_1(r) + u_2(r),$$

where

$$u_1(r) = 2\pi \int_0^r \int_0^\pi \frac{1}{4\pi|\mathbf{r} - \mathbf{r}'|} \varrho_0 r'^2 \sin\theta \, d\theta dr',$$

$$u_2(r) = 2\pi \int_r^{r_0} \int_0^\pi \frac{1}{4\pi|\mathbf{r} - \mathbf{r}'|} \varrho_0 r'^2 \sin\theta \, d\theta dr'.$$

For $u_1(r)$, $r > r'$ so

$$u_1(r) = \frac{1}{2}\varrho_0 \int_0^r \frac{1}{r} 2r'^2 dr' = \frac{1}{r}\varrho_0 \frac{1}{3}r^3 = \frac{1}{3}\varrho_0 r^2.$$

For $u_2(r)$, $r < r'$

$$u_2(r) = \frac{1}{2}\varrho_0 \int_r^{r_0} \int_0^\pi \frac{1}{r'} \sum_{l=0}^\infty \left(\frac{r}{r'}\right)^l P_l(\cos\theta) r'^2 \sin\theta \, d\theta dr'$$

$$= \varrho_0 \int_r^{r_0} r' dr' = \frac{1}{2}\varrho_0 (r_0^2 - r^2).$$

Thus for $r < r_0$,

$$u(r) = \frac{1}{3}\varrho_0 r^2 + \frac{1}{2}\varrho_0 (r_0^2 - r^2) = \frac{1}{2}\varrho_0 r_0^2 - \frac{1}{6}\varrho_0 r^2.$$

6.5.2 Green's Function for Boundary Value Problems

The Green's function derived in the last section (fundamental solution) enables us to obtain the solution $u(\mathbf{r})$ of Poisson's equation which goes to zero at infinity. Often the solution of the Poisson's equation is required to satisfy additional or other boundary conditions. For example, suppose we have an electric charge distribution near a grounded conducting sphere. The electrostatic potential, in addition to satisfying the Poisson's equation, must vanish at the surface of the sphere. If we can find a Green's function that also satisfies this boundary condition, then we can still use the integral (6.31) to find the potential.

Thus, we want to find a $G(\mathbf{r}, \mathbf{r}')$ that satisfy the equation $\nabla^2 G(\mathbf{r}, \mathbf{r}') - \delta(\mathbf{r} - \mathbf{r}')$ and at the same time vanishes at certain boundaries. In general, it is difficult to find such a function directly. However, we note that if $F(\mathbf{r}, \mathbf{r}')$ satisfies the Laplace's equation $\nabla^2 F(\mathbf{r}, \mathbf{r}') = 0$, then

$$G(\mathbf{r}, \mathbf{r}') = -\frac{1}{4\pi|\mathbf{r} - \mathbf{r}'|} + F(\mathbf{r}, \mathbf{r}') \tag{6.32}$$

satisfies the equation that defines the Green's function

$$\nabla^2 G(\mathbf{r}, \mathbf{r}') = \nabla^2 \left[-\frac{1}{4\pi|\mathbf{r} - \mathbf{r}'|}\right] + \nabla^2 F(\mathbf{r}, \mathbf{r}') = \nabla^2 \left[-\frac{1}{4\pi|\mathbf{r} - \mathbf{r}'|}\right] = \delta(\mathbf{r} - \mathbf{r}').$$

It might be possible to adjust the constants in the solution of the Laplace's equation $F(\mathbf{r}, \mathbf{r}')$ in such a way that (6.32) is equal to zero at the boundaries. For boundaries of simple geometry, such as planes, spheres, circular cylinders, this can indeed be done. We will illustrate this procedure in the following example.

Example 6.5.3. Find the Green's function $G(\mathbf{r}, \mathbf{r}')$ for solving Poisson's equation outside of a grounded sphere of radius R centered at the origin.

Solution 6.5.3. We try to find $G(\mathbf{r}, \mathbf{r}')$ in the form of

$$G(\mathbf{r}, \mathbf{r}') = -\frac{1}{4\pi|\mathbf{r} - \mathbf{r}'|} + F(\mathbf{r}, \mathbf{r}'),$$

where $F(\mathbf{r}, \mathbf{r}')$ satisfies the Laplace's equation

$$\nabla^2 F(\mathbf{r}, \mathbf{r}') = 0,$$

where ∇^2 is acting on $\mathbf{r}$ (not on $\mathbf{r}'$). With $\mathbf{r} = \mathbf{r}(r, \theta)$ and $\mathbf{r}'$ as the z-axis, $F(\mathbf{r}, \mathbf{r}')$ is given by

$$F(\mathbf{r}, \mathbf{r}') = \sum_{l=0}^{\infty} \left(a_l r^l + b_l \frac{1}{r^{l+1}} \right) P_l(\cos\theta).$$

Since $F(\mathbf{r}, \mathbf{r}')$ is required to go to zero as $r \to \infty$, all a_l must be set to zero. Therefore

$$F(\mathbf{r}, \mathbf{r}') = \sum_{l=0}^{\infty} b_l \frac{1}{r^{l+1}} P_l(\cos\theta).$$

If $\mathbf{r}$ is a point on the surface of the sphere (or as a radial vector from the center to a point on the surface of the sphere, i.e., $|\mathbf{r}| = r = R$), then

$$F(\mathbf{r}, \mathbf{r}')|_{r=R} = \sum_{l=0}^{\infty} b_l \frac{1}{R^{l+1}} P_l(\cos\theta),$$

where $\mathbf{r}$, $\mathbf{r}'$, and θ are shown in Fig. 6.11.

With this configuration,

$$\frac{1}{|\mathbf{r} - \mathbf{r}'|} = \frac{1}{(R^2 - 2Rr'\cos\theta + r'^2)^{1/2}} = \frac{1}{r'} \sum_{l=0}^{\infty} \left(\frac{R}{r'} \right)^l P_l(\cos\theta).$$

The condition $G(\mathbf{r}, \mathbf{r}')|_{r=R} = 0$ requires

$$-\frac{1}{4\pi} \frac{1}{r'} \sum_{l=0}^{\infty} \left(\frac{R}{r'} \right)^l P_l(\cos\theta) + \sum_{l=0}^{\infty} b_l \frac{1}{R^{l+1}} P_l(\cos\theta) = 0.$$

Clearly

$$\frac{1}{4\pi} \frac{R^l}{r'^{l+1}} = \frac{b_l}{R^{l+1}}$$

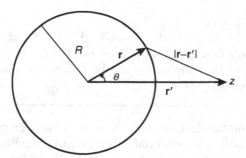

Fig. 6.11. The $\mathbf{r}, \mathbf{r}'$, and θ in the configuration of the Green's function. When $\mathbf{r}$ is a point on the surface of the sphere, then $|\mathbf{r}| = R$

or

$$b_l = \frac{R^{2l+1}}{4\pi r'^{l+1}}.$$

Therefore

$$F(\mathbf{r}, \mathbf{r}') = \frac{1}{4\pi} \sum_{l=0}^{\infty} \frac{R^{2l+1}}{r'^{l+1}} \frac{1}{r^{l+1}} P_l(\cos\theta).$$

It is interesting to write the last expression in the form of

$$\sum_{l=0}^{\infty} \frac{R^{2l+1}}{r'^{l+1}} \frac{1}{r^{l+1}} P_l(\cos\theta) = \frac{R}{r'r} \sum_{l=0}^{\infty} \left(\frac{R^2}{r'r}\right)^l P_l(\cos\theta)$$

$$= \frac{R}{r'r} \frac{1}{\left[1 - 2\frac{R^2}{r'r}\cos\theta + \left(\frac{R^2}{r'r}\right)^2\right]^{1/2}} = \frac{R}{r'} \frac{1}{\left[r^2 - 2r\frac{R^2}{r'}\cos\theta + \left(\frac{R^2}{r'}\right)^2\right]^{1/2}}.$$

With the help of the following figure and the law of cosine, we see that

$$\left[r^2 - 2r\frac{R^2}{r'}\cos\theta + \left(\frac{R^2}{r'}\right)^2\right]^{1/2} = \left|\mathbf{r} - \frac{R^2}{r'^2}\mathbf{r}'\right|.$$

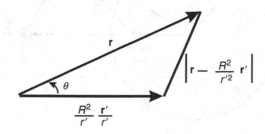

Therefore the Green's function in the presence of a grounded sphere is given by

$$G(\mathbf{r}, \mathbf{r}') = -\frac{1}{4\pi}\frac{1}{|\mathbf{r}-\mathbf{r}'|} + \frac{1}{4\pi}\frac{R/r'}{|\mathbf{r}-(R^2/r'^2)\mathbf{r}'|}.$$

This kind of Green's function is sometimes called Dirichlet Green's function. (When the values of the solution on the boundary are specified, it is called Dirichlet boundary condition.)

Example 6.5.4. Find the electrostatic potential outside of a grounded conducting sphere of radius R. The potential is produced by a point charge q located at a distance a from the center of the sphere, and $a > R$.

Solution 6.5.4. The potential is given by the solution of the Poisson's equation

$$\nabla^2 u(\mathbf{r}) = -\varrho(\mathbf{r})$$

with the boundary condition

$$u(\mathbf{r})|_{r=R} = 0.$$

Let the line joining the center of the sphere and the charge coincide with the z-axis and $\mathbf{k}$ be a unit vector in the z direction, as shown in Fig. 6.12. So the charge distribution can be expressed as

$$\varrho(\mathbf{r}) = q\delta(\mathbf{r} - a\mathbf{k}).$$

In terms of the Dirichlet Green's function, the potential is given by

$$u(\mathbf{r}) = -\iiint \varrho(\mathbf{r}')G(\mathbf{r}, \mathbf{r}')d^3\tau'$$

$$= \frac{1}{4\pi}\iiint q\delta(\mathbf{r}' - a\mathbf{k})\left[\frac{1}{|\mathbf{r}-\mathbf{r}'|} - \frac{R/r'}{|\mathbf{r}-(R^2/r'^2)\mathbf{r}'|}\right]d^3\tau'.$$

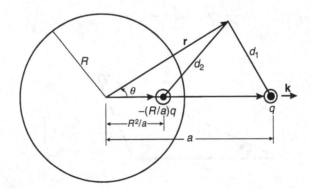

Fig. 6.12. Method of image. A point charge q is placed at a distance a from the center of a grounded sphere of radius R. The unit vector $\mathbf{k}$ is the direction of the line joining the center and the charge

The effect of the delta function is to replace $\mathbf{r}'$ by $a\mathbf{k}$ and r' by a in the Green's function. Thus

$$u(\mathbf{r}) = \frac{1}{4\pi}\left[\frac{q}{|\mathbf{r}-a\mathbf{k}|} - \frac{(R/a)q}{|\mathbf{r} - (R^2/a)\mathbf{k}|}\right] = \frac{1}{4\pi}\frac{q}{d_1} - \frac{1}{4\pi}\frac{(R/a)q}{d_2}$$

$$= \frac{1}{4\pi}\left[\frac{q}{[r^2 - 2ra\cos\theta + a^2]^{1/2}} - \frac{(R/a)q}{[r^2 - 2r(R^2/a)\cos\theta + (R^2/a)^2]^{1/2}}\right].$$

The first term is simply the Coulomb potential due to the point charge. The second term is caused by the presence of the sphere. It is interesting to note that if we replace the sphere by a charge $-(R/a)q$ placed at $(R^2/a)\mathbf{k}$, we would get the same potential for $r > R$. This is shown in Fig. 6.12. In fact, this is routinely done in electromagnetic theory. It is known as the "method of images."

Exercises

1. Find the steady-state distribution of temperature in a sector of a circular plate of radius 10 and angle $\pi/4$ if the temperature is maintained at $0°$ along the radii and at $100°$ along the curved edge.

   Ans. $u = \dfrac{400}{\pi}\sum_{n\ odd}\dfrac{1}{n}\left(\dfrac{\rho}{10}\right)^{4n}\sin 4n\varphi.$

2. Find the steady-state distribution of temperature in a circular annulus of inner radius 1 and outer radius 2 if the inner circle is held at $0°$ and the outer circle has half its circumference at $0°$ and half at $100°$.

   Ans. $u = \dfrac{50\ln\rho}{\ln 2} + \dfrac{200}{\pi}\sum_{n\ odd}\dfrac{1}{n}\dfrac{\rho^n - \rho^{-n}}{(2^n - 2^{-n})}\sin n\varphi.$

3. Find the solution the following boundary value problem inside the circular annulus

$$\nabla^2 u(\rho, \varphi) = 0, \quad 1 \le \rho \le 2,$$
$$u(1, \varphi) = \sin\varphi, \quad u(2, \varphi) = \sin\varphi.$$

   Ans. $u = \left(\dfrac{1}{3}\rho + \dfrac{2}{3}\dfrac{1}{\rho}\right)\sin\varphi.$

4. A particle of mass m is contained in a right circular cylindrical box of radius R and height H. The particle is described by a wavefunction satisfying the Schrödinger wave equation

$$-\frac{\hbar^2}{2m}\nabla^2\psi(r, \theta, z) = E\psi(r, \theta, z)$$

and the condition that the wavefunction go to zero over the surface of the box. Find the lowest permitted energy.

Ans. $E_{\min} = \dfrac{\hbar^2}{2m} \left[\left(\dfrac{2.4048}{R} \right)^2 + \left(\dfrac{\pi}{H} \right)^2 \right].$

5. A membrane is stretched over a fixed circular frame at $r = c$. Its displacement $z(r, \theta, t)$ satisfies the wave equation

$$\frac{\partial^2 z}{\partial t^2} = a^2 \nabla^2 z$$

and the boundary condition

$$z(c, \theta, t) = 0.$$

If the initial displacement of the membrane is

$$z(r, \theta, 0) = f(r),$$

show that z has no θ dependence and

$$z(r, \theta, t) = \frac{2}{c^2} \sum_{j=1}^{\infty} \frac{J_0(\lambda_j r) \cos \lambda_j at}{[J_1(\lambda_j c)]^2} \int_0^c r J_0(\lambda_j r) f(r) dr,$$

where λ_j are the roots of $J_0(\lambda c) = 0$.

6. In previous problem, if

$$z(r, \theta, 0) = A J_0(\lambda_k r),$$

where λ_k is some root of $J_0(\lambda c) = 0$, show that

$$z(r, \theta, t) = A J_0(\lambda_k r) \cos(\lambda_k at).$$

Observe that these displacements are periodic in t, thus the membrane gives a musical note.

7. Consider a solid circular cylinder of radius c and length L. Find the steady-state distribution of temperature inside the cylinder if the temperature at both the curved and top surfaces are kept a $0°$ and the bottom surface is specified by a given function $g(\rho)$.

Ans. $u(\rho, z) = \sum_{g=1} A_j J_0(k_{0j}\rho) \sinh k_{0j}(L - z),$

$$A_j = \frac{2}{c^2 J_1^2(k_{0j}c) \sinh k_{0j}L} \int_0^c g(\rho') J_0(k_{0j}\rho')\rho' d\rho'.$$

8. Starting with the heat conduction equation

$$\nabla^2 u = \frac{1}{a^2}\frac{\partial u}{\partial t},$$

derive the following formula for the temperature in an infinitely long right-angled cylindrical wedge bounded by a surface $r = c$ and the planes $\theta = 0$ and $\theta = \pi/2$, when these surfaces are kept at temperature zero and the initial temperature is $f(r, \theta)$

$$u(r, \theta, t) = \sum_{n=1}^{\infty}\sum_{j=1}^{\infty} A_{nj} J_{2n}(\lambda_{nj} r) \sin 2n\theta e^{-(\lambda_{nj}a)^2 t},$$

where λ_{nj} are the positive roots of $J_{2n}(\lambda c) = 0$ and A_{nj} are given by

$$\pi c^2 [J_{2n+1}(\lambda_{nj}c)]^2 A_{nj} = 8 \int_0^{\pi/2}\int_0^c \sin 2n\theta r J_{2n}(\lambda_{nj} r) f(r, \theta) dr d\theta.$$

9. Let $u(r, t)$ be the temperature in a thin circular plate whose edge, $r = 1$, is kept at temperature zero and whose initial temperature is 1. If there is surface heat transfer from the faces of the plate, the heat equation has the form

$$\frac{\partial u}{\partial t} = \frac{\partial^2 u}{\partial r^2} + \frac{1}{r}\frac{\partial u}{\partial r} - hu, \quad h > 0.$$

Show that

$$u(r, t) = 2e^{-ht}\sum_{j=1}^{\infty}\frac{J_0(\lambda_j r)}{\lambda_j J_1(\lambda_j)}e^{-\lambda_j^2 t},$$

where λ_j are the positive roots of $J_0(\lambda) = 0$.

10. *Shrunken fitting.* Over a long solid cylinder $r \le 1$, initially at uniform temperature A, is tightly fitted a long hollow cylinder $1 \le r \le 2$ of the same material at temperature B. The outer surface $r = 2$ is then kept at temperature B. Show that the temperature in the cylinder of radius 2 so formed is

$$u(r, t) = B + \frac{A - B}{2}\sum_{j=1}^{\infty}\frac{J_1(\lambda_j)}{\lambda_j[J_1(2\lambda_j)]^2}J_0(\lambda_j r)\exp(-\lambda_j^2 a^2 t),$$

where λ_j are the positive roots of $J_0(2\lambda) = 0$ and a^2 is the constant in the heat conduction equation.

Hint: Let $u(r, t) = U(r, t) + B$, find the differential equation, boundary conditions and initial condition of $U(r, t)$. Then solve for $U(r, t)$.

11. A two-dimensional region having the shape of a quarter of a circle of radius b is initially at the uniform temperature of $100°$. At $t = 0$, the temperature around the entire boundary is reduced to $0°$ and maintained thereafter at that value. Find the temperature at any point of the region at any subsequent time.

Ans. $u(r, \theta, t) = \sum_m \sum_n A_{nm} J_{2n}(\lambda_{nm} r) \sin 2n\theta \exp(-\lambda_{nm}^2 a^2 t)$, where λ_{nm} are the roots of $J_{2n}(\lambda_{nm} b) = 0$, and A_{nm} is given by

$$\frac{b^2}{2} J_{2n+1}(\lambda_{nm} b) A_{nm} = \frac{400}{n\pi} \int_0^b r J_{2n}(\lambda_{nm} r) dr.$$

12. A very long cylinder of radius c is split lengthwise into two halves. The surfaces of the halves of the cylinder are kept at temperature of $+u_0$ and $-u_0$, respectively. Find the steady-state temperature everywhere inside the cylinder.

Ans. $u(r, \theta) = \frac{4u_0}{\pi} \sum_{n=1} \frac{1}{2n-1} \left(\frac{r}{c}\right)^{2n-1} \sin(2n-1)\theta$.

13. A two-dimensional region having the shape of a semicircle of radius b is initially at the uniform temperature of $100°$. At $t = 0$, the temperature around the bounding diameter is reduced to $0°$ and maintained thereafter at that temperature. The curved boundary is maintained at $100°$. Find the temperature at any point of the region at any subsequent time.

Ans. $u(r, \theta, t) = \frac{400}{\pi} \sum_{n(\text{odd})} \frac{1}{n} \left(\frac{r}{b}\right)^n \sin n\theta +$
$\sum_{n(\text{odd})} \sum_m B_{nm} J_n(\lambda_{nm} r) \sin n\theta \exp(-\lambda_{nm}^2 a^2 t)$, where B_{nm} is given by

$B_{nm} = \frac{2}{b^2 J_{nm}^2(\lambda_{n+1} b)} \frac{400}{n\pi} \int_0^b \left[1 - \left(\frac{r}{b}\right)^n\right] J_n(\lambda_{nm} r) r \, dr$, and λ_{nm} are the roots of $J_n(\lambda b) = 0$.

Hint: Let $u(r, \theta, t) = V(r, \theta) + U(r, \theta, t)$, and choose the boundary conditions for $V(r, \theta)$ to be $V(r, 0) = V(r, \pi) = 0$ and $V(b, \theta) = 100$. Find the boundary and initial conditions for $U(r, \theta, t)$. Then solve for V and U.

14. Find the steady-state temperature at any point in a spherical shell of inner radius b_1 and outer radius b_2 if the temperature distributions $u(b_1, \theta) = f_1(\theta)$ and $u(b_2, \theta) = f_2(\theta)$ are maintained over the inner and outer surfaces, respectively.

Ans. $u(r, \theta) = \sum_{n=1} \left(A_n r^n + B_n \frac{1}{r^{n+1}}\right) P_n(\cos \theta)$ with A_n and B_n given by

$$A_n b_1^n + B_n \frac{1}{b_1^{n+1}} = \frac{2n+1}{2} \int_0^\pi f_1(\theta) P_n(\cos \theta) \sin \theta \, d\theta,$$

$$A_n b_2^n + B_n \frac{1}{b_2^{n+1}} = \frac{2n+1}{2} \int_0^\pi f_2(\theta) P_n(\cos \theta) \sin \theta \, d\theta.$$

15. The temperature distribution $u(r, \theta) = f(\theta)$ is maintained over the curved surface of a hemisphere of radius b. The plane boundary of the hemisphere is kept at the temperature zero. Find the steady-state temperature at any point in the hemisphere.

Ans. $u(r, \theta) = \sum_{n \text{ odd}} \dfrac{2n+1}{b^n} r^n P_n(\cos\theta) \displaystyle\int_0^{\pi/2} f(\theta') P_n(\cos\theta') \sin\theta' d\theta'$.

16. If V satisfies the Laplace's equation $\nabla^2 V = 0$ throughout the domains $r < c$ and $r > c$, and if $V \to 0$ as $r \to \infty$, and $V = 1$ on the spherical surface $r = c$, show that $V = 1$ when $r \le c$ and $V = c/r$ when $r \ge c$.

17. Write a formula for the steady temperature $u(r, \theta)$ in a solid sphere $r \le 1$ if, for all ϕ, $u(1, \theta) = 1$ when $0 < \theta < \frac{1}{2}\pi$ and $u(1, \theta) = 0$ when $\frac{1}{2}\pi < \theta < \pi$.

Ans. $u(r, \theta) = \frac{1}{2} + \frac{1}{2}\sum_{n=0}^{\infty}[P_{2n}(0) - P_{2n+2}(0)]r^{2n+1}P_{2n+1}(\cos\theta)$.

18. Let $u(r, \theta)$ denote steady temperatures in a hollow sphere $a \le r \le b$ when $u(a, \theta) = f(\cos\theta)$ and $u(b, \theta) = 0$, $0 < \theta < \pi$. Derive the formula

$$u(r, \theta) = \sum_{n=0}^{\infty} A_n \frac{b^{2n+1} - r^{2n+1}}{b^{2n+1} - a^{2n+1}} \left(\frac{a}{r}\right)^{n+1} P_n(\cos\theta),$$

$$A_n = \frac{2n+1}{2}\int_{-1}^{1} f(\cos\theta) P_n(\cos\theta) d\cos\theta.$$

19. Find the electrostatic potential u, which satisfies the Poisson's equation

$$\nabla^2 u(\mathbf{r}) = -\varrho(\mathbf{r}),$$

due to the charge distribution

$$\varrho(\mathbf{r}) = \begin{cases} Ar\cos\theta & \text{for } 0 \le r < R, \\ 0 & \text{for } \quad r > R. \end{cases}$$

Ans. $u(r, \theta) = \begin{cases} \dfrac{A}{15}\dfrac{R^5}{r^2}\cos\theta & \text{for } r > R, \\ A\left(\dfrac{1}{6}R^2 r - \dfrac{1}{10}r^3\right)\cos\theta & \text{for } r < R. \end{cases}$

20. If electric charges are distributed over the surface of a sphere of radius R with surface charge densities

$$\varrho(\mathbf{r}) = C\cos\theta,$$

where θ is the angle between $\mathbf{r}$ and the z-axis, and C is a constant. Let z be a point on the z-axis and $z > R$. Use the Green's function to show that the electrostatic potential u is given by

$$u(z) = \frac{1}{4\pi}\int_0^{\pi} \frac{2\pi\varrho(\theta')R^2 \sin\theta'}{\sqrt{R^2 + z^2 - 2Rz\cos\theta'}} d\theta'.$$

Evaluate this integral. Find the potential $u(r, \theta)$ for $r > R$.

Hint: Use the generating function of the Legendre polynomial, and note that $P_1(\cos \theta) = \cos \theta$.

Ans. $u(z) = \dfrac{1}{z^2} \left(\dfrac{1}{3} R^3 C \right), \quad u(r, \theta) = \dfrac{1}{r^2} \left(\dfrac{1}{3} R^3 C \right) \cos \theta.$

Part IV

Variational Methods

7

Calculus of Variation

A basic problem in calculus is to find the value of x that maximizes or minimizes a given function $y = f(x)$. The calculus of variation extends this problem to finding a function that maximizes or minimizes a definite integral.

Consider the integral

$$I = \int_{x_1}^{x_2} F(y, y', x)dx, \qquad (7.1)$$

where F depends not only on x, but also on y which is a function of x, and on y', the derivative of y with respect to x. The form of F is fixed by the physics of the problem. The only thing that can be changed in the attempt to make I larger or smaller is the form of the function $y(x)$. In this sense, the integral is a function of y. The common terminology is that I is a functional of the curve $y(x)$.

The calculus of variation provides a method for finding the function $y(x)$ that makes the integral stationary, i.e., the function that makes the value of the integral a local maximum or minimum.

Calculus of variation is one of the oldest problems in mathematical physics, developed soon after the invention of calculus. At first it was used to solve mathematically interesting problems, such as finding the shape of a hanging rope. Later it was found that many principles in classical physics, ranging from optics to mechanics, can be stated in the form that certain integrals have stationary values. In modern physics, calculus of variation was found to be equally useful in finding the eigenvalues and eigenfunctions of quantum systems.

In this chapter, y' is used to denote dy/dx, unless stated otherwise. We will also assume that all functions we need to deal with are sufficiently smooth and differentiable.

7.1 The Euler–Lagrange Equation

7.1.1 Stationary Value of a Functional

If a given form of the function $y = y(x)$ gives the integral in (7.1) a minimum value, any neighboring function must give the integral a value equal to or greater than the minimum. This is illustrated in Fig. 7.1. The solid line is the curve $y = y(x)$, along which the integral is a minimum, the broken lines represent small variations of this path. This family of curves is conveniently represented by

$$Y(x) = y(x) + \alpha\eta(x), \tag{7.2}$$

where α is a small parameter, and $\eta(x)$ is an arbitrary function of x that is bounded, continuous, and has a continuous first derivative. If the two end points are fixed as shown in Fig. 7.1, then $\eta(x)$ has to satisfy the boundary conditions $\eta(x_1) = \eta(x_2) = 0$. Replacing y with Y in (7.1), we have

$$I(\alpha) = \int_{x_1}^{x_2} F(Y, Y', x)\mathrm{d}x. \tag{7.3}$$

Now we have the values of the integral along a family of curves passing through the two end points, each of the curves is labeled by the variable α. The curve which makes I stationary has the label $\alpha = 0$.

A necessary, but not sufficient, condition for minimum is the vanishing of the first derivative. Thus we require

$$\left.\frac{\mathrm{d}I}{\mathrm{d}\alpha}\right|_{\alpha=0} = 0, \quad \text{for all } \eta(x). \tag{7.4}$$

Since α does not depend on x, differentiation can be carried out under the integral sign

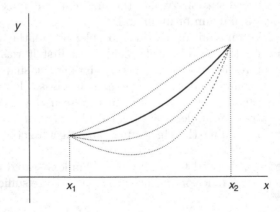

Fig. 7.1. The *solid line* is the curve $y(x)$ along which the integral is stationary. The *broken lines* are curves $y(x) + \alpha\eta(x)$ representing small variations from the solid path. They all pass the two end points

$$\frac{\mathrm{d}I}{\mathrm{d}\alpha} = \int_{x_1}^{x_2} \left[\frac{\partial F}{\partial Y} \frac{\mathrm{d}Y}{\mathrm{d}\alpha} + \frac{\partial F}{\partial Y'} \frac{\mathrm{d}Y'}{\mathrm{d}\alpha} \right] \mathrm{d}x. \tag{7.5}$$

It is clear from (7.2) that

$$Y'(x) = y'(x) + \alpha \eta'(x),$$

$$\frac{\mathrm{d}Y}{\mathrm{d}\alpha} = \eta(x), \quad \frac{\mathrm{d}Y'}{\mathrm{d}\alpha} = \eta'(x).$$

Setting $\alpha = 0$ is equivalent to setting $Y(x) = y(x)$, $Y'(x) = y'(x)$. Therefore after α is set to zero, (7.5) becomes

$$\left. \frac{\mathrm{d}I}{\mathrm{d}\alpha} \right|_{\alpha=0} = \int_{x_1}^{x_2} \left[\frac{\partial F}{\partial y} \eta(x) + \frac{\partial F}{\partial y'} \eta'(x) \right] \mathrm{d}x = 0. \tag{7.6}$$

The second term can be integrated by parts

$$\int_{x_1}^{x_2} \frac{\partial F}{\partial y'} \eta'(x) \mathrm{d}x = \int_{x_1}^{x_2} \frac{\partial F}{\partial y'} \frac{\mathrm{d}\eta}{\mathrm{d}x} \mathrm{d}x = \int_{x_1}^{x_2} \frac{\partial F}{\partial y'} \mathrm{d}\eta$$

$$= \left. \frac{\partial F}{\partial y'} \eta(x) \right|_{x_1}^{x_2} - \int_{x_1}^{x_2} \frac{\mathrm{d}}{\mathrm{d}x} \left(\frac{\partial F}{\partial y'} \right) \eta(x) \mathrm{d}x.$$

The integrated term is zero because $\eta(x_2) = \eta(x_1) = 0$. Therefore we have

$$\left. \frac{\mathrm{d}I}{\mathrm{d}\alpha} \right|_{\alpha=0} = \int_{x_1}^{x_2} \left[\frac{\partial F}{\partial y} - \frac{\mathrm{d}}{\mathrm{d}x} \left(\frac{\partial F}{\partial y'} \right) \right] \eta(x) \mathrm{d}x = 0.$$

Since $\eta(x)$ is an arbitrary function, our intuition tells us that

$$\frac{\partial F}{\partial y} - \frac{\mathrm{d}}{\mathrm{d}x} \left(\frac{\partial F}{\partial y'} \right) = 0. \tag{7.7}$$

This equation was derived by Euler in 1744. It is known as Euler–Lagrange equation, because it is also the basis for Lagrange's formulation of classical mechanics.

If we expand the total derivative with respect to x,

$$\frac{\mathrm{d}}{\mathrm{d}x} \left(\frac{\partial F}{\partial y'} \right) = \frac{\partial}{\partial x} \left(\frac{\partial F}{\partial y'} \right) + \frac{\partial}{\partial y} \left(\frac{\partial F}{\partial y'} \right) \frac{\mathrm{d}y}{\mathrm{d}x} + \frac{\partial}{\partial y'} \left(\frac{\partial F}{\partial y'} \right) \frac{\mathrm{d}y'}{\mathrm{d}x}$$

$$= \frac{\partial^2 F}{\partial x \partial y'} + \frac{\partial^2 F}{\partial y \partial y'} \frac{\mathrm{d}y}{\mathrm{d}x} + \frac{\partial^2 F}{\partial y'^2} \frac{\mathrm{d}y'}{\mathrm{d}x},$$

the Euler–Lagrange equation becomes

$$\frac{\partial F}{\partial y} - \frac{\partial^2 F}{\partial x \partial y'} - \frac{\partial^2 F}{\partial y \partial y'} y' - \frac{\partial^2 F}{\partial y'^2} y'' = 0$$

a second-order differential equation. Since the form of F is given, this equation can be solved to give the desired extremum function $y(x)$.

The condition of (7.4) is only a necessary condition for a minimum, the solution $y(x)$ could also produce a maximum or even a point of inflection of the function $I(\alpha)$ at $\alpha = 0$. To mathematically decide the nature of the extremum, one has to investigate the sign of higher derivatives of $I(\alpha)$. While this can be done, but it is rather complicated. Fortunately, for most applications, Euler–Lagrange equation by itself is enough to give a complete solution of the problem, because the existence and the nature of an extremum   are often clear from the physical or geometrical meaning of the problem.

7.1.2 Fundamental Theorem of Variational Calculus

The Euler–Lagrange equation is the center piece of calculus of variation. One can establish it with mathematical rigor by the following theorem, known as the fundamental theorem of variational calculus.

Theorem 7.1.1. *If $f(x)$ is a continuous function on the interval (x_1, x_2), and if*

$$\int_{x_1}^{x_2} f(x)\eta(x)\mathrm{d}x = 0$$

for every continuously differentiable function $\eta(x)$ that satisfies the boundary conditions $\eta(x_1) = \eta(x_2) = 0$, then $f(x) = 0$ in the interval (x_1, x_2).

Proof. Let us assume that at some point ξ in the interval (x_1, x_2), $f(\xi) \neq 0$. Assume $f(\xi) > 0$. Since $f(x)$ is continuous, there must be a region around ξ, within which $f(x) > 0$. This region is a subinterval in (x_1, x_2). This means that we can find two numbers a and b in (x_1, x_2) such that inside $a < x < b$, $f(x) > 0$. Now it is readily verified that the function $\eta(x)$ represented by

$$\eta(x) = \begin{cases} 0 & x_1 \leq x \leq a \\ (x-a)^2(x-b)^2 & a \leq x \leq b \\ 0 & b \leq x \leq x_2 \end{cases}$$

is continuously differentiable on (x_1, x_2) and satisfies the boundary conditions $\eta(x_1) = \eta(x_2) = 0$. For this particular $\eta(x)$, we have

$$\int_{x_1}^{x_2} f(x)\eta(x)dx = \int_a^b f(x)(x-a)^2(x-b)^2\mathrm{d}x > 0$$

which contradicts the given condition. This eliminates the possibility that $f(\xi) > 0$ at some ξ inside (x_1, x_2). Similar argument shows that it is also not possible for $f(\xi) < 0$ at some ξ inside (x_1, x_2). Thus the theorem is established.

$\square$

This shows that for I to be stationary, F must satisfy the Euler–Lagrange equation. Since F is a given function, the Euler–Lagrange equation is a differential equation for $y(x)$.

Before we proceed further, let us illustrate how to use the Euler–Lagrange equation with a simple example.

Example 7.1.1. Shortest distance between two points in a plane.
Find the equation $y = y(x)$ of a curve joining two points (x_1, y_1) and (x_2, y_2) in the plane so that the distance between them measured along the curve is a minimum.

Solution 7.1.1. Let ds be the length of a small arc in a plane, then

$$ds^2 = dx^2 + dy^2$$

or

$$ds = \sqrt{dx^2 + dy^2} = \sqrt{1 + \left(\frac{dy}{dx}\right)^2}\, dx.$$

The total length of any curve going between the two points is

$$I = \int_{x_1}^{x_2} ds = \int_{x_1}^{x_2} \sqrt{1 + \left(\frac{dy}{dx}\right)^2}\, dx = \int_{x_1}^{x_2} \sqrt{1 + y'^2}\, dx.$$

This equation is in the form of (7.1) with

$$F = \sqrt{1 + y'^2}.$$

The condition that the curve be the shortest path is that I be a minimum. Thus F must satisfy the Euler–Lagrange equation. Substituting it into (7.7) with

$$\frac{\partial F}{\partial y} = 0, \quad \frac{\partial F}{\partial y'} = \frac{y'}{\sqrt{1 + y'^2}},$$

we have

$$\frac{d}{dx} \frac{y'}{\sqrt{1 + y'^2}} = 0,$$

or

$$\frac{y'}{\sqrt{1 + y'^2}} = c,$$

where c is a integration constant. Squaring the equation and solving for y', we get

$$y' = \frac{c}{\sqrt{1 - c^2}}.$$

In fact, by inspection we can conclude directly that

$$y' = a,$$

where a is another constant. But this is clearly the equation of a straight line

$$y = ax + b,$$

where b is another constant of integration. The constants a and b are determined by the condition that the curve passes through the two end points, (x_1, y_1), (x_2, y_2).

Strictly speaking, the straight line has only been proved to be an extremum path, but for this problem it is obviously also a minimum.

7.1.3 Variational Notation

In literature on calculus of variation, the symbol δ is often found as a differential operator. It is defined in the following way.

Expanding (7.3) as a Taylor series in α, we have

$$I(\alpha) = I(0) + \left. \frac{\mathrm{d}I}{\mathrm{d}\alpha} \right|_{\alpha=0} \alpha + O(\alpha^2).$$

The variation of I due to the variation of $y(x)$ expressed in (7.2) is simply

$$I(\alpha) - I(0) = \left. \frac{\mathrm{d}I}{\mathrm{d}\alpha} \right|_{\alpha=0} \alpha + O(\alpha^2).$$

The first-order variation of I is denoted as δI, which is just the first term on the right-hand side of this equation

$$\delta I = \left. \frac{\mathrm{d}I}{\mathrm{d}\alpha} \right|_{\alpha=0} \alpha.$$

Using (7.6), we have

$$\delta I = \alpha \int_{x_1}^{x_2} \left[\frac{\partial F}{\partial y} \eta(x) + \frac{\partial F}{\partial y'} \eta'(x) \right] \mathrm{d}x. \tag{7.8}$$

With the family of curves defined in (7.2), the variations of $y(x)$ and $y'(x)$ are given by

$$\delta y(x) = Y(x) - y(x) = \alpha \eta(x),$$
$$\delta y'(x) = Y'(x) - y'(x) = \alpha \eta'(x).$$

Expanding F in a Taylor series

$$F(y + \delta y, y' + \delta y', x) = F(y, y', x) + \frac{\partial F}{\partial y}\delta y + \frac{\partial F}{\partial y'}\delta y' + \cdots,$$

we see that the first-order variation of F is given by

$$\delta F = \frac{\partial F}{\partial y}\delta y + \frac{\partial F}{\partial y'}\delta y'.$$

These equations enable us to write δI as

$$\delta I = \delta \int_{x_1}^{x_2} F\,\mathrm{d}x = \int_{x_1}^{x_2} \delta F\,\mathrm{d}x$$

$$= \int_{x_1}^{x_2} \left[\frac{\partial F}{\partial y}\delta y + \frac{\partial F}{\partial y'}\delta y'\right]\mathrm{d}x$$

$$= \alpha \int_{x_1}^{x_2} \left[\frac{\partial F}{\partial y}\eta(x) + \frac{\partial F}{\partial y'}\eta'(x)\right]\mathrm{d}x,$$

which is identical to (7.8).

Since α, as a parameter, cannot be identically equal to zero, the condition

$$\left.\frac{\mathrm{d}I}{\mathrm{d}\alpha}\right|_{\alpha=0} = 0$$

can be replaced by the statement

$$\delta I = 0.$$

Although the δ symbol is not much used any more in mathematics, but it does enable us to operate on functionals in a formal way. Therefore it still appears frequently in applications.

7.1.4 Special Cases

In certain special cases, the Euler–Lagrange equation can be reduced to a first-order differential equation.

Integrand Does Not Depend on y Explicitly

In this case

$$\frac{\partial F}{\partial y} = 0.$$

The Euler–Lagrange equation is reduced to

$$\frac{\mathrm{d}}{\mathrm{d}x}\left(\frac{\partial F}{\partial y'}\right) = 0.$$

Therefore

$$\frac{\partial F}{\partial y'} = c,$$

where c is a constant. This is a first-order differential equation which does not depend on y. Solving for y', we obtain an equation of the form

$$y' = f(x, c)$$

from which $y(x)$ can be found.

Example 7.1.2. Find the curve $y = y(x)$ passing through $(1, 0)$ and $(2, 1)$ that renders the following functional stationary:

$$I = \int_1^2 \frac{1}{x}\sqrt{1 + y'^2}dx.$$

Solution 7.1.2. The integrand does not contain y, therefore

$$\frac{\partial F}{\partial y'} = c,$$

or

$$\frac{\partial}{\partial y'}\left(\frac{1}{x}\sqrt{1 + y'^2}\right) = \frac{y'}{x\sqrt{1 + y'^2}} = c,$$

so that

$$y' = cx\sqrt{1 + y'^2}.$$

Squaring and solving for y', we get

$$y'^2 = c^2 x^2(1 + y'^2),$$

or

$$y' = \frac{cx}{\sqrt{1 - c^2 x^2}}$$

from which it follows that:

$$y = \int \frac{cx\,dx}{\sqrt{1 - c^2 x^2}} = -\frac{1}{c}\sqrt{1 - c^2 x^2} + c',$$

where c' is another constant of integration. Hence

$$(y - c')^2 = \frac{1}{c^2} - x^2.$$

Since the curve passes through $(1, 0)$ and $(2, 1)$, so we have

$$c'^2 = \frac{1}{c^2} - 1, \quad (1 - c')^2 = \frac{1}{c^2} - 4.$$

It follows that:
$$c' = 2, \quad c = \frac{1}{\sqrt{5}}.$$

Thus the curve is given by
$$x^2 + (y - 2)^2 = 5.$$

Integrand Does Not Depend on x Explicitly

In this case
$$\frac{\partial F}{\partial x} = 0$$

and F is a function of y and y'. Since $y = y(x)$, so F must still implicitly depend on x through y and y', i.e.,

$$\frac{\mathrm{d}F}{\mathrm{d}x} = \frac{\partial F}{\partial x} + \frac{\partial F}{\partial y}\frac{\mathrm{d}y}{\mathrm{d}x} + \frac{\partial F}{\partial y'}\frac{\mathrm{d}y'}{\mathrm{d}x}$$

$$= \frac{\partial F}{\partial y}y' + \frac{\partial F}{\partial y'}y''.$$

Furthermore,

$$\frac{\mathrm{d}}{\mathrm{d}x}\left(y'\frac{\partial F}{\partial y'}\right) = y''\frac{\partial F}{\partial y'} + y'\frac{\mathrm{d}}{\mathrm{d}x}\left(\frac{\partial F}{\partial y'}\right).$$

Subtracting one from the other, we have

$$\frac{\mathrm{d}F}{\mathrm{d}x} - \frac{\mathrm{d}}{\mathrm{d}x}\left(y'\frac{\partial F}{\partial y'}\right) = \frac{\partial F}{\partial y}y' - y'\frac{\mathrm{d}}{\mathrm{d}x}\left(\frac{\partial F}{\partial y'}\right),$$

or

$$\frac{\mathrm{d}}{\mathrm{d}x}\left(F - y'\frac{\partial F}{\partial y'}\right) = \left[\frac{\partial F}{\partial y} - \frac{\mathrm{d}}{\mathrm{d}x}\left(\frac{\partial F}{\partial y'}\right)\right]y'.$$

The quantity in the bracket of the right-hand side is equal to zero because of the Euler–Lagrange equation, therefore

$$\frac{\mathrm{d}}{\mathrm{d}x}\left(F - y'\frac{\partial F}{\partial y'}\right) = 0.$$

Thus

$$F - y'\frac{\partial F}{\partial y'} = c, \tag{7.9}$$

where c is constant. This is a first-order differential equation which can be solved for $y(x)$.

Example 7.1.3. Find $y(x)$ so that the following integral is stationary

$$I = \int_{x_1}^{x_2} \frac{\sqrt{1 + y'^2}}{1 + y} \, dx.$$

Solution 7.1.3. The integrand does not depend on x, hence

$$\frac{\sqrt{1 + y'^2}}{1 + y} - y' \frac{\partial}{\partial y'} \frac{\sqrt{1 + y'^2}}{1 + y} = c$$

or

$$\frac{\sqrt{1 + y'^2}}{1 + y} - \frac{y'^2}{(1 + y)\sqrt{1 + y'^2}} = \frac{1}{(1 + y)\sqrt{1 + y'^2}} = c.$$

Thus

$$(1 + y)^2 (1 + y'^2) = \frac{1}{c^2}.$$

It follows that:

$$y'^2 = \frac{1}{c^2(1 + y)^2} - 1 = \frac{1 - c^2(1 + y)^2}{c^2(1 + y)^2}$$

and

$$y' = \frac{\sqrt{1 - c^2(1 + y)^2}}{c(1 + y)}.$$

Thus

$$\frac{c(1 + y)}{\sqrt{1 - c^2(1 + y)^2}} dy = dx.$$

Integrating both sides, we get

$$-\frac{1}{c}\sqrt{1 - c^2(1 + y)^2} = x + c'$$

or

$$1 - c^2(1 + y)^2 = c^2(x + c')^2.$$

So the solution is given by the circle

$$(x + c')^2 + (1 + y)^2 = \frac{1}{c^2},$$

where c and c' are two constants.

7.2 Constrained Variation

Very often we want to find the curve $y = y(x)$ that not only renders the integral

$$I = \int_{x_1}^{x_2} F(y, y', x) \mathrm{d}x \qquad (7.10)$$

an extremum, but also makes the second integral

$$J = \int_{x_1}^{x_2} G(y, y', x) \mathrm{d}x \qquad (7.11)$$

equal to a prescribed value. The curve is required to pass through the two end points (x_1, y_1) and (x_2, y_2), and the given functions F and G are supposed to be twice differentiable.

In essence we follow the same procedure as before by letting $y(x)$ denote the actual extremizing function and introducing a family of "neighboring" functions $Y(x)$ with respect to which we carry out the extremization. We cannot, however, express $Y(x)$ as functions depending on a single parameter as shown in (7.2), because the constant value of J would determine that parameter, which would, in turn, determine I. That would make it impossible to extremize I. For this reason we introduce a family of two parameters

$$Y(x) = y(x) + \alpha_1 \eta_1(x) + \alpha_2 \eta_2(x), \qquad (7.12)$$

where $\eta_1(x)$ and $\eta_2(x)$ are arbitrary differentiable functions for which

$$\eta_1(x_1) = \eta_1(x_2) = 0, \qquad (7.13)$$
$$\eta_2(x_1) = \eta_2(x_2) = 0. \qquad (7.14)$$

These conditions ensure that every curve in the family passes through (x_1, y_1) and (x_2, y_2) for all values of α_1 and α_2.

We replace $y(x)$ by $Y(x)$ in (7.10) and (7.11) so as to form

$$I(\alpha_1, \alpha_2) = \int_{x_1}^{x_2} F(Y, Y', x) \mathrm{d}x,$$

$$J(\alpha_1, \alpha_2) = \int_{x_1}^{x_2} G(Y, Y', x) \mathrm{d}x.$$

Clearly the parameters α_1 and α_2 are not independent, because J is to be maintained at a constant value.

We proceed by forming a combination

$$K(\alpha_1, \alpha_2) = I(\alpha_1, \alpha_2) + \lambda J(\alpha_1, \alpha_2) = \int_{x_1}^{x_2} H(Y, Y', x) \mathrm{d}x,$$

where
$$H = F + \lambda G.$$

The yet undetermined constant λ is known as the Lagrange multiplier. Now if I is stationary, and J is constant, then K must also be stationary. The conditions for K to be stationary are

$$\frac{\partial K}{\partial \alpha_1} = 0, \quad \frac{\partial K}{\partial \alpha_2} = 0,$$

where the partial derivatives are to be evaluated at $\alpha_1 = 0$ and $\alpha_2 = 0$.

Calculating the two partial derivatives, we have

$$\frac{\partial K}{\partial \alpha_i} = \int_{x_1}^{x_2} \left[\frac{\partial H}{\partial Y} \frac{\partial Y}{\partial \alpha_i} + \frac{\partial H}{\partial Y'} \frac{\partial Y'}{\partial \alpha_i} \right] dx, \quad i = 1, 2.$$

It is clear from (7.12) that

$$\frac{\partial Y}{\partial \alpha_i} = \eta_i(x), \quad \frac{\partial Y'}{\partial \alpha_i} = \eta_i'(x).$$

Thus

$$\frac{\partial K}{\partial \alpha_i} = \int_{x_1}^{x_2} \left[\frac{\partial H}{\partial Y} \eta_i + \frac{\partial H}{\partial Y'} \eta_i' \right] dx, \quad i = 1, 2.$$

Integrating by parts the second term of the integrand, we get

$$\frac{\partial K}{\partial \alpha_i} = \eta_i(x) \frac{\partial H}{\partial Y'} \Big|_{x_1}^{x_2} + \int_{x_1}^{x_2} \left[\frac{\partial H}{\partial Y} - \frac{d}{dx} \left(\frac{\partial H}{\partial Y'} \right) \right] \eta_i dx, \quad i = 1, 2.$$

The integrated term can be dropped because of the boundary conditions (7.13) and (7.14). Now we set $\alpha_1 = 0$ and $\alpha_2 = 0$, so Y and Y' are replaced by y and y'. For the two partial derivatives of K to vanish, we must have

$$\int_{x_1}^{x_2} \left[\frac{\partial H}{\partial y} - \frac{d}{dx} \left(\frac{\partial H}{\partial y'} \right) \right] \eta_i(x) dx = 0, \quad i = 1, 2.$$

Because $\eta_1(x)$ and $\eta_2(x)$ are both arbitrary functions, the two relations embodied in this equation are essentially one. By the fundamental theorem of calculus of variation, we conclude that

$$\frac{\partial H}{\partial y} - \frac{d}{dx} \left(\frac{\partial H}{\partial y'} \right) = 0.$$

This is the same Euler–Lagrange equation except with F replaced by H which is equal to $F + \lambda G$. It is a second-order differential equation that $y(x)$ must satisfy in order to keep J at a constant value and render I stationary.

Solution of this equation yields $y(x)$ that has three undetermined quantities: two constants of integration and the Lagrange multiplier λ. These quantities can be fixed by the boundary conditions at the two end points and by the requirement that J must be kept at its prescribed value.

Example 7.2.1. A curve of length L passes through x_1 and x_2 on the x-axis. Find the shape of the curve so that the area enclosed by the curve and the x-axis is the largest possible.

Solution 7.2.1. The area is given by

$$I = \int_{x_1}^{x_2} y \, dx$$

and the length of the curve is

$$J = \int_{x_1}^{x_2} ds = \int_{x_1}^{x_2} \sqrt{1 + y'^2} dx.$$

We want to maximize I subject to the condition that J is equal to the constant L. So we solve the Euler–Lagrange equation

$$\frac{d}{dx} \frac{\partial H}{\partial y'} - \frac{\partial H}{\partial y} = 0,$$

where

$$H = y + \lambda \sqrt{1 + y'^2}.$$

Since H does not explicitly depend on x, so

$$y' \frac{\partial H}{\partial y'} - H = c_1$$

or

$$\frac{\lambda y'^2}{\sqrt{1 + y'^2}} - y - \lambda \sqrt{1 + y'^2} = c_1.$$

This can be simplified to

$$(c_1 + y)\sqrt{1 + y'^2} = -\lambda,$$

or

$$1 + y'^2 = \frac{\lambda^2}{(c_1 + y)^2}$$

from which we have

$$y' = \frac{\sqrt{\lambda^2 - (c_1 + y)^2}}{c_1 + y}.$$

Thus

$$\frac{(c_1 + y)dy}{\sqrt{\lambda^2 - (c_1 + y)^2}} = dx.$$

Let $z = \lambda^2 - (c_1 + y)^2$, $\mathrm{d}z = -2(c_1 + y)\mathrm{d}y$, so

$$\int \frac{(c_1 + y)\mathrm{d}y}{\sqrt{\lambda^2 - (c_1 + y)^2}} = -\int \frac{1}{2} z^{-1/2}\mathrm{d}z = -\int \mathrm{d}z^{1/2} = -\int \mathrm{d}\sqrt{\lambda^2 - (c_1 + y)^2}.$$

Hence

$$-\sqrt{\lambda^2 - (c_1 + y)^2} = x + c_2,$$

where c_1 and c_2 are two constants of integration. Squaring both sides, we get

$$(x + c_2)^2 + (c_1 + y)^2 = \lambda^2.$$

So the curve should be an arc of a circle passing through the two given points. The constants c_1, c_2, and λ may be fixed by requiring that the curve passes through the appropriate end points, and has the required length between these points.

7.3 Solutions to Some Famous Problems

7.3.1 The Brachistochrone Problem

Suppose that a bead is sliding down a wire without friction as shown in Fig. 7.2. We learnt in physics class that at the point (x, y), the kinetic energy of the bead is $\frac{1}{2}mv^2$ and the potential energy is $-mgy$, if we take $y = 0$ as our reference level. Because of the conservation of energy, the sum of these two energies must be equal to zero

$$\frac{1}{2}mv^2 - mhy = 0,$$

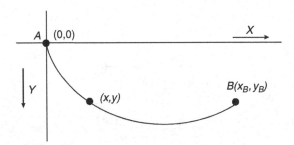

Fig. 7.2. The Brachistochrone problem. A bead sliding down on a wire from A to B without friction, the problem is to find the shape of the wire so that it takes the least amount of time

since initially at $(0,0)$, both kinetic and potential energies are equal to zero. Therefore the velocity at that point is

$$v = \sqrt{2gy}.$$

Thus the time it takes from A to B is

$$T = \int dt = \int \frac{ds}{v} = \int \frac{\sqrt{(dx)^2 + (dy)^2}}{v}$$

$$= \frac{1}{\sqrt{2g}} \int_0^{x_B} \frac{\sqrt{1+y'^2}}{\sqrt{y}}dx. \tag{7.15}$$

Now the question is: what should be the shape of the wire so that the time it takes from A to B is the shortest? This famous problem is known as the Brachistochrone problem (from the Greek words meaning shortest time).

In 1696, Johann Bernoulli proposed this problem and he addressed it "to the shrewdest mathematicians in all the world" and allowed 6 months for anyone to come up with a solution. This marks the beginning of general interest in the calculus of variation. Five correct solutions were submitted – by Newton, Leibniz, L'Hospital, himself, and his brother Jakob Bernoulli. They independently with different methods arrived at the same answer. The required shape is a cycloid, the curve traced by a point on the rim of a wheel as it rolls on a horizontal surface.

We can answer the question by minimizing the integral in (7.15). Since the integrand does not explicitly depend on x, so the Euler–Lagrange equation can be written in the form of (7.9)

$$F - y'\frac{\partial F}{\partial y'} = c$$

with

$$F = \frac{\sqrt{1+y'^2}}{\sqrt{y}}.$$

Thus

$$\frac{\sqrt{1+y'^2}}{\sqrt{y}} - \frac{y'^2}{\sqrt{y}\sqrt{1+y'^2}} = c$$

or

$$\frac{1}{\sqrt{y}\sqrt{1+y'^2}} = c.$$

Squaring both sides, we have

$$y(1+y'^2) = \frac{1}{c^2}.$$

It follows that:

$$y' = \sqrt{\frac{1 - c^2 y}{c^2 y}}.$$

With $y' = \dfrac{dy}{dx}$, this equation can be written as

$$\sqrt{\frac{c^2 y}{1 - c^2 y}}\, dy = dx.$$

To find $y(x)$, we have to integrate both sides of this equation. We can carry out the integration by a change of variable. Let

$$c^2 y = \frac{1}{2}(1 - \cos\theta) \qquad (7.16)$$

so

$$1 - c^2 y = \frac{1}{2}(1 + \cos\theta)$$

and

$$dy = \frac{1}{2c^2}\sin\theta\, d\theta.$$

Since

$$\cos\theta = \cos 2\frac{\theta}{2} = \cos^2\frac{\theta}{2} - \sin^2\frac{\theta}{2},$$

$$\sin\theta = \sin 2\frac{\theta}{2} = 2\sin\frac{\theta}{2}\cos\frac{\theta}{2},$$

we can write

$$1 - \cos\theta = 2\sin^2\frac{\theta}{2},$$

$$1 + \cos\theta = 2\cos^2\frac{\theta}{2},$$

$$dy = \frac{1}{c^2}\sin\frac{\theta}{2}\cos\frac{\theta}{2}d\theta.$$

Therefore

$$\sqrt{\frac{c^2 y}{1 - c^2 y}}\, dy = \frac{1}{c^2}\frac{\sin(\theta/2)}{\cos(\theta/2)}\sin\frac{\theta}{2}\cos\frac{\theta}{2}d\theta = \frac{1}{c^2}\sin^2\frac{\theta}{2}d\theta$$

$$= \frac{1}{2c^2}(1 - \cos\theta)d\theta.$$

Hence

$$\int \sqrt{\frac{c^2 y}{1 - c^2 y}}\, dy = \int dx$$

becomes

$$\frac{1}{2c^2} \int (1 - \cos\theta)d\theta = \int dx,$$

or

$$\frac{1}{2c^2}(\theta - \sin\theta) = x + c'.$$

Now the curve goes through $(0,0)$, i.e., when $x = 0$, y must also be zero. But, by (7.16), $y = (1 - \cos\theta)/2c^2$, so when $y = 0$, θ must be equal to zero. Therefore $x = \theta = 0$ must satisfy the above equation. Thus $c' = 0$. The remaining constant is fixed by the condition that the curve should pass the other end point.

As a result, the required curve is given by the parametric equations

$$x = \frac{1}{2c^2}(\theta - \sin\theta),$$

$$y = \frac{1}{2c^2}(1 - \cos\theta).$$

These are the equations of a cycloid that can be seen as follows. Let a circle of radius r rolls along the x-axis as shown in Fig. 7.3. The origin is so chosen that the point P makes contact with x-axis at the origin. When the circle has revolved through an angle of θ radians, it will have rolled a distance $OC = r\theta$ from the origin as shown in Fig. 7.3. Hence the center of the circle will be at the point $(r\theta,\ r)$. It is clear from the figure that the x coordinate of P is

$$x = r\theta + r\cos\phi = r\theta + r\cos\left(\frac{3\pi}{2} - \theta\right)$$

$$= r\theta - r\sin\theta = r(\theta - \sin\theta)$$

and the y coordinate of P is

$$y = r + r\sin\phi = r + r\sin\left(\frac{3\pi}{2} - \theta\right)$$

$$= r - r\cos\theta = r(1 - \cos\theta).$$

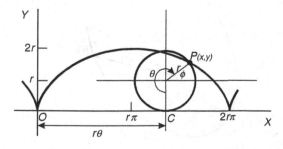

Fig. 7.3. Cycloid. The *curve* traced out by the point P on a *circle* while the *circle* is rolling on the x-axis

Therefore the solution of the Brachistochrone problem is a cycloid, since the parametric equations for the required curve are identical to the last two equations with $r = 1/2c^2$. Note that, since we have taken the downward direction of the y-axis as positive, the circle which generates the cycloid rolls along the under side of the x-axis.

The correct minimum path is shown in Fig. 7.2. It is somewhat surprising that going to the bottom of the curve and then back up to B actually takes less time than sliding on the straight line from A to B.

7.3.2 Isoperimetric Problems

The word isoperimetric means same perimeter. The most famous isoperimetric problem is to find the plane curve of given length which encloses the greatest possible area.

Let C be a closed plane curve as shown in Fig. 7.4. The area inside C can be found as follows. The shaded infinitesimal area is given by half of the product of its base and height

$$dA = \frac{1}{2}rh = \frac{1}{2}r\, dr \sin\theta = \frac{1}{2}\left|\mathbf{r} \times d\mathbf{r}\right|.$$

Let $\mathbf{r} = x\mathbf{i} + y\mathbf{j}$, so $d\mathbf{r} = dx\mathbf{i} + dy\mathbf{j}$, and

$$\mathbf{r} \times d\mathbf{r} = \begin{vmatrix} \mathbf{i} & \mathbf{j} & \mathbf{k} \\ x & y & 0 \\ dx & dy & 0 \end{vmatrix} = (x\, dy - y\, dx)\mathbf{k}.$$

Thus

$$dA = \frac{1}{2}\left|\mathbf{r} \times d\mathbf{r}\right| = \frac{1}{2}(x\, dy - y\, dx).$$

It follows that the area enclosed by C is:

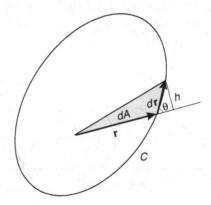

Fig. 7.4. Area inside a *closed* curve C. The *infinitesimal shaded area* is given by $\left|\mathbf{r} \times d\mathbf{r}\right|/2 = (x\, dy - y\, dx)/2$

$$A = \oint_C dA = \oint_C \frac{1}{2}(x\,dy - y\,dx) = \oint_C \frac{1}{2}(xy' - y)dx.$$

The length of the curve is

$$L = \oint_C ds = \oint_C (dx^2 + dy^2)^{1/2} = \oint_C (1 + y'^2)^{1/2}dx.$$

Now we want to maximize A and keep L constant. According to the theory of constrained variation, we have to solve the Euler–Lagrange equation

$$\frac{d}{dx}\frac{\partial H}{\partial y'} - \frac{\partial H}{\partial y} = 0$$

with

$$H - \frac{1}{2}(xy' - y) + \lambda(1 + y'^2)^{1/2}.$$

Since

$$\frac{\partial H}{\partial y'} = \frac{1}{2}x + \frac{\lambda y'}{(1 + y'^2)^{1/2}} \quad \text{and} \quad \frac{\partial H}{\partial y} = -\frac{1}{2},$$

we have

$$\frac{d}{dx}\left[\frac{1}{2}x + \frac{\lambda y'}{(1 + y'^2)^{1/2}}\right] + \frac{1}{2} = 0$$

or

$$\frac{d}{dx}\left[\frac{\lambda y'}{(1 + y'^2)^{1/2}}\right] = -1.$$

Integrating once

$$\frac{\lambda y'}{(1 + y'^2)^{1/2}} = -x + c_1.$$

Squaring

$$(\lambda y')^2 = (x - c_1)^2(1 + y'^2),$$

and solving for y',

$$y' = \frac{\pm(x - c_1)}{[\lambda^2 - (x - c_1)^2]^{1/2}}.$$

Since

$$\frac{d}{dx}[\lambda^2 - (x - c_1)^2]^{1/2} = -\frac{(x - c_1)}{[\lambda^2 - (x - c_1)^2]^{1/2}},$$

so we have

$$y' = \frac{dy}{dx} = \mp\frac{d}{dx}[\lambda^2 - (x - c_1)^2]^{1/2}.$$

Integrating again

$$y = \mp[\lambda^2 - (x - c_1)^2]^{1/2} + c_2$$

or

$$(y - c_2)^2 = \lambda^2 - (x - c_1)^2.$$

Clearly this is a circle of radius λ centered at (c_1, c_2)

$$(x - c_1)^2 + (y - c_2)^2 = \lambda^2.$$

The area enclosed

$$A = \pi\lambda^2 = \pi \left(\frac{L}{2\pi}\right)^2 = \frac{L^2}{4\pi}$$

must be the maximum, since the minimum area is obviously zero when the curve is squeezed into a line.

7.3.3 The Catenary

The Latin word Catena means a chain. The catenary is a problem of hanging chain. Its history is parallel to that of Brachistochrone. In 1690, Jakob Bernoulli proposed the problem: "to find the curve assumed by a loose rope hung freely from two fixed points." One year later, three correct solutions were submitted – by Christian Huygens, Leibnitz, and Johann Bernoulli.

The rope will assume a shape for which its potential energy is a minimum. If ϱ is the mass per unit length of the rope, then the potential energy of an infinitesimal section of length ds due to gravity is $\varrho\,\mathrm{d}s g y$, where g is the gravitational constant. Let the two fixed points be A and B, the total potential energy of the rope is given by the functional

$$I = \int_A^B \varrho g y \, \mathrm{d}s = \varrho g \int_A^B y \, \mathrm{d}s.$$

On using $\mathrm{d}s = \sqrt{(\mathrm{d}x)^2 + (\mathrm{d}y)^2} = \sqrt{1 + y'^2}\mathrm{d}x$, and ignoring the constant factor ϱg, we have to minimize the functional

$$I = \int_A^B y\sqrt{1 + y'^2}\mathrm{d}x = \int_A^B F(y, y') \, \mathrm{d}x. \qquad (7.17)$$

Since F is independent of x, the Euler–Lagrange equation is given by

$$F - y'\frac{\partial F}{\partial y'} = c \qquad (7.18)$$

or

$$y\sqrt{1 + y'^2} - y'\frac{yy'}{\sqrt{1 + y'^2}} = c.$$

Thus

$$y\left(1 + y'^2\right) - yy'^2 = c\sqrt{1 + y'^2},$$

or

$$y = c\sqrt{1 + y'^2}.$$

It follows that:

$$y' = \sqrt{\frac{y^2}{c^2} - 1},$$

so

$$\frac{dy}{\sqrt{(y/c)^2 - 1}} = dx. \tag{7.19}$$

Recall

$$\cosh x = \frac{1}{2}(e^x + e^{-x}), \quad \sinh x = \frac{1}{2}(e^x - e^{-x}),$$

$$\cosh^2 x - \sinh^2 x = 1, \quad \frac{d}{dx}\cosh x = \sinh x.$$

With the substitution

$$\frac{y}{c} = \cosh z$$

and

$$dy = c\sinh z\, dz, \quad \sqrt{(y/c)^2 - 1} = \sqrt{\cosh^2 z - 1} = \sinh z,$$

we can write (7.19) as

$$c\, dz = dx.$$

Integrating once again, we have

$$cz = x + b,$$

$$z = \frac{x + b}{c}.$$

Hence

$$y = c\cosh\frac{x + b}{c}. \tag{7.20}$$

The constants c and b can be determined if the coordinates of the fixed points are known. For example, if the coordinates of A and B are $(-x_0, y_0)$ and (x_0, y_0), respectively, then

$$y_0 = c\cosh\frac{-x_0 + b}{c} = c\cosh\frac{x_0 + b}{c}.$$

Since $\cosh(-x) = \cosh(x)$, it follows that $b = 0$. In this case,

$$y = c\cosh\frac{x}{c}. \tag{7.21}$$

This is the equation of the catenary. The shape of the catenary is shown in Fig. 7.5

It should be noted that if the length of the rope L is given

$$L = \int_A^B ds = \int_A^B \sqrt{1 + y'^2}dx,$$

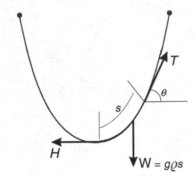

Fig. 7.5. The catenary. The shape of a hanging chain is given by $y = c\cosh(x/c)$, known as the catenary

then L must be kept constant. According the theory of constrained variation, we have to minimize the functional

$$K = \int_A^B \varrho g y \sqrt{1 + y'^2} dx + \lambda \int_A^B \sqrt{1 + y'^2} dx.$$

In this case, F in (7.18) is given by

$$F = \varrho g y \sqrt{1 + y'^2} + \lambda \sqrt{1 + y'^2}.$$

Following the same procedure as above, one can show that

$$y = \frac{c_1}{\varrho g} \cosh \frac{\varrho g (x + c_2)}{c_1} - \frac{\lambda}{\varrho g}.$$

The constants c_1, c_2, and λ are to be determined by the length of the rope L and the coordinates of A and B. As we see that the character of the catenary is not changed.

This problem can also be solved with the "ordinary calculus." Let $x = 0$ be the lowest point of the rope, and H be the tension at that point. Clearly, H acts horizontally. Let T be the tension at the other end of the section of length s, as shown in Fig. 7.5. The weight of rope of that section is $w = \varrho g s$. Since it is in equilibrium, the forces must be balanced in both x and y directions,

$$T \sin \theta = \varrho g s,$$
$$T \cos \theta = H.$$

The ratio of these two equations is

$$\tan \theta = \frac{\varrho g}{H} s,$$

which is the slope of the curve. That is

$$y' = \frac{dy}{dx} = \tan\theta = \frac{\varrho g}{H}s.$$

It follows:

$$\frac{d}{dx}y' = \frac{\varrho g}{H}\frac{ds}{dx}.$$

Since $ds = \sqrt{dx^2 + dy^2} = \sqrt{1 + y'^2}dx$,

$$\frac{ds}{dx} = \sqrt{1 + y'^2}.$$

Thus

$$\frac{d}{dx}y' = \frac{\varrho g}{H}\sqrt{1 + y'^2}$$

or

$$\frac{dy'}{\sqrt{1 + y'^2}} = \frac{\varrho g}{H}dx.$$

Let

$$y' = \sinh z,$$

the last equation becomes

$$\frac{d\sinh z}{\sqrt{1 + \sinh^2 z}} = \frac{\cosh z\, dz}{\cosh z} = dz = \frac{\varrho g}{H}dx.$$

Integrating once

$$z = \frac{\varrho g}{H}x + c'$$

or

$$y' = \sinh\left(\frac{\varrho g}{H}x + c'\right).$$

At $x = 0$, $y' = 0$, therefore $c' = 0$, since $\sinh(0) = 0$. Thus

$$\frac{dy}{dx} = \sinh\left(\frac{\varrho g}{H}x\right)$$

or

$$dy = \sinh\left(\frac{\varrho g}{H}x\right)dx.$$

Integrating again, we get

$$y = \frac{H}{\varrho g}\cosh\left(\frac{\varrho g}{H}x\right) + b.$$

If we adjust the y-axis scale in such a way that $y(0) = H/\varrho g$, then $b = 0$ since $\cosh(0) = 1$. Thus

$$y(x) = \frac{H}{\varrho g}\cosh\left(\frac{\varrho g}{H}x\right),$$

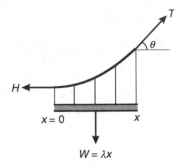

Fig. 7.6. The cable of a suspension bridge. The cable of a suspension bridge is in the form of parabola, which looks like a catenary but is not

which is in the same form as (7.21). In addition, we have given a physical meaning to the constant c of (7.21).

It is interesting to note that the shape of the cable of a suspension bridge is not a catenary, although it looks like one. Suppose that the cable is supporting a uniform roadway, and the weight of the cable is negligible compared to the weight of the roadway. A section of the cable is shown in Fig. 7.6. We take $x = 0$, $y = 0$ at the center of the span. This section supports a portion of the roadway whose weight W is proportional to the distance x, since the roadway is assumed uniform, i.e.,

$$W = \lambda x,$$

where λ is the weight per unit length. It is clear from Fig. 7.6 that

$$T \sin \theta = \lambda x,$$
$$T \cos \theta = H.$$

Thus

$$\tan \theta = \frac{\lambda}{H} x.$$

Since $\tan \theta$ is the slope of the cable

$$\frac{\mathrm{d}y}{\mathrm{d}x} = \tan \theta = \frac{\lambda}{H} x$$

or

$$\mathrm{d}y = \frac{\lambda}{H} x \, \mathrm{d}x.$$

Integrating, we get

$$y = \frac{1}{2} \frac{\lambda}{H} x^2 + c.$$

This is a parabola. While the catenary certainly looks like a parabola, we see that they are fundamentally different. The parabola is algebraic, while the catenary is transcendental.

For a long time, the shape of a hanging chain was thought as a parabola. In 1646, Huygens (at age 17) proved that it could not possibly be a parabola. But it was not until 1691, after the invention of calculus, the catenary was correctly described as a hyperbolic cosine function.

7.3.4 Minimum Surface of Revolution

The catenary is also the solution of the problem of the minimum surface of revolution passing through two given points A and B.

The area of the surface of revolution generated by rotating the curve $y = y(x)$ about the x-axis is

$$I = 2\pi \int_A^B y\,ds = 2\pi \int_A^B y\sqrt{1+y'^2}dx.$$

The integral we seek to minimize is the same as (7.17). Therefore the required curve is a catenary given by (7.20)

$$y(x) = c\cosh\frac{x+b}{c}.$$

The surface generated by the rotation of the catenary is called *catenoid*. The values of the arbitrary constants c and b are determined by the conditions

$$y(x_A) = y_A, \quad y(x_B) = y_B.$$

Unfortunately this is only an incomplete solution, because the two points A and B must satisfy certain condition in order to have a curve of the form (7.17) passing through them. In other words, if the condition is not satisfied, there is no surface in the class of smooth surfaces of revolution which achieves the minimum area.

We will find this condition for the following problem. Let the coordinates of A be $(-x_0, y_0)$ and of B be (x_0, y_0). As shown in (7.21), in this case the catenary can be written as

$$y = c\cosh\frac{x}{c}.$$

The constant c is determined by the ratio y_0/x_0. Now if we define

$$u = \frac{x}{c}, \quad v = \frac{y}{c}$$

and

$$u_0 = \frac{x_0}{c}, \quad v_0 = \frac{y_0}{c},$$

we can write (7.21) as

$$v = \cosh u. \tag{7.22}$$

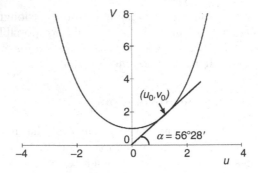

Fig. 7.7. To have a minimum surface of revolution the curve $v = \cosh u$ and $v = \frac{y_0}{x_0} u$ must intersect

On the other hand, we can take another curve

$$\frac{v}{u} = \frac{y}{x} = \frac{y_0}{x_0}$$

or

$$v = \frac{y_0}{x_0} u \qquad (7.23)$$

which is a straight line. Since (u_0, v_0) satisfies both (7.22) and (7.23), it must be the point of intersection the two curves expressed by these equations. Figure 7.7 shows the curve $v = \cosh u$ and a straight line $v = (\tan \alpha)u$ that is tangent to this curve. It is clear that if $(y_0/x_0) < \tan \alpha$, then the two curves will not intersect, and no catenary can be drawn from A to B.

The angle α can be found by noting that at the point u_0 where the straight line is the tangent to the curve, we have the following relationship:

$$\frac{v}{u} = \frac{dv}{du}.$$

Since $v = \cosh u$ and $dv/du = \sinh u$, so we have

$$\frac{\cosh u}{u} - \sinh u = 0.$$

This equation can be solved numerically. For example, this book is written with the computer software "Scientific WorkPlace." It came with a computer algebra system "MuPAD." After this equation is typed in the math mode, click the "compute" and "solve" button, the program will return with an answer:

$$u = 1.1997.$$

This means that $(u_0, v_0) = (1.1997,\ \cosh 1.1997)$. It follows that:

$$\alpha = \tan^{-1}\left(\frac{\cosh 1.1997}{1.1997}\right) = 0.9885 \text{ radians } (56°28').$$

Thus, if

$$\frac{y_0}{x_0} < \tan \alpha = 1.5089$$

the straight line $v = \dfrac{y_0}{x_0}u$ and the curve $v = \cosh u$ will not meet, and there is no twice-differentiable minimizing curve.

This limitation can be illustrated with a soap film experiment. Because of surface tension, a soap film will form a surface of minimum energy, which happens to be the surface of minimum area with the frame as the boundary.

A soap film will form a catenoid between two parallel rings of radius y_0 with their centers $2x_0$ apart on an axis perpendicular to the rings, as shown in Fig. 7.8, if y_0/x_0 is greater than 1.5089. We can increase x_0. When y_0/x_0 becomes less than 1.5089, the catenoid will no longer be formed, and the soap film will cover only the two rings to give a surface area of $2\pi y_0^2$. Clearly the solution is discontinuous and beyond the scope of the variational theory.

Example 7.3.1. Find the area of the minimum surface of revolution shown in Fig. 7.8 with $x_0 = 1$ and $y_0 = 2$.

Solution 7.3.1. The area of the catenoid is given by

$$A = 2\pi \int_{-1}^{1} y\sqrt{1 + y'^2}\,dx,$$

where

$$y = c \cosh \frac{x}{c}.$$

Since

$$y\sqrt{1 + y'^2} = c \cosh \frac{x}{c}\sqrt{1 + \sinh^2 \frac{x}{c}} = c \cosh^2 \frac{x}{c},$$

the area can be written as

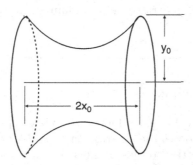

Fig. 7.8. The catenoid. A minimum surface of revolution can be formed between the two rings if y_0 is greater than $1.5089x_0$

$$A = 2\pi c \int_{-1}^{1} \cosh^2 \frac{x}{c} dx = \pi c \int_{-1}^{1} \left(1 + \cosh \frac{2x}{c}\right) dx$$

$$= \pi c^2 \left(\frac{2}{c} + \sinh \frac{2}{c}\right).$$

Since
$$\frac{y_0}{x_0} = 2 > 1.5089,$$

there are two intersection points of

$$v = \cosh u \quad \text{and} \quad v = 2u.$$

The equation
$$\cosh u - 2u = 0$$

can be solved numerically to give

$$u_0 = \begin{cases} 0.5894, \\ 2.1268. \end{cases}$$

Recall $u_0 = x_0/c$, so

$$c = \frac{x_0}{u_0} = \begin{cases} \frac{1}{0.5894} = 1.6967, \\ \frac{1}{2.1268} = 0.4702. \end{cases}$$

Therefore

$$A = \begin{cases} \pi (1.6967)^2 \left(\frac{2}{1.6967} + \sinh \frac{2}{1.6967}\right) = 23.968, \\ \pi (0.4702)^2 \left(\frac{2}{0.4702} + \sinh \frac{2}{0.4702}\right) = 27.382. \end{cases}$$

Thus, rotating
$$y = 1.6967 \cosh \frac{x}{1.6967},$$

around the x-axis will generate the minimum surface of revolution with an area of 23.968.

7.3.5 Fermat's Principle

In the 1650s Pierre de Fermat adopted the view espoused by Aristotelians that nature always chooses the shortest path, and formulated a "principle of least time" for geometrical optics. It says that a ray of light traveling from one point to another takes the path which requires the shortest time.

If the velocity of the ray of light in a medium is v, then the time it takes from A to B is

$$T = \int \frac{ds}{v} = \int \frac{1}{v}\sqrt{(dx)^2 + (dy)^2} = \int_A^B \frac{1}{v}(1 + y'^2)^{1/2}dx,$$

where $y(x)$ is the path of the ray. In optics, a very useful quantity is the index of refraction n, defined as

$$n = \frac{c}{v},$$

where c is the speed of light in vacuum, which is a constant. To have the shortest time is to require the following integral to be stationary:

$$I = \int_A^B n(1 + y'^2)^{1/2}dx.$$

Let us assume that n is not a function of x. In that case, the integrand

$$F = n(1 + y'^2)^{1/2}$$

does not explicitly contain the independent variable x. Therefore

$$F - y'\frac{\partial}{\partial y'}F = k,$$

where k is a constant. Carrying out the differentiation, we have

$$n(1 + y'^2)^{1/2} - y'n\frac{y'}{(1 + y'^2)^{1/2}} = k$$

or

$$n(1 + y'^2) - ny'^2 = k(1 + y'^2)^{1/2}.$$

Thus

$$n = k(1 + y'^2)^{1/2}.$$

Since y' is the slope of $y(x)$, so $y' = \tan\phi$ where ϕ is the angle between the instantaneous direction of the ray and the x-axis. Thus the last equation can be written as

$$n = k(1 + \tan^2\phi)^{1/2} = k\left(1 + \frac{\sin^2\phi}{\cos^2\phi}\right)^{1/2} = k\frac{1}{\cos\phi}.$$

Therefore the general result is in the form of

$$n\cos\phi = k.$$

If n is not changing in space, then ϕ must be a constant since k is a constant. This means that the ray of light is traveling in a straight line. This is certainly the case.

Suppose that the ray travels from one medium with index of refraction n_1 to another medium with index of refraction n_2, and the interface between

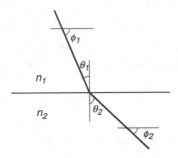

Fig. 7.9. Snell's law: $n_1 \sin \theta_1 = n_2 \sin \theta_2$. If $n_1 > n_2$, then $\theta_1 < \theta_2$

them is a plane as shown in Fig. 7.9. Since k is constant along the whole path, we must have

$$n_1 \cos \phi_1 = n_2 \cos \phi_2.$$

This is the well-known Snell's law. Usually the Snell's law is written as

$$n_1 \sin \theta_1 = n_2 \sin \theta_2,$$

where θ_1 and θ_2 are the angles between the ray and the normal of the interface as shown in Fig. 7.9. Since $\theta_1 + \phi_1 = \theta_2 + \phi_2 = \pi/2$, the last two equations are identical.

If the ray is going from medium 1 to medium 2, θ_1 is known as the angle of incidence θ_i and θ_2, the angle of transmission (refraction) θ_t. If $\theta_t = \pi/2$, the incident angle is known as the critical angle θ_c

$$\theta_c = \sin^{-1} \frac{n_2}{n_1}.$$

If the angle of incidence is greater than the critical angle, the ray is reflected back. The angle between the reflected ray and the normal is known as the angle of reflection θ_r. In that case, $n_1 \sin \theta_i = n_1 \sin \theta_r$. Therefore

$$\theta_i = \theta_r,$$

another well-known fact in geometrical optics.

In fact, the entire geometrical optics can be derived from the Fermat principle.

Suppose that the light is going through a series of mediums as shown in Fig. 7.10. If $n_1 > n_2 > n_3 > n_4$, then $\theta_1 < \theta_2 < \theta_3$. If θ_3 is greater than the critical angle, then the ray will be reflected back as shown in the figure. It is clear that if the index of refraction is decreasing continuously (this is equivalent to say that if the velocity of the light v is increasing continuously), the path of the light will become a continuous curve. If v is proportional to $\sqrt{y}$ (with the positive y-axis directed downward), the path of the light will become the Brachistochrone curve shown in Fig. 7.2. In fact, Johann Bernoulli first solved the Brachistochrone problem with this optical analogy.

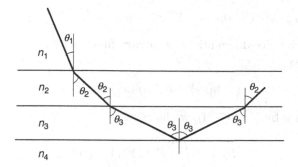

Fig. 7.10. Light going through a series of mediums with increasing index of refraction

This is also the reason for the mirage that one often sees while driving in hot roads. One sees "water" on the road, but when one gets there, it is dry. The explanation is this. The air is very hot just above the road and is cooler up higher. Light travels faster in the hot region because the air is more expanded and therefore thinner. So the light the sky, heading for the road, is going faster and faster. As a consequence, it follows a curved path, like the Brachistochrone curve shown in Fig. 7.2. When it ended in our eyes, we thought it was reflected from the water on the road.

7.4 Some Extensions

Often the functionals contain higher derivatives, or several independent or dependent variables. The Euler–Lagrange equations for these problems can be derived in a similar way.

7.4.1 Functionals with Higher Derivatives

Consider the functional

$$I = \int_{x_1}^{x_2} F(y, y', y'', x) \mathrm{d}x, \tag{7.24}$$

where the values of y and y' are specified at the end points

$$y(x_1) = A_0, \quad y'(x_1) = A_1,$$
$$y(x_2) = B_0, \quad y'(x_2) = B_1.$$

Among all possible functions that satisfy these boundary conditions, we want to find the function $y(x)$ for which the functional I has an extremum.

To solve this problem, we follow the previous procedure. We define a family of curves that satisfies these boundary conditions

$$Y(x) = y(x) + \alpha\eta(x), \quad Y'(x) = y' + \alpha\eta', \quad Y'' = y'' + \alpha\eta'',$$

where $\eta(x)$ is a twice-differentiable arbitrary function that satisfies the boundary conditions

$$\eta(x_1) = \eta(x_2) = 0, \quad \eta'(x_1) = \eta'(x_2) = 0.$$

Replacing y by Y in (7.24), we have

$$I(\alpha) = \int_{x_1}^{x_2} F(Y, Y', Y'', x)dx.$$

A necessary condition for it to be an extremum is that

$$\left.\frac{dI}{d\alpha}\right|_{\alpha=0} = 0.$$

Carrying the differentiation inside the integral, we have

$$\left.\frac{dI}{d\alpha}\right|_{\alpha=0} = \int_{x_1}^{x_2} \left[\frac{\partial F}{\partial y}\eta + \frac{\partial F}{\partial y'}\eta' + \frac{\partial F}{\partial y''}\eta''\right] dx.$$

We have already shown that

$$\int_{x_1}^{x_2} \frac{\partial F}{\partial y'}\eta'dx = -\int_{x_1}^{x_2} \frac{d}{dx}\left[\frac{\partial F}{\partial y'}\right] \eta\, dx.$$

Similarly, with integration by parts

$$\int_{x_1}^{x_2} \frac{\partial F}{\partial y''}\eta''dx = \int_{x_1}^{x_2} \frac{\partial F}{\partial y''}\frac{d\eta'}{dx}dx = \int_{x_1}^{x_2} \frac{\partial F}{\partial y''}d\eta'$$

$$= \left.\frac{\partial F}{\partial y''}\eta'\right|_{x_1}^{x_2} - \int_{x_1}^{x_2} \frac{d}{dx}\left(\frac{\partial F}{\partial y''}\right)\eta'dx.$$

The integrated part is zero because of the boundary conditions of $\eta'(x)$. With integration by parts again, the last term becomes

$$-\int_{x_1}^{x_2} \frac{d}{dx}\left(\frac{\partial F}{\partial y''}\right)\eta'dx = -\left.\frac{d}{dx}\left(\frac{\partial F}{\partial y''}\right)\eta\right|_{x_1}^{x_2} + \int_{x_1}^{x_2} \frac{d^2}{dx^2}\left(\frac{\partial F}{\partial y''}\right)\eta\, dx.$$

Again the integrated part vanishes because of the boundary conditions of $\eta(x)$. Therefore

$$\left.\frac{dI}{d\alpha}\right|_{\alpha=0} = \int_{x_1}^{x_2} \left[\frac{\partial F}{\partial y} - \frac{d}{dx}\left(\frac{\partial F}{\partial y'}\right) + \frac{d^2}{dx^2}\left(\frac{\partial F}{\partial y''}\right)\right]\eta\, dx = 0.$$

It follows that the function $y(x)$, for which I is stationary, must satisfy the differential equation:

$$\frac{\partial F}{\partial y} - \frac{d}{dx}\left(\frac{\partial F}{\partial y'}\right) + \frac{d^2}{dx^2}\left(\frac{\partial F}{\partial y''}\right) = 0.$$

Notice the alternating sign in this equation. Clearly, the function $y(x)$ that minimizes the functional

$$I = \int_{x_1}^{x_2} F(y, y', y'', \ldots, y^n, x)dx$$

is the solution of

$$\frac{\partial F}{\partial y} - \frac{d}{dx}\left(\frac{\partial F}{\partial y'}\right) + \cdots (-1)^n \frac{d^n}{dx^n}\left(\frac{\partial F}{\partial y^n}\right) = 0.$$

7.4.2 Several Dependent Variables

Consider the integral

$$I = \int_{t_1}^{t_2} F(x, y, x', y', t)dt,$$

where x and y are twice-differentiable functions of the independent variable t. Their derivatives with respect to t are, respectively, x' and y'. The values of $x(t_1)$, $y(t_1)$ and $x(t_2)$, $y(t_2)$ are specified. We want to find the differential equations that x and y must satisfy so that the value of I is stationary. We can solve this problem by the same procedure as in the case of one dependent variable.

Let $x(t)$ and $y(t)$ be the actual curve along which I is stationary. We denote the family of curves that go through the two fixed points at t_1 and t_2 as

$$X(t) = x(t) + \alpha\varepsilon(t), \qquad Y(t) = y(t) + \alpha\eta(t),$$

where $\varepsilon(t)$ and $\eta(t)$ are arbitrary differentiable functions for which

$$\varepsilon(t_1) = \varepsilon(t_2) = 0, \quad \eta(t_1) = \eta(t_2) = 0.$$

These boundary conditions ensure that every curve in the family goes through the two end points. The parameter α specifies each individual curve and the actual curve that minimizes I is labeled as $\alpha = 0$. As a consequence

$$X' = x' + \alpha\varepsilon', \quad Y' = y' + \alpha\eta'.$$

Replacing x and y by X and Y, respectively, in the integral I, it becomes a function of α

$$I(\alpha) = \int_{t_1}^{t_2} F(X, Y, X', Y', t)dt.$$

A necessary condition for I to be stationary is

$$\left.\frac{\mathrm{d}I}{\mathrm{d}\alpha}\right|_{\alpha=0} = 0.$$

Since α does not depend on t, the differentiation can carry out inside the integral,

$$\frac{\mathrm{d}I}{\mathrm{d}\alpha} = \int_{t_1}^{t_2} \left[\frac{\partial F}{\partial X}\frac{\mathrm{d}X}{\mathrm{d}\alpha} + \frac{\partial F}{\partial Y}\frac{\mathrm{d}Y}{\mathrm{d}\alpha} + \frac{\partial F}{\partial X'}\frac{\mathrm{d}X'}{\mathrm{d}\alpha} + \frac{\partial F}{\partial Y'}\frac{\mathrm{d}Y'}{\mathrm{d}\alpha}\right]\mathrm{d}t$$

$$= \int_{t_1}^{t_2} \left[\frac{\partial F}{\partial X}\varepsilon + \frac{\partial F}{\partial Y}\eta + \frac{\partial F}{\partial X'}\varepsilon' + \frac{\partial F}{\partial Y'}\eta'\right]\mathrm{d}t.$$

Setting $\alpha = 0$ is equivalent to replace X and Y by x and y. Thus

$$\left.\frac{\mathrm{d}I}{\mathrm{d}\alpha}\right|_{\alpha=0} = \int_{t_1}^{t_2} \left[\frac{\partial F}{\partial x}\varepsilon + \frac{\partial F}{\partial y}\eta + \frac{\partial F}{\partial x'}\varepsilon' + \frac{\partial F}{\partial y'}\eta'\right]\mathrm{d}t.$$

This relation must hold for all possible choices of $\varepsilon(t)$ and $\eta(t)$, as long as they satisfy the boundary conditions. In particular, it holds for the special choice in which $\varepsilon(t)$ is identically equal to zero and $\eta(t)$ is still arbitrary. For this choice, the last equation becomes

$$\left.\frac{\mathrm{d}I}{\mathrm{d}\alpha}\right|_{\alpha=0} = \int_{t_1}^{t_2} \left[\frac{\partial F}{\partial y}\eta + \frac{\partial F}{\partial y'}\eta'\right]\mathrm{d}t.$$

This equation is identical to (7.6) with the independent variable x replaced by t. Following the same procedure, we obtain

$$\frac{\partial F}{\partial y} - \frac{\mathrm{d}}{\mathrm{d}t}\frac{\partial F}{\partial y'} = 0.$$

Similarly,

$$\frac{\partial F}{\partial x} - \frac{\mathrm{d}}{\mathrm{d}t}\frac{\partial F}{\partial x'} = 0.$$

Therefore for this system, we have two separate but simultaneous Euler–Lagrange equations for $x(t)$ and $y(t)$. Clearly, if we have n dependent variables, the analysis will lead to n separate but simultaneous equations.

This method is easily generalized to cases with more than one constraint. If we wish to find the stationary value of the integral I, which has n dependent variables, subject to multiple constraints that the values of the integrals J_j be held constant for $i = 1, 2, \ldots m$, then we simply find the unconstrained stationary value of the new integral K,

$$K = I + \sum_{j=1}^{m} \lambda_j J_j.$$

With

$$I = \int_{t_1}^{t_2} F(x_1, \ldots, x_n, x'_1, \ldots, x'_n, t)dt,$$

$$J_j = \int_{t_1}^{t_2} G_j(x_1, \ldots, x_n, x'_1, \cdot\cdot, x'_n, t)dt, \quad i = 1, 2, \ldots, m.$$

Following the same procedure, one can obtain the set of Euler–Lagrange equations

$$\frac{d}{dt}\frac{\partial H}{\partial x'_i} - \frac{\partial H}{\partial x_i} = 0, \quad i = 1, 2, \ldots, n,$$

where

$$H = F + \sum_{j=1}^{m} \lambda_i G_j.$$

This is a set of n coupled differential equations.

7.4.3 Several Independent Variables

For problems in more than one dimension, we need to consider functionals that depend on more than one independent variables. Let us consider the following double integral of x and y over some region R

$$I = \iint_R F(u, u'_x, u'_y, x, y)dxdy, \tag{7.25}$$

where u is a function of x and y, and

$$u'_x = \frac{\partial u(x, y)}{\partial x}, \quad u'_y = \frac{\partial u(x, y)}{\partial y}.$$

Let the region R be bounded by the curve C. The values of $u(x, y)$ are specified on C. We assume F is continuous and twice differentiable. We wish to determine the function $u(x, y)$ for which I is stationary with respect to small changes of u.

Procedures analogous to the one for one-dimensional problems can be used to solve this two-dimensional problem. Let $u(x, y)$ be the function for which the integral I is stationary, and the trial functions be of the form

$$U(x, y) = u(x, y) + \alpha\eta(x, y),$$

where $\eta(x, y) = 0$ on C. We will use the following notations to express partial derivatives

$$U'_x = \frac{\partial U}{\partial x} = \frac{\partial u}{\partial x} + \alpha\frac{\partial \eta}{\partial x} = u'_x + \alpha\eta'_x,$$

$$U'_y = \frac{\partial U}{\partial y} = \frac{\partial u}{\partial y} + \alpha\frac{\partial \eta}{\partial y} = u'_y + \alpha\eta'_y.$$

It follows that:

$$\frac{\mathrm{d}U}{\mathrm{d}\alpha} = \eta, \quad \frac{\mathrm{d}U'_x}{\mathrm{d}\alpha} = \eta'_x, \quad \frac{\mathrm{d}U'_y}{\mathrm{d}\alpha} = \eta'_y.$$

Replacing $u(x, y)$ with $U(x, y)$ in (7.25), we have

$$I(\alpha) = I = \iint_R F(U, U'_x, U'_y, x, y)\mathrm{d}x\mathrm{d}y.$$

A necessary condition for $u(x, y)$ to be the function for which I is an extremum is that the derivative of I must vanish at $\alpha = 0$

$$\frac{\mathrm{d}}{\mathrm{d}\alpha}I(\alpha)\bigg|_{\alpha=0} = 0.$$

Since α does not depend on x or y, the differentiation can be carried out under the integral sign

$$\frac{\mathrm{d}}{\mathrm{d}\alpha}I(\alpha) = \iint_R \left[\frac{\partial F}{\partial U}\frac{\mathrm{d}U}{\mathrm{d}\alpha} + \frac{\partial F}{\partial U'_x}\frac{\mathrm{d}U'_x}{\mathrm{d}\alpha} + \frac{\partial F}{\partial U'_y}\frac{\mathrm{d}U'_y}{\mathrm{d}\alpha}\right]\mathrm{d}x\mathrm{d}y$$

$$= \iint_R \left[\frac{\partial F}{\partial U}\eta + \frac{\partial F}{\partial U'_x}\eta'_x + \frac{\partial F}{\partial U'_y}\eta'_y\right]\mathrm{d}x\mathrm{d}y.$$

In the limit of $\alpha \to 0$, we have

$$\frac{\mathrm{d}}{\mathrm{d}\alpha}I(\alpha)\bigg|_{\alpha=0} = \iint_R \left[\frac{\partial F}{\partial u}\eta + \frac{\partial F}{\partial u'_x}\eta'_x + \frac{\partial F}{\partial u'_y}\eta'_y\right]\mathrm{d}x\mathrm{d}y.$$

The second term on the right-hand side can be written as

$$\iint_R \frac{\partial F}{\partial u'_x}\eta'_x\mathrm{d}x\mathrm{d}y = \int_{y_1}^{y_2}\left[\int_{x=C_1(y)}^{x=C_2(y)}\frac{\partial F}{\partial u'_x}\frac{\partial \eta}{\partial x}\mathrm{d}x\right]\mathrm{d}y,$$

where $C_1(y)$ and $C_2(y)$ are shown in Fig. 7.11.

With a given y, we can use integration by parts to write

$$\int_{x=C_1(y)}^{x=C_2(y)}\frac{\partial F}{\partial u'_x}\frac{\partial \eta}{\partial x}\mathrm{d}x = \left[\frac{\partial F}{\partial u'_x}\eta\right]_{x=C_1(y)}^{x=C_2(y)} - \int_{x=C_1(y)}^{x=C_2(y)}\eta\frac{\partial}{\partial x}\left(\frac{\partial F}{\partial u'_x}\right)\mathrm{d}x.$$

The integrated part is equal to zero because on the boundary $\eta(x, y) = 0$. Therefore

$$\iint_R \frac{\partial F}{\partial u'_x}\eta'_x\mathrm{d}x\mathrm{d}y = -\int_{y_1}^{y_2}\left[\int_{x=C_1(y)}^{x=C_2(y)}\eta\frac{\partial}{\partial x}\left(\frac{\partial F}{\partial u'_x}\right)\mathrm{d}x\right]\mathrm{d}y$$

$$= -\iint_R \eta\frac{\partial}{\partial x}\left(\frac{\partial F}{\partial u'_x}\right)\mathrm{d}x\mathrm{d}y.$$

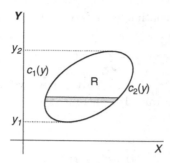

Fig. 7.11. Double integral. A double integral over the region R can be carried out by first integrating x from the left boundary $x_1 = C_1(y)$ to the right boundary $x_2 = C_2(y)$ with a fixed y, then integrating y from y_1 to y_2

Similarly,

$$\iint_R \frac{\partial F}{\partial u'_y} \eta'_y \, dx\, dy = -\iint_R \eta \frac{\partial}{\partial y}\left(\frac{\partial F}{\partial u'_y}\right) dx\, dy.$$

Thus,

$$\frac{d}{d\alpha} I(\alpha)\bigg|_{\alpha=0} = \iint_R \left[\frac{\partial F}{\partial u} - \frac{\partial}{\partial x}\left(\frac{\partial F}{\partial u'_x}\right) - \frac{\partial}{\partial y}\left(\frac{\partial F}{\partial u'_y}\right)\right] \eta(x,y) \, dx\, dy = 0.$$

Since $\eta(x,y)$ is arbitrary except at the boundary, we conclude that the term in the bracket must be equal to zero,

$$\frac{\partial F}{\partial u} - \frac{\partial}{\partial x}\left(\frac{\partial F}{\partial u'_x}\right) - \frac{\partial}{\partial y}\left(\frac{\partial F}{\partial u'_y}\right) = 0.$$

This is the Euler–Lagrange equation in two dimensions. Extension to three and higher dimensions is straightforward.

7.5 Sturm–Liouville Problems and Variational Principles

7.5.1 Variational Formulation of Sturm–Liouville Problems

Suppose we seek a function $y = y(x)$ in the range of $x_1 \leq x \leq x_2$ which satisfies the boundary condition

$$y(x_1) = 0, \quad y(x_2) = 0$$

and makes the value of the following integral stationary:

$$I = \int_{x_1}^{x_2} \left[p(x)y'^2 - q(x)y^2\right] dx, \tag{7.26}$$

where $p(x)$ and $q(x)$ are continuous differential functions of x. In addition, we require the integral

$$J = \int_{x_1}^{x_2} w(x)y^2 \mathrm{d}x \tag{7.27}$$

to equal to a prescribed value with a given positive function $w(x)$.

According to the constrained variational theory, the answer is given by the Euler–Lagrange equation

$$\frac{\mathrm{d}}{\mathrm{d}x}\frac{\partial H}{\partial y'} - \frac{\partial H}{\partial y} = 0$$

with

$$H = \left[p(x)y'^2 - q(x)y^2 \right] - \lambda w(x)y^2.$$

(The sign of λ is immaterial, since it is an undetermined multiplier. We use $-\lambda$ instead of positive to conform with the sign convention of the Sturm–Liouville problem, see below.) Since

$$\frac{\mathrm{d}}{\mathrm{d}x}\frac{\partial H}{\partial y'} = 2\frac{\mathrm{d}}{\mathrm{d}x}(p(x)y'),$$

$$\frac{\partial H}{\partial y} = -2q(x)y - 2\lambda w(x)y,$$

the Euler–Lagrange equation becomes

$$\frac{\mathrm{d}}{\mathrm{d}x}(p(x)y') + q(x)y + \lambda w(x)y = 0. \tag{7.28}$$

This is a Sturm–Liouville equation with eigenvalue λ.

This opens up a relation between the calculus of variation and eigenvalue problems.

Note that J of (7.27) is just a normalization integral. If we want $y(x)$ to be normalized to one with respect to the weight function $w(x)$, then $y(x)$ should be replaced by $y(x)J^{-1/2}$. Replace $y(x)$ in (7.26) by $y(x)J^{-1/2}$, it becomes

$$K\left[y(x)\right] = \frac{\displaystyle\int_{x_1}^{x_2} \left[p(x)y'^2 - q(x)y^2 \right] \mathrm{d}x}{\displaystyle\int_{x_1}^{x_2} w(x)y^2 \mathrm{d}x} = \frac{I}{J}. \tag{7.29}$$

Since the denominator J is constant, the stationary value of I corresponds to the stationary value of K. That is, the solution of the Sturm–Liouville equation (7.28) is still the function that minimizes the functional $K[y(x)]$.

Integrating the first term in the numerator of $K[y(x)]$ by parts, we get

$$\int_{x_1}^{x_2} p(x)y'^2 \mathrm{d}x = \int_{x_1}^{x_2} p(x)y'\frac{\mathrm{d}y}{\mathrm{d}x}\mathrm{d}x = \int_{x_1}^{x_2} p(x)y'\mathrm{d}y$$

$$= p(x)y'(x)y(x)\big|_{x=x_1}^{x=x_2} - \int_{x_1}^{x_2} y\,\mathrm{d}\left[p(x)y'\right].$$

The integrated part is equal to zero because of the boundary conditions of $y(x)$. Thus

$$\int_{x_1}^{x_2} p(x)y'^2 \mathrm{d}x = -\int_{x_1}^{x_2} y\frac{\mathrm{d}}{\mathrm{d}x}\left[p(x)y'\right]\mathrm{d}x.$$

It follows that:

$$K[y(x)] = \frac{-\int_{x_1}^{x_2} y\left\{\frac{\mathrm{d}}{\mathrm{d}x}\left[p(x)y'\right] + q(x)y\right\}\mathrm{d}x}{\int_{x_1}^{x_2} w(x)y^2\mathrm{d}x}. \qquad (7.30)$$

If $y(x)$ is the ith eigenfunction of (7.28), then

$$\frac{\mathrm{d}}{\mathrm{d}x}\left[p(x)y_i'\right] + q(x)y_i = -\lambda w(x)y_i. \qquad (7.31)$$

Substituting it into (7.30), we get

$$K\left[y_i(x)\right] = \frac{I}{J} = \frac{\lambda_i \int_{x_1}^{x_2} y_i w(x)y_i \mathrm{d}x}{\int_{x_1}^{x_2} w(x)y_i^2\mathrm{d}x} = \lambda_i. \qquad (7.32)$$

Thus, the eigenvalue λ, introduced originally as the undetermined multiplier, is the stationary value of the functional $K[y(x)]$. The function $y(x)$ that minimizes $K[y(x)]$ is the corresponding eigenfunction.

7.5.2 Variational Calculations of Eigenvalues and Eigenfunctions

The advantage of the variational formulation of the Sturm–Liouville equation is that one can use (7.29) or (7.30) to make systematic estimates of the eigenvalues and eigenfunctions of such equations.

The value of the functional $K[y(x)]$ can be calculated for any function of $y(x)$. There is a theorem which says that the functional $K[\phi(x)]$ of (7.29) evaluated with any function $\phi(x)$ that satisfies the same boundary conditions as given in the eigenvalue problem will be greater or equal to the smallest eigenvalue.

Let $\{y_i(x)\}$ be the set of eigenfunctions of the Sturm–Liouville problem. We may not know what they are, but we know that they are orthogonal and can be made orthonormal

$$\int_{x_1}^{x_2} y_i(x)y_j(x)w(x)\mathrm{d}x = \delta_{ij}, \qquad (7.33)$$

and they form a complete set. Therefore $\phi(x)$ can be expressed as

$$\phi(x) = \sum_i c_i y_i(x).$$

Substituting this expression into the functional of (7.30)

$$K[\phi(x)] = \frac{-\int_{x_1}^{x_2} \phi \left\{ \frac{d}{dx} \left[p(x)\phi' \right] + q(x)\phi \right\} dx}{\int_{x_1}^{x_2} w(x)\phi^2 dx},$$

and using (7.31) and (7.33), we have

$$K[\phi(x)] = \frac{\sum_i c_i^2 \lambda_i}{\sum_i c_i^2}.$$

Let λ_1 be the smallest eigenvalue, then

$$K[\phi(x)] - \lambda_1 = \frac{\sum_i c_i^2 \lambda_i}{\sum_i c_i^2} - \lambda_1 = \frac{\sum_i c_i^2 (\lambda_i - \lambda_1)}{\sum_i c_i^2}.$$

Since every λ_i, $i \neq 1$, is greater than λ_1, therefore $K[\phi(x)] - \lambda_1 > 0$, and

$$K[\phi(x)] \geq \lambda_1.$$

The equal sign holds only if $\phi(x) = y_1(x)$, the ground state eigenfunction. (The ground state is the state with the smallest eigenvalue.) This is often called the Rayleigh–Ritz variational principle.

Now we can approximate $y_1(x)$ with any reasonable trial function $\phi(x)$ that satisfies the boundary conditions. The eigenvalue obtained from (7.29)

$$\lambda_u = K[\phi(x)] = \frac{\int_{x_1}^{x_2} \left[p(x)\phi'^2 - q(x)\phi^2 \right] dx}{\int_{x_1}^{x_2} w(x)\phi^2 dx} \tag{7.34}$$

is always greater than or equal to λ_1. We can include parameters in $\phi(x)$, these parameters can be varied to minimize $K[\phi(x)]$ and thereby improve the estimate of the ground state eigenvalue.

As an illustration, let us consider the equation

$$y'' + \lambda y = 0$$

with the boundary conditions

$$y(0) = 0, \quad y(1) = 0.$$

This is a Sturm–Livouville problem with $p(x) = 1$, $q(x) = 0$, and $w(x) = 1$. This problem is simple enough for all of us to know the exact solutions

$$y = \sin \sqrt{\lambda} x,$$
$$\lambda = n^2 \pi^2, \quad n = 1, 2, \dots .$$

Therefore the lowest eigenvalue is

$$\lambda_1 = \pi^2 = 9.8696.$$

Now let us use the Rayleigh–Ritz method to approximate λ_1. We may use the simple function

$$\phi(x) = x(1-x)$$

as our trial function, since it satisfies the boundary conditions $\phi(0) = \phi(1) = 0$. Substituting this function into (7.34), with $\phi'(x) = 1-2x$, $p(x) = 1$, $q(x) = 0$, and $w(x) = 1$, we have

$$\lambda_u = K[\phi(x)] = \frac{\displaystyle\int_{x_1}^{x_2}(1-2x)^2\mathrm{d}x}{\displaystyle\int_{x_1}^{x_2}x^2(1-x)^2\mathrm{d}x} = \frac{1/3}{1/30} = 10,$$

which is only 1.3% in error compared to the exact value of π^2.

To calculate the eigenvalue, it is not necessary to use a normalized trial function, because of the denominator in $K[\phi(x)]$. However, it should be kept in mind that the trial function is an approximation of the eigenfunction only within a multiplicative constant. The comparison between the normalized trial function $\sqrt{30}x(1-x)$ and the normalized exact eigenfunction $\sqrt{2}\sin\pi x$ is shown in Fig. 7.12.

The result can be improved by introducing more terms with parameters. These parameters can be adjusted to minimize $K[\phi(x)]$, since no matter what these parameters are, the results are always upper bounds to λ_1. For example, we may use

$$\phi_1(x) = x(1-x) + cx^2(1-x)^2$$

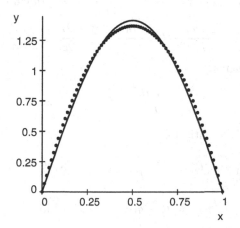

Fig. 7.12. Comparison between the normalized exact eigenfunction $\sqrt{2}\sin\pi x$ *(solid line)* and the normalized trial function $\sqrt{30}x(1-x)$ *(dotted line)*

as the trial function. As a consequence, $K\left[\phi_1(x)\right]$ becomes a function of c. Take the derivative of $K\left[\phi_1(x)\right]$ with respect to c and set it to zero, we find $c = 1.1353$. Use this value of c, we find

$$\lambda_u = K\left[\phi_1(x)\right] = 9.8697,$$

which is very close to the exact value of $\pi^2 = 9.8696$. When normalized, this trial function becomes $4.404x(1-x) + 4.990x^2(1-x)^2$. Plotted against x, this function is indistinguishable from the exact eigenfunction in the scale shown in Fig. 7.12. This suggests that if the eigenvalue calculated from the trial function is very good, the trial function is probably also a good approximation of the eigenfunction.

This method can be extended to the second and higher eigenvalues by imposing additional restrictions of the trial functions to only those that are orthogonal to the eigenfunctions corresponding to the lower eigenvalues.

For example, we may use a trial function in the form of

$$\phi(x) = c_1 f_1(x) + c_2 f_2(x),$$

where $f_1(x)$ and $f_2(x)$ are known as basis. We will show that in minimizing the functional

$$K\left[\phi(x)\right] = \frac{I\left[\phi(x)\right]}{J\left[\phi(x)\right]},$$

we will obtain two "eigenvalues." If $f_2(x)$ is orthogonal to $f_1(x)$

$$\int_{x_1}^{x_2} f_1(x) f_2(x) w(x) \mathrm{d}x = 0,$$

then they may approximate the first two true eigenvalues of the Sturm–Liouville problem.

To minimize $K[\phi(x)]$, we need to set both $\partial K/\partial c_1$ and $\partial K/\partial c_2$ to zero. Since

$$\frac{\partial K}{\partial c_i} = \frac{\partial I/\partial c_i}{J} - I\frac{\partial J/\partial c_i}{J^2} = \frac{1}{J}\left[\frac{\partial I}{\partial c_i} - \frac{I}{J}\frac{\partial J}{\partial c_i}\right] = 0$$

and $J > 0$, this means

$$\frac{\partial I}{\partial c_i} - \frac{I}{J}\frac{\partial J}{\partial c_i} = 0, \quad i = 1, 2. \tag{7.35}$$

Carrying out the differentiation

$$\frac{\partial I}{\partial c_1} = \frac{\partial}{\partial c_1}I[c_1 f_1 + c_2 f_2] = \frac{\partial}{\partial c_1}\int_{x_1}^{x_2}[p(c_1 f_1' + c_2 f_2')^2 - q(c_1 f_1 + c_2 f_2)^2]\mathrm{d}x$$

$$= 2\int_{x_1}^{x_2}[p(c_1 f_1' + c_2 f_2')f_1' - q(c_1 f_1 + c_2 f_2)f_1]\mathrm{d}x$$

$$= 2c_1\int_{x_1}^{x_2}[pf_1'f_1' - qf_1 f_1]\mathrm{d}x + 2c_2\int_{x_1}^{x_2}[pf_2'f_1' - qf_2 f_1]\mathrm{d}x.$$

Clearly we can write both derivatives as

$$\frac{\partial I}{\partial c_i} = 2 \sum_{j=1}^{2} c_j a_{ji}, \quad i = 1, 2,$$

$$a_{ji} = \int_{x_1}^{x_2} [p f'_j f'_i - q f_j f_i] dx.$$

Similarly,

$$\frac{\partial J}{\partial c_i} = 2 \sum_{j=1}^{2} c_j b_{ji}, \quad k = 1, 2,$$

$$b_{ji} = \int_{x_1}^{x_2} f_j f_i w \, dx.$$

According to (7.32), our "approximate eigenvalue" is given by

$$K = \frac{I}{J} = \lambda.$$

Therefore (7.35) can be written as

$$\sum_{j=1}^{2} c_j (a_{ji} - \lambda b_{ji}) = 0, \quad i = 1, 2.$$

Since f_1 and f_2 are orthogonal, so $b_{ij} = 0$ for $i \neq j$. Thus we have

$$(a_{11} - \lambda b_{11}) c_1 + a_{21} c_2 = 0,$$
$$a_{12} c_1 + (a_{22} - \lambda b_{22}) c_2 = 0.$$

For a nonzero solution, λ must satisfy the secular equation

$$\begin{vmatrix} a_{11} - \lambda b_{11} & a_{21} \\ a_{12} & a_{22} - \lambda b_{22} \end{vmatrix} = 0.$$

This is a quadratic equation, λ will have two roots. They may approximate the first two eigenvalues of the problem.

To illustrate the procedure, let us try to approximate the first two eigenvalues of the previous problem

$$y'' + \lambda y = 0, \quad y(0) = y(1) = 0.$$

We choose the following trial function:

$$\phi(x) = c_1 f_1(x) + c_2 f_2(x)$$

with

$$f_1(x) = x(1 - x),$$
$$f_2(x) = x(1 - x)(1 + ax).$$

Both $f_1(x)$ and $f_2(x)$ satisfy the boundary conditions. The constant a is determined from the orthogonal condition

$$\int_0^1 f_1(x)f_2(x)\mathrm{d}x = 0$$

to be -2. With $f_1' = 1 - 2x$, $f_2' = 1 - 6x + 6x^2$, one can readily find

$$a_{11} = \int_0^1 f_1'^2\mathrm{d}x = \frac{1}{3}, \quad a_{22} = \int_0^1 f_2'^2\mathrm{d}x = \frac{1}{5},$$

$$a_{12} = a_{21} = \int_0^1 f_1'f_2'\mathrm{d}x = 0,$$

$$b_{11} = \int_0^1 f_1^2\mathrm{d}x = \frac{1}{30}, \quad b_{22} = \int_0^1 f_2^2\mathrm{d}x = \frac{1}{210}.$$

Therefore the secular equation is

$$\begin{vmatrix} \frac{1}{3} - \frac{1}{30}\lambda & 0 \\ 0 & \frac{1}{5} - \frac{1}{210}\lambda \end{vmatrix} = 0$$

which has two roots

$$\lambda_1 = 10, \quad \lambda_2 = 42.$$

These two roots are to be compared with the first two exact eigenvalues

$$\lambda_1 = \pi^2 = 9.8696, \quad \lambda_2 = 4\pi^2 = 39.48.$$

7.6 Rayleigh–Ritz Methods for Partial Differential Equations

The Euler–Lagrange equations for functionals with more than one independent variables are partial differential equations. The minimizing function of the functional will be the solution of the corresponding partial differential equations.

Consider the functional

$$I = \int_{x_1}^{x_2} \int_{y_1}^{y_2} F(u, u_x', u_x'', u_y', u_y'', x, y)\mathrm{d}x\mathrm{d}y,$$

where x and y are two independent variables and

$$u = u(x, y),$$

$$u'_x = \frac{\partial u}{\partial x}, \quad u''_x = \frac{\partial^2 u}{\partial x^2},$$

$$u'_y = \frac{\partial u}{\partial y}, \quad u''_y = \frac{\partial^2 u}{\partial y^2}.$$

With the methods developed for functionals with two independent variables and with higher derivatives, one can easily show that the Euler–Lagrange equation for the functional is

$$\frac{\partial F}{\partial u} - \frac{\partial}{\partial x}\left(\frac{\partial F}{\partial u'_x}\right) + \frac{\partial^2}{\partial x^2}\left(\frac{\partial F}{\partial u''_x}\right)$$

$$-\frac{\partial}{\partial y}\left(\frac{\partial F}{\partial u'_y}\right) + \frac{\partial^2}{\partial y^2}\left(\frac{\partial F}{\partial u''_y}\right) = 0.$$

Many important partial differential equations in mathematical physics can be put in this form. In what follows, we will interpret these partial differential equations as the Euler–Lagrange equations of some functionals, then use the Rayleigh–Ritz method to approximate the minimizing functions of these functionals. The minimizing functions will then be the solutions of these partial differential equations.

7.6.1 Laplace's Equation

To find the Euler–Lagrange equation for the following two-dimensional functional

$$I = \int_{x_1}^{x_2} \int_{y_1}^{y_2}\left[\left(\frac{\partial u}{\partial x}\right)^2 + \left(\frac{\partial u}{\partial y}\right)^2\right] dxdy,$$

we can write the integrand as

$$F = u'^2_x + u'^2_y.$$

Thus

$$\frac{\partial F}{\partial u} = 0, \quad \frac{\partial F}{\partial u'_x} = 2u'_x = 2\frac{\partial u}{\partial x}, \quad \frac{\partial}{\partial x}\left(\frac{\partial F}{\partial u'_x}\right) = 2\frac{\partial^2 u}{\partial x^2},$$

$$\frac{\partial F}{\partial u'_y} = 2u'_y = 2\frac{\partial u}{\partial y}, \quad \frac{\partial}{\partial y}\left(\frac{\partial F}{\partial u'_y}\right) = 2\frac{\partial^2 u}{\partial y^2}.$$

Therefore the Euler–Lagrange equation for this functional

$$\frac{\partial F}{\partial u} - \frac{\partial}{\partial x}\left(\frac{\partial F}{\partial u'_x}\right) - \frac{\partial}{\partial y}\left(\frac{\partial F}{\partial u'_y}\right) = 0$$

is the Laplace equation

$$\frac{\partial^2 u}{\partial x^2} + \frac{\partial^2 u}{\partial y^2} = 0.$$

We have developed this relation in two dimensions, the extension to three dimension is obvious. Now we will use the notation

$$\nabla^2 u = \frac{\partial^2 u}{\partial x^2} + \frac{\partial^2 u}{\partial y^2},$$

$$|\nabla u|^2 = \nabla u \cdot \nabla u = \left(\mathbf{i}\frac{\partial u}{\partial x} + \mathbf{j}\frac{\partial u}{\partial y}\right) \cdot \left(\mathbf{i}\frac{\partial u}{\partial x} + \mathbf{j}\frac{\partial u}{\partial y}\right) = \left(\frac{\partial u}{\partial x}\right)^2 + \left(\frac{\partial u}{\partial y}\right)^2,$$

so that all the results will be automatically valid for three dimensions.

What we have shown is that the function u that minimizes the functional

$$I = \int\int |\nabla u|^2 \, dx dy$$

will also be the solution of the Laplace equation

$$\nabla^2 u = 0.$$

Now we turn it around, saying that to solve the Laplace equation with some boundary conditions is to find the function satisfying the same boundary conditions that minimizes the functional.

One way of finding the minimizing function is first to approximate it with a trial function with many terms

$$u(x, y) = f_0(x, y) + c_1 f_1(x, y) + \ldots + c_n f_n(x, y). \tag{7.36}$$

Then adjust the coefficients $c_1, c_2, \ldots, c_n$ so that the functional is as small as possible. Note that, as long as the trial function satisfies the boundary conditions, one additional term will make it closer to the true minimizing function. This is because the trial function that has the term $c_{n+1} f_{n+1}(x, y)$ automatically includes all the previous terms. If the minimizing process cannot make the functional smaller than the previous minimum, it will make $c_{n+1} = 0$ and settle with the previous minimum. Therefore by including more and more nonzero terms, the trial function will get closer and closer to the true solution.

To illustrate this process, let us take three terms

$$u(x, y) = f_0(x, y) + c_1 f_1(x, y) + c_2 f_2(x, y).$$

Putting it in the functional, we have

$$I = \int\int F \, dx dy,$$

$$F = \nabla(f_0 + c_1 f_1 + c_2 f_2) \cdot \nabla(f_0 + c_1 f_1 + c_2 f_2)$$

$$= |\nabla f_0|^2 + c_1^2 |\nabla f_1|^2 + c_2^2 |\nabla f_2|^2 + 2c_1 \nabla f_0 \cdot \nabla f_1$$

$$+ 2c_2 \nabla f_0 \cdot \nabla f_1 + 2c_1 c_2 \nabla f_1 \cdot \nabla f_2.$$

To minimize it, we have to set the following derivatives to zero.

$$\frac{\partial I}{\partial c_1} = \int \int \left\{ 2c_1 |\nabla f_1|^2 + 2\nabla f_0 \cdot \nabla f_1 + 2c_2 \nabla f_1 \cdot \nabla f_2 \right\} dxdy = 0,$$

$$\frac{\partial I}{\partial c_2} = \int \int \left\{ 2c_2 |\nabla f_2|^2 + 2\nabla f_0 \cdot \nabla f_2 + 2c_1 \nabla f_1 \cdot \nabla f_2 \right\} dxdy = 0.$$

These two equations can be put in the form of

$$a_{11}c_1 + a_{12}c_2 = b_1,$$
$$a_{21}c_1 + a_{22}c_2 = b_2$$

with

$$a_{11} = \int \int |\nabla f_1|^2 \, dxdy, \qquad a_{12} = \int \int \nabla f_1 \cdot \nabla f_2 dxdy,$$

$$a_{21} = \int \int \nabla f_1 \cdot \nabla f_2 dxdy, \qquad a_{22} = \int \int |\nabla f_2|^2 \, dxdy,$$

$$b_1 = -\int \int \nabla f_1 \cdot \nabla f_0 dxdy, \qquad b_2 = -\int \int \nabla f_0 \cdot \nabla f_2 dxdy.$$

Therefore

$$c_1 = \frac{\begin{vmatrix} b_1 & a_{12} \\ b_2 & a_{22} \end{vmatrix}}{\begin{vmatrix} a_{11} & a_{12} \\ a_{21} & a_{22} \end{vmatrix}}, \quad c_2 = \frac{\begin{vmatrix} a_{11} & b_1 \\ a_{21} & b_2 \end{vmatrix}}{\begin{vmatrix} a_{11} & a_{12} \\ a_{21} & a_{22} \end{vmatrix}}.$$

It is clear that with the trial function given by (7.36), the coefficients c_i are determined by the system of n linear equations

$$\sum_{j=1}^{n} a_{ij}c_j = b_i, \quad i = 1, 2, \ldots, n,$$

where

$$a_{ij} = \int \int \nabla f_i \cdot \nabla f_j dxdy, \quad b_i = -\int \int \nabla f_0 \cdot \nabla f_i dxdy.$$

The true solution can be approached by ever larger n. With a computer, such calculations are not difficult to carry out.

Example 7.6.1. Find a three term approximation to the solution of the Laplace's equation with the boundary conditions in the region shown in the following figure.

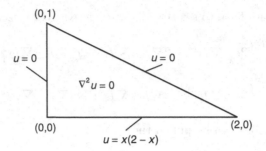

(0,1)

$u = 0$

$u = 0$

$\nabla^2 u = 0$

(0,0)

(2,0)

$u = x(2 - x)$

Solution 7.6.1. The equation for the straight line going through $(0,1)$ and $(2,0)$ is $x = 2 - 2y$. Therefore the region is bounded by the lines $x = 0$, $y = 0$ and $x = 2 - 2y$. the boundary conditions are

$$u(0,y) = 0, \quad u(x,0) = x(2 - x), \quad u(2 - 2y, y) = 0.$$

It is readily seen that the following simple function satisfies these boundary conditions:

$$f_0(x,y) = x(2 - x - 2y).$$

It is clear that functions in the form of

$$u(x,y) = x(2 - x - 2y)(1 + c_1 y + c_2 y^2)$$

will also satisfy the boundary conditions. Writing it in the form of

$$u(x,y) = f_0(x,y) + c_1 f_1(x,y) + c_2 f_2(x,y),$$

we have

$$f_1(x,y) = yx(2 - x - 2y), \quad f_2(x,y) = y^2 x(2 - x - 2y).$$

Carrying out the integration, we find

$$a_{11} = \int_0^1 \int_0^{2-2y} |\nabla f_1|^2 \, dxdy = \frac{2}{9},$$

$$a_{12} = a_{21} = \int_0^1 \int_0^{2-2y} \nabla f_1 \cdot \nabla f_2 dxdy = \frac{22}{315},$$

$$a_{22} = \int_0^1 \int_0^{2-2y} |\nabla f_2|^2 \, dxdy = \frac{11}{315},$$

$$b_1 = -\int_0^1 \int_0^{2-2y} \nabla f_0 \cdot \nabla f_1 dxdy = -\frac{2}{15},$$

$$b_2 = -\int_0^1 \int_0^{2-2y} \nabla f_0 \cdot \nabla f_2 dxdy = -\frac{28}{143}.$$

Therefore

$$c_1 = \frac{\begin{vmatrix} -\frac{2}{15} & \frac{22}{315} \\ -\frac{28}{143} & \frac{11}{315} \end{vmatrix}}{\begin{vmatrix} \frac{2}{9} & \frac{22}{315} \\ \frac{22}{315} & \frac{11}{315} \end{vmatrix}} = -\frac{7}{13}, \quad c_2 = \frac{\begin{vmatrix} \frac{2}{9} & -\frac{2}{15} \\ \frac{22}{315} & -\frac{28}{143} \end{vmatrix}}{\begin{vmatrix} \frac{2}{9} & \frac{22}{315} \\ \frac{22}{315} & \frac{11}{315} \end{vmatrix}} = -\frac{28}{143}.$$

Thus, the three term approximation to the solution of the Laplace equation is

$$u(x,y) = x(2 - x - 2y)\left(1 - \frac{7}{13}y - \frac{28}{143}y^2\right).$$

7.6.2 Poisson's Equation

It is easy to show that the Poisson's equation

$$\nabla^2 u = \rho$$

is the Euler–Lagrange equation of the functional

$$I = \int\int \left[|\nabla u|^2 + 2u\rho\right] dxdy. \tag{7.37}$$

Since the integrand of this functional is

$$F = u_x'^2 + u_y'^2 + 2u\rho,$$

the Euler–Lagrange equation

$$\frac{\partial F}{\partial u} - \frac{\partial}{\partial x}\left(\frac{\partial F}{\partial u_x'}\right) - \frac{\partial}{\partial y}\left(\frac{\partial F}{\partial u_y'}\right) = 0$$

is clearly

$$2\rho - 2\frac{\partial^2 u}{\partial x^2} - 2\frac{\partial^2 u}{\partial y^2} = 0$$

which is identical to the Poisson's equation $\nabla^2 u = \rho$.

Therefore to solve this Poisson's equation with some boundary conditions is equivalent to finding the function that satisfies the same boundary conditions and minimizes the corresponding functional (7.37).

Again we can approximate the solution with a trial function

$$u(x,y) = c_1 f_1(x,y) + c_2 f_2(x,y) + \cdots + c_n f_n(x,y). \tag{7.38}$$

The same method as we used in solving the Laplace's equation can be used to determine the coefficients $c_1, c_2, \ldots, c_n$. However, There is one difference.

Note that in (7.38) there is no $f_0(x, y)$ term with $c_0 = 1$. This is because any constant times the solution of a Laplace equation is still a solution. In other words, the coefficients in (7.36) are determined only up to a multiplicative constant. Therefore we can arbitrarily assign the coefficient of $f_0(x, y)$ to be one. We have no such freedom in solving the Poisson's equation because of the presence of the nonhomogeneous term $\rho(x, y)$. Therefore the trial function for the Poisson's equation has to start with the term $c_1 f_1(x, y)$.

For example, suppose we want to solve the following problem:

$$\nabla^2 u = \rho(x, y), \quad 0 \le x \le 1,\ 0 \le y \le 1,$$
$$u = 0, \quad \text{on the boundary of the square.}$$

We may choose a trial function in the form of

$$u(x, y) = xy(1 - x)(1 - y)(c_1 + c_2 x + c_3 y + c_4 x^2 + \cdots),$$

which clearly satisfies the boundary conditions. This function can be written in the form (7.38) with

$$f_1 = xy(1 - x)(1 - y), \quad f_2 = x f_1, \quad f_3 = y f_1, \quad f_4 = x^2 f_1, \ldots.$$

Put it into the functional

$$I = \int_0^1 \int_0^1 \left[\left| \nabla \sum_{j=1}^{n} c_j f_j \right|^2 + 2 \left(\sum_{j=1}^{n} c_j f_j \right) \rho \right] dx dy.$$

To minimize it, we set the derivatives with respect to c_j to zero

$$\frac{\partial I}{\partial c_1} = 0, \frac{\partial I}{\partial c_2} = 0, \ldots, \frac{\partial I}{\partial c_n} = 0.$$

The result is a system of n linear equations

$$\sum_{j=1}^{n} a_{ij} c_j = b_i, \quad i = 1, 2, \ldots, n,$$

where

$$a_{ij} = \int_0^1 \int_0^1 \nabla f_i \cdot \nabla f_j dx dy, \quad b_i = - \int_0^1 \int_0^1 f_i(x, y) \rho(x, y) dx dy.$$

This matrix equation can be solved for the set of coefficients $c_1, c_2, \ldots, c_n$. One strategy would be to calculate the functional with larger and larger n, until it is stabilized. This way we can get the approximation as close to the true solution as we want. Such calculations would have to be carried out on a computer.

7.6.3 Helmholtz's Equation

The Helmholtz's equation with boundary conditions is an eigenvalue problem. The Rayleigh–Ritz method developed for Sturm–Liouville problems can be used to obtain the solution. Consider the two-dimensional problem

$$\frac{\partial^2 u}{\partial x^2} + \frac{\partial^2 u}{\partial y^2} = \lambda u. \tag{7.39}$$

Multiplying both sides by u from the left and integrating, we have

$$\int \int u \left(\frac{\partial^2}{\partial x^2} + \frac{\partial^2}{\partial y^2} \right) u \, dxdy = \int \int \lambda u^2 dxdy.$$

If u is not an eigenfunction, we can show that

$$\lambda = \frac{\int \int u \left(\frac{\partial^2 u}{\partial x^2} + \frac{\partial^2 u}{\partial y^2} \right) dxdy}{\int \int u^2 dxdy} = \frac{\int \int u \nabla^2 u \, dxdy}{\int \int u^2 dxdy} \tag{7.40}$$

is an upper bound of the lowest eigenvalue by expanding u in terms of the eigenfunctions, as we did in the one-dimensional case.

In the context of variational principle, $\lambda[u]$ is a functional. One can show that to minimize $\lambda[u]$ is equivalent to minimizing the following functional (see exercise 11):

$$K[u] = \int \int \left[u \left(\frac{\partial^2 u}{\partial x^2} + \frac{\partial^2 u}{\partial y^2} \right) - \lambda u^2 \right] dxdy.$$

The integrand F of this functional is

$$F = u(u_x'' + u_y'') - \lambda u^2,$$

so

$$\frac{\partial F}{\partial u} = (u_x'' + u_y'') - 2\lambda u,$$

$$\frac{\partial F}{\partial u_x'} = \frac{\partial F}{\partial u_y'} = 0, \quad \frac{\partial F}{\partial u_x''} = \frac{\partial F}{\partial u_y''} = u.$$

Therefore the Euler–Lagrange equation

$$\frac{\partial F}{\partial u} - \frac{\partial}{\partial x} \left(\frac{\partial F}{\partial u_x'} \right) + \frac{\partial^2}{\partial x^2} \left(\frac{\partial F}{\partial u_x''} \right) - \frac{\partial}{\partial y} \left(\frac{\partial F}{\partial u_y'} \right) + \frac{\partial^2}{\partial y^2} \left(\frac{\partial F}{\partial u_y''} \right) = 0$$

for this functional becomes

$$(u_x'' + u_y'') - 2\lambda u + u_x'' + u_y'' = 0,$$

which is identical to the original equation

$$\nabla^2 u = \lambda u.$$

It is interesting to note that the Euler–Lagrange equation for another functional

$$K[u] = \int \int \left[\left(\frac{\partial u}{\partial x} \right)^2 + \left(\frac{\partial u}{\partial y} \right)^2 + \lambda u^2 \right] dx dy \qquad (7.41)$$

is also the Helmholtz's equation (7.39). The integrand F of this functional is

$$F = (u'_x)^2 + (u'_x)^2 + \lambda u^2.$$

Since

$$\frac{\partial F}{\partial u} = 2\lambda u, \quad \frac{\partial}{\partial x} \left(\frac{\partial F}{\partial u'_x} \right) = 2u''_x, \quad \frac{\partial}{\partial y} \left(\frac{\partial F}{\partial u'_y} \right) = 2u''_y,$$

the Euler–Lagrange equation becomes

$$2\lambda u - 2u''_x - 2u''_y = 0,$$

which is identical to (7.39).

Since to minimize the functional of (7.41) is equivalent to minimizing

$$\lambda = \frac{- \int \int \left[\left(\frac{\partial u}{\partial x} \right)^2 + \left(\frac{\partial u}{\partial y} \right)^2 \right] dx dy}{\int \int u^2 dx dy} = \frac{- \int \int |\nabla u|^2 \, dx dy}{\int \int u^2 dx dy}, \qquad (7.42)$$

we can approximate the ground state energy by either (7.40) or (7.42).

This is not surprising. When the boundary values of u are specified, one can use divergence theorem to show that $\int \int u \nabla^2 u \, dx dy$ and $- \int \int |\nabla u|^2 \, dx dy$ differ at most by a constant. Since

$$\int \int \nabla \cdot (u \nabla u) dx dy = \oint u \nabla u \cdot \mathbf{n} \, dl = \text{constant},$$

where the line integral is along the boundary of the area. But

$$\nabla \cdot (u \nabla u) = \nabla u \cdot \nabla u + u \nabla^2 u.$$

Therefore

$$\int \int u \nabla^2 u \, dx dy = - \int \int |\nabla u|^2 \, dx dy + \text{constant}.$$

Thus if u minimizes (7.40), it must also minimize (7.42). Usually it is (7.42) that is more convenient.

Example 7.6.2. Use the variational method to estimate the lowest vibrational frequency of circular membrane of radius c.

Solution 7.6.2. As we have learned in Chap. 6 that the vibration is governed by the wave equation

$$\nabla^2 \phi = \frac{1}{a^2} \frac{\partial^2}{\partial t^2} \phi.$$

The time-dependent part can be separated out to give $T(t) = \cos \omega t$, where $\omega = 2\pi \nu$ and ν is the frequency. The space part is then governed by the Helmholtz's equation

$$\nabla^2 u(x, y) = -\frac{\omega^2}{a^2} u(x, y).$$

Therefore the frequency of the vibration of any normal mode is determined by the eigenvalue of this equation. The boundary condition of u is that it is zero on the rim of the circular membrane. According to (7.42),

$$\frac{\omega^2}{a^2} = \frac{\int \int |\nabla u|^2 \, dx dy}{\int \int u^2 dx dy}.$$

Any u, as long as it satisfies the boundary condition will give an upper limit to the lowest frequency. For a circular membrane, clearly it is more convenient to do the integration in the polar coordinates. Written in polar coordinates, the boundary condition of $u(r, \theta)$ is

$$u(c, \theta) = 0.$$

The simplest trial function satisfying this boundary condition is

$$u(r, \theta) = r - c.$$

In polar coordinates

$$|\nabla u|^2 = \left(\frac{\partial}{\partial r} (r - c) \right)^2 = 1.$$

Thus

$$\int \int |\nabla u|^2 \, dx dy = \int_0^c \int_0^{2\pi} 1 \cdot r \, d\theta dr = \pi c^2,$$

$$\int \int u^2 dx dy = \int_0^c \int_0^{2\pi} (r - c)^2 r \, d\theta dr = \frac{1}{6} \pi c^4.$$

Hence

$$\frac{\omega^2}{a^2} = \frac{6}{c^2}, \quad \omega = 2.449 \frac{a}{c}.$$

We have shown in Chap. 6, the exact value of numerical factor is given by the first zero of $J_0(x)$ which is 2.405. It is seen that even with such a simple trial function, we still can get a reasonable estimate.

7.7 Hamilton's Principle

The Euler–Lagrange equation for the functional

$$I = \int_{t_1}^{t_2} F(y, y', t)\mathrm{d}t = \int_{t_1}^{t_2} (py'^2 - qy^2)\mathrm{d}t \qquad (7.43)$$

is

$$\frac{\mathrm{d}}{\mathrm{d}t}\frac{\partial F}{\partial y'} - \frac{\partial F}{\partial y} = 2py'' + 2qy = 0. \qquad (7.44)$$

Let

$$p = \frac{1}{2}m, \quad q = \frac{1}{2}k,$$

then (7.44) becomes

$$my'' + ky = 0,$$

which we recognize as the equation of a harmonic oscillator with mass m and spring constant k. Put the same values of p and q into (7.43), we find

$$I = \int_{t_1}^{t_2} \left(\frac{1}{2}my'^2 - \frac{1}{2}ky^2\right)\mathrm{d}t.$$

It is readily seen the first term is the kinetic energy T and the second term is the potential energy V of the harmonic oscillator

$$T = \frac{1}{2}my'^2, \quad V = \frac{1}{2}ky^2.$$

Thus the functional can be written as

$$I = \int_{t_1}^{t_2} (T - V)\mathrm{d}t. \qquad (7.45)$$

Therefore for an harmonic oscillator, we have found that the Newton's equation of motion is identical with the Euler–Lagrange equation for the functional of (7.45). This is just a special case of a general principle known as Hamilton's principle. It was first announced in 1834 by the brilliant Irish mathematician William Rowan Hamilton.

The difference between kinetic energy and potential energy $T - V$ is denoted by L and is called the Lagrangian. Hamilton's principle states that the motion of a system from t_1 to t_2 is such that the time integral of the Lagrangian L, (7.45), known as "action," has a stationary value. The Lagrangian is specified by a set of "generalized" coordinates $q_1, q_2, \ldots, q_n$ and their time derivatives $\dot{q}_1, \dot{q}_2, \ldots, \dot{q}_n$. From here on, we will follow the convention in mechanics, a dot on top means derivative with respect to time (Newton's notation). The Euler–Lagrange equations for the action functional (7.45) are usually called simply Lagrangian equations,

$$\frac{\partial L}{\partial q_i} - \frac{d}{dt}\left(\frac{\partial L}{\partial \dot{q}_i}\right) = 0, \quad i = 1, 2, \ldots, n.$$

These are a set of n simultaneous second-order differential equations. These equations were independently developed by Joseph L. Lagrange (1736–1816). They are equivalent to Newton's equations of motion. The Lagrange equations deal only with scalar variables, whereas Newton's equations are intrinsically vector equations. In many situations it is much easier to generate the correct differential equations of motion from the Lagrangian equations than from Newton's equation.

For example, suppose that a particle of mass m is moving in an arbitrary potential field. Then

$$T = \frac{1}{2}m(\dot{x}_1^2 + \dot{x}_2^2 + \dot{x}_3^2), \quad V = V(x_1, x_2, x_3)$$

and

$$L = \frac{1}{2}m(\dot{x}_1^2 + \dot{x}_2^2 + \dot{x}_3^2) - V(x_1, x_2, x_3).$$

The Lagrangian equations give us

$$\frac{\partial L}{\partial x_i} - \frac{d}{dt}\left(\frac{\partial L}{\partial \dot{x}_i}\right) = -\frac{\partial V}{\partial x_i} - \frac{d}{dt}(m\dot{x}_i) = 0, \quad i = 1, 2, 3$$

or

$$m\ddot{x}_i = -\frac{\partial V}{\partial x_i}.$$

Since $-\partial V/\partial x_i$ is the force on the particle in the x_i direction, this is simply Newton's second law which may be written in vector form as

$$m\ddot{\mathbf{r}} = \mathbf{F}.$$

From the following examples, we will see that Hamilton's principle is equally valid for system of continuum, and the "generalized" q_is need not be any standard coordinate set. They can be selected to match the conditions of the physical problem.

Example 7.7.1. Use Hamilton's principle to derive the wave equation for small transverse oscillations of a taut string.

Solution 7.7.1. Let ρ and τ be the linear density and tension of the oscillating string shown in Fig. 7.13.

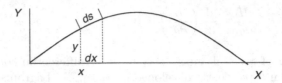

Fig. 7.13. Small oscillations of a taut string

Since the transverse speed of any part of the string is $\partial y/\partial t$, we can easily determine the kinetic energy of the vibration. It is

$$T = \int_0^L \frac{1}{2}\rho \left(\frac{\partial y}{\partial t}\right)^2 dx.$$

The potential energy V is found by considering the increase of length of the element dx. This element has increased its length from dx to ds. We have therefore done an amount of work $\tau(ds - dx)$. Since the potential energy is equal to the work done, summing all the work done along the line, we have

$$V = \int_0^L \tau(ds - dx).$$

But

$$ds - dx = [(dx)^2 + (dy)^2]^{1/2} - dx = \left[1 + \left(\frac{dy}{dx}\right)^2\right]^{1/2} dx - dx,$$

and

$$\left[1 + \left(\frac{dy}{dx}\right)^2\right]^{1/2} = 1 + \frac{1}{2}\left(\frac{dy}{dx}\right)^2 + \cdots.$$

Since it is a small oscillation, we will take the first two terms. Thus

$$ds - dx = \frac{1}{2}\left(\frac{dy}{dx}\right)^2 dx$$

and

$$V = \int_0^L \frac{1}{2}\tau \left(\frac{dy}{dx}\right)^2 dx.$$

Therefore the Lagrangian is given by

$$L = T - V = \int_0^L \left[\frac{1}{2}\rho\left(\frac{\partial y}{\partial t}\right)^2 - \frac{1}{2}\tau\left(\frac{dy}{dx}\right)^2\right] dx$$

and the action integral becomes

$$I = \int_{t_1}^{t_2} \int_0^L \left[\frac{1}{2}\rho\left(\frac{\partial y}{\partial t}\right)^2 - \frac{1}{2}\tau\left(\frac{dy}{dx}\right)^2\right] dxdt.$$

This is a two-dimensional functional. The independent variables are x and t. The integrand L can be written in the form of

$$L = \frac{1}{2}\rho y_t'^2 - \frac{1}{2}\tau y_x'^2.$$

The Lagrangian equation is

$$\frac{\partial L}{\partial y} - \frac{\partial}{\partial t}\left(\frac{\partial L}{\partial y_t'}\right) - \frac{\partial}{\partial x}\left(\frac{\partial L}{\partial y_x'}\right) = 0,$$

which becomes

$$-\rho\frac{\partial}{\partial t}y_t' + \tau\frac{\partial}{\partial x}y_x' = -\rho\frac{\partial}{\partial t}\frac{\partial y}{\partial t} + \tau\frac{\partial}{\partial x}\frac{\partial y}{\partial x} = 0.$$

This is exactly the same wave equation we derived before

$$\frac{\partial^2 y}{\partial x^2} = \frac{\rho}{\tau}\frac{\partial^2 y}{\partial t^2}.$$

Example 7.7.2. (a) Find the angular acceleration of a pendulum of length l. (b) A bead of mass m slides freely on a frictionless circular wire of radius r that rotates in a horizontal plane about a point on the circular wire with a constant angular velocity ω. Show that the bead oscillates as a pendulum of length $l = g/\omega^2$ about the line joining the center of rotation and the center of the circle.

Solution 7.7.2. (a) It is clear from Fig. 7.14a that the coordinates of m are given by

$$x = l\cos\theta, \quad y = l\sin\theta.$$

The kinetic energy is

$$T = \frac{1}{2}m(\dot{x}^2 + \dot{y}^2) = \frac{1}{2}ml^2(\sin^2\theta + \cos^2\theta)\dot{\theta}^2 = \frac{1}{2}ml^2\dot{\theta}^2.$$

Choosing the reference level for potential energy at distance l below the point of suspension, we have

$$V = mgl(1 - \cos\theta).$$

Thus the Lagrangian L is

$$L = T - V = \frac{1}{2}ml^2\dot{\theta}^2 - mgl(1 - \cos\theta).$$

So we have only one independent variable θ, which is our "generalized" coordinate. Hence we have the Lagrangian equation

$$\frac{\partial L}{\partial \theta} - \frac{d}{dt}\left(\frac{\partial L}{\partial \dot{\theta}}\right) = 0$$

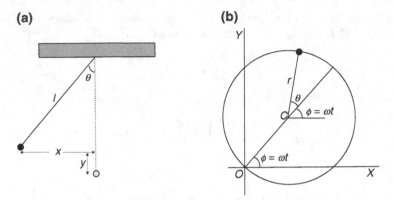

Fig. 7.14. Motions of pendulums

or
$$-mgl\sin\theta - \frac{\mathrm{d}}{\mathrm{d}t}\left(ml^2\dot\theta\right) = 0.$$

Therefore
$$\ddot\theta = -\frac{g}{l}\sin\theta.$$

(b) Let C be the center of the circular wire, and the angles θ and ϕ are indicated in the Fig. 7.14b. As the wire rotates counterclockwise with an angular velocity ω, so $\phi = \omega t$. The coordinates x and y of the bead are seen to be

$$x = r\cos\omega t + r\cos(\theta + \omega t),$$
$$y = r\sin\omega t + r\sin(\theta + \omega t).$$

Since the motion is taken place in a horizontal plan, the potential energy can be taken as zero. The kinetic energy is

$$T = \frac{1}{2}m\left(\dot x^2 + \dot y^2\right)$$
$$= \frac{1}{2}mr^2\left[\omega^2 + \left(\dot\theta + \omega\right)^2 + 2\omega\left(\dot\theta + \omega\right)\cos\theta\right],$$

which is also the Lagrangian L, since $V = 0$. Thus

$$\frac{\partial L}{\partial\theta} - \frac{\mathrm{d}}{\mathrm{d}t}\left(\frac{\partial L}{\partial\dot\theta}\right) = 0$$

becomes
$$mr^2\left[-\omega\left(\dot\theta + \omega\right)\sin\theta - \ddot\theta + \omega\sin\theta\dot\theta\right] = 0,$$

or
$$\ddot\theta = -\omega^2\sin\theta.$$

Thus we see that the bead oscillates about the line joining the center of rotation and the center of the circular wire like a pendulum of length $l = g/\omega^2$.

Hamilton's principle and Fermat's principle are only examples showing that the physical universe follows paths through space and time based on extrema principles. Almost in all branches of physics, one can find such a principle. Why the nature operates in accordance with this principle of economy is a question for philosophers and theologians. As scientists, we can just enjoy the elegance of the theory. However this is not to say that variation principle is merely a device to provide an alternative derivation of known results. In fact its impact on the development of science cannot be overemphasized. When the basic physics is not yet known, a postulated variational principle can be very useful. A shining example is Richard Feynman's formulation of quantum electrodynamics which is based on the principle of least action. For his achievements, he was awarded a 1965 Nobel prize in physics.

Variational principle as a computation tool is also very important. With variational methods, energy levels of all kinds of molecules can now be calculated to a high degree of accuracy. John Pople codified such calculations in a computer program known as GAUSSIAN. He was awarded a 1998 Nobel prize in chemistry.

Exercises

1. Find the Euler–Lagrange equation for

$$\text{(a) } F = x^2 y^2 - y'^2, \quad \text{(b) } F = \sqrt{xy} + y'^2.$$

Ans. (a) $y'' + x^2 y = 0$, (b) $\dfrac{1}{4}\sqrt{\dfrac{x}{y}} - y'' = 0$.

2. Find the curve $y(x)$ that will make the following functional stationary

$$\text{(a) } I = \int_a^b (y^2 + y'^2 + 2ye^x)dx,$$

$$\text{(b) } I = \int_a^b \frac{y'^2}{x^3}dx.$$

Ans. (a) $y = \frac{1}{2}xe^x + c_1 e^x + c_2 e^{-x}$, (b) $y = c_1 x^4 + c_2$.

3. Find the function $y(x)$ that passes through the points $(0,0)$ and $(1,1)$ and minimizes

$$I(y) = \int_0^1 (y^2 + y'^2)dx.$$

Ans. $y(x) = 0.42e^x - 0.42e^{-x}$.

4. Find the function $y(x)$ that passes through the points $(0,0)$ and $(\pi/2, 1)$ and minimizes

$$I(y) = \int_0^1 (y'^2 - y^2)\mathrm{d}x.$$

Ans. $y(x) = \sin x$.

5. What would be the functional corresponding to the following problem:

$$\frac{\partial^2 u}{\partial x^2} + \frac{\partial^2 u}{\partial y^2} = 1, \quad 0 < x < 1,\ 0 < y < 1,$$

$$u = 0, \quad \text{on the boundary.}$$

Ans. $I(u) = \int_0^1 \int_0^1 \left[\left(\frac{\partial u}{\partial x}\right)^2 + \left(\frac{\partial u}{\partial y}\right)^2 + 2u \right] \mathrm{d}x\mathrm{d}y$.

6. Show that if the integrand of the following integral:

$$I = \int_{t_1}^{t_2} F(x, y, x', y')\mathrm{d}t$$

does not explicitly contain the independent variable t, then the Euler–Lagrange equations lead to

$$F - x'\frac{\partial F}{\partial x'} - y'\frac{\partial F}{\partial y'} = C,$$

where C is a constant.

7. Find the Euler–Lagrange equation for the functional

$$I = \int_0^1 (yy'' + 4y)\mathrm{d}x.$$

Ans. $y'' + 2 = 0$.

8. Find the Euler–Lagrange equation for the functional

$$I = \int_0^1 (-y'^2 + 4y)\mathrm{d}x.$$

Ans. $y'' + 2 = 0$.

9. Show that the Euler–Lagrange equation for the three-dimensional functional

$$I = \iiint \left[\left(\frac{\partial u}{\partial x}\right)^2 + \left(\frac{\partial u}{\partial y}\right)^2 + \left(\frac{\partial u}{\partial z}\right)^2 \right] \mathrm{d}x\mathrm{d}y\mathrm{d}z$$

is given by the Laplace's equation

$$\frac{\partial^2 u}{\partial x^2} + \frac{\partial^2 u}{\partial y^2} + \frac{\partial^2 u}{\partial z^2} = 0.$$

10. Estimate the lowest vibrational frequency of a circular drum-head with radius a, using the functional

$$\frac{\omega^2}{v^2} = \frac{-\int\int u\nabla^2 u\,dxdy}{\int\int u^2 dxdy}$$

and the trial function

$$u(r) = r - a.$$

Ans. $\omega = 2.449v/a$.

11. If $I[u]$ and $J[u]$ are both two-dimensional functionals and

$$\lambda[u] = \frac{I[u]}{J[u]},$$

show that to minimize $\lambda[u]$ is equivalent to minimizing the functional $K[u]$

$$K[u] = I[u] - \lambda J[u].$$

Hint: Replace $u(x,y)$ by $U(x,y) + \alpha\eta(x,y)$, and show that $\left.\dfrac{d\lambda}{d\alpha}\right|_{\alpha=0} = 0$

leads to $\left[\dfrac{dI}{d\alpha} - \lambda\dfrac{dJ}{d\alpha}\right]_{\alpha=0} = 0.$

12. Find the Euler–Lagrange equation for the functional

$$I = \int_0^1 xy'^2 dx$$

subject to the constraint

$$\int_0^1 xy^2 dx = 1.$$

Ans. $xy'' + y' - \lambda xy = 0$.

13. Find the Euler–Lagrange equation for the functional

$$I = \int_0^1 (py'^2 - qy^2)dx$$

subject to the constraint

$$\int_0^1 ry^2 dx = 1.$$

Ans. $\frac{d}{dx}(py') + (q - \lambda r)y = 0$.

14. Show the equivalence of the following two forms of Euler–Lagrange equations:

$$\frac{\partial F}{\partial y} - \frac{d}{dx}\left(\frac{\partial F}{\partial y'}\right) = 0,$$

$$\frac{\partial F}{\partial x} - \frac{d}{dx}\left(F - y'\frac{\partial F}{\partial y'}\right) = 0.$$

15. Approximate the solution of the problem

$$y'' + \left(\frac{\pi}{2}\right)^2 y = 0,$$
$$y(0) = 1, \quad y(1) = 0$$

with a trial function

$$y = 1 - x^2.$$

With this trial function, find the eigenvalue and compare it with the exact value.

Ans. $\lambda = 2.5$, $\lambda/\lambda_{exat} = 1.013$.

16. In the previous problem, use a trial function

$$y = 1 - x^n.$$

Find the optimum value of n. With that n, what is λ/λ_{exat}?

Ans. $n = 1.7247$, $\lambda/\lambda_{exat} = 1.003$.

17. Find the function $y(x)$ that will extremize the integral

$$I = \int_0^a y'^2 dx$$

subject to the constraint

$$\int_0^a y^2 dx = 1, \quad y(0) = 0, \quad y(a) = 0.$$

Ans. $y(x) = \left(\frac{2}{a}\right)^{1/2} \sin\frac{n\pi}{a}x$.

18. Use the Fermat principle to find the path followed by a light ray if the index of refraction is proportional to

(a) y^{-1}, (b) y.

Ans. (a) $(x - c_1)^2 + y^2 = c_2^2$, (b) $y = c_1 \cosh\dfrac{x - c_2}{c_1}$.

19. Use a trial function of the form

$$u = (r - c) + b(r - c)^2$$

to calculate the lowest frequency of the vibration of a circular membrane of radius c.

Ans. $\omega = 2.4203\ a/c$.

20. **Conservation of energy.** If

$$T = \frac{1}{2}\sum_{i=1}^{n} m_i \dot{q}_i^2, \quad V = V(q_1, q_2, \ldots, q_n)$$

use Hamilton's principle to show that

$$T + V = \text{constant}.$$

Hint: From the fact that the independent variable t does not appear explicitly in the integrand, show that

$$L - \sum_{i=1}^{n} \dot{q}_i \frac{\partial L}{\partial \dot{q}_i} = \text{constant}.$$

21. Derive Lagrangian equation of motion for a particle in a gravitation field constraint to be on a circle of radius c in a fixed vertical plane.

Ans. $\dfrac{d}{dt}(mc^2\dot{\theta}) + mgc\cos\theta = 0$.

References

This bibliograph includes the references cited in the text and a few other books and tables that might be useful.

1. M. Abramowitz, I.A. Stegun: *Handbook of Mathematical Functions* (Dover, New York 1970)
2. G.B. Arfken, H.J. Weber: *Mathematical Methods for Physicists,* 5th edn. (Academic Press, San Diego, 2001)
3. M. L. Boas: *Mathematical Methods in the Physical Sciences,* 3rd edn. (Wiley, New York 2006)
4. T.C. Bradbury: *Mathematical Methods with Applications to Problems in the Physical Sciences* (Wiley, New York 1984)
5. E.O. Brigham: *The Fast Fourier Transform and Its Applications* (Prentice Hall, Upper Saddle River 1988)
6. E. Butkov: *Mathematical Physics* (Addison-Wesley, Reading 1968)
7. F.W. Byron, Jr., R.W. Fuller: *Mathematics of Classical and Quantum Physics* (Dover, New York 1992)
8. T.L. Chow: *Mathematical Methods for Physicists: A Concise Introduction* (Cambridge University Press, Cambridge 2000)
9. R.V. Churchill: *Fourier Series and Boundary Value Problems,* 2nd edn. (McGraw-Hill, New York 1963)
10. H. Cohen: *Mathematics for Scientists and Engineeers* (Prentice-Hall, Englewood Cliffs 1992)
11. R.E. Collins: *Mathematical Methods for Physicists and Engineers* (Reinhold, New York 1968)
12. R. Courant, D. Hillbert: *Methods of Mathematical Physics* (Wiley, New York 1989)
13. C.H. Edwards Jr., D.E. Penney: *Differential Equations and Boundary Value Problems* (Prentice-Hall, Englewood Cliffs 1996)
14. A. Erdélyi, W. Magnus, F. Oberhettinger, F. Tricomi: *Tables of Integral Transforms, Vol. 1* (McGraw-Hill, New York 1954)
15. R.P. Feynman, R.B. Leighton, M. Sands: *The Feynman Lectures on Physics, Vol. I*, Chapter 50 (Addison-Wesley, Reading 1963)
16. I.S. Gradshteyn, I.M. Ryzhik: *Table of Integrals, Series and Products* (Academic Press, Orlando 1980)
17. D.W. Hardy, C.L. Walker: *Doing Mathematics with Scientific WorkPlace and Scientific Notebook,* Version 5 (MacKichan, Poulsbo 2003)

18. S. Hasssani: *Mathematical Methods: For Students of Physics and Related Fields* (Springer, New York 2000)
19. F.B. Hilderbrand: *Advanced Calculus for Applications,* 2nd edn. (Prentice-Hall, Englewood Cliffs 1976)
20. H. Jeffreys, B.S. Jeffreys: *Mathematical Physics* (Cambridge University Press, Cambridge 1962)
21. D.E. Johnson and J.R. Johnson: *Mathematical Methods in Engineering Physics* (Prentice-Hall, Upper Sadddle River 1982)
22. D.W. Jordan, P. Smith: *Mathematical Techniques: An Introduction for the Engineering, Physical, and Mathematical Sciences,* 3rd edn. (Oxford University Press, Oxford 2002)
23. E. Kreyszig: *Advanced Engineering Mathematics,* 8th edn. (Wiley, New York 1999)
24. B.R. Kusse, E.A. Westwig: *Mathematical Physics: Applied Mathematics for Scientists and Engineers,* 2nd edn. (Wiley, New York 2006)
25. S.M. Lea: *Mathematics for Physicists* (Brooks/Cole, Belmont 2004)
26. M.J. Lighthill: *Introduction to Fourier Analysis and Generalised Functions* (Cambridge University Press, Cambridge 1958)
27. W. Magnus, F. Oberhattinger, R.S. Soni: *Formulas and Theorems for the Special Functions of Mathematical Physics* (Springer, New York, 1966)
28. H. Margenau, G.M. Murphy: *Methods of Mathematical Physics* (Van Nostrand, Princeton 1956)
29. J. Mathew, R.L. Walker: *Mathematical Methods of Physics,* 2nd edn. (Benjamin, New York 1970)
30. N.W. Mclachlan: *Bessel Functions for Engineers,* 2nd edn. (Oxford University Press, Oxford 1955)
31. D.A. McQuarrie: *Mathematical Methods for Scientists and Engineers* (University Science Books, Sausalito 2003)
32. P.M. Morse, H. Feshbach: *Methods of Theoretical Physics* (McGraw-Hill, New York 1953)
33. J.M.H. Olmsted: *Advanced Calculus* (Prentice Hall, Englewood Cliffs 1961)
34. M.C. Potter, J.L. Goldber, E.F. Aboufadel: *Advanced Engineering Mathematics,* 3rd edn. (Oxford University Press, New York 2005)
35. D.L. Powers: *Boundary Value Problems* (Academic Press, New York 1972)
36. W.H. Press, S.A. Teukolsky, W.T. Vettering, B.P. Flannery: *Numerical Recipes,* 2nd edn. (Cambridge University Press, Cambridge 1992)
37. K.F. Riley, M.P. Hobson, S.J. Bence: *Mathematical Methods for Physics and Engineering,* 2nd edn. (Cambridge University Press, Cambridge 2002)
38. H. Sagan: *Introduction to the Calculus of Variations* (Dover, New York 1992)
39. K.A. Stroud, D.J. Booth: *Advanced Engineering Mathematics,* 4th edn. (Industrial Press, New York 2003)
40. N.M. Temme: *Special Functions: An Introduction to the Classical Functions of Mathematical Physics* (Wiley, New York 1996)
41. G.P. Tolstov: *Fourier Series* (Dover, New York 1976)
42. R. Winstock: *Calculus of Variations, with Applications to Physics and Engineering* (Dover, New York 1974)
43. C.R. Wylie, L.C. Barrett: *Advanced Engineering Mathematics,* 5th edn. (McGraw-Hill, New York 1982)
44. E. Zauderer: *Partial Differential Equations of Applied Mathematics* (Wiley, New York 1983)
45. S. Zhang, J. Jin: *Computation of Special Functions* (Wiley, New York 1996)

Index

434   Index